Measuring the Natural Environment, Second Edition

Measurements of natural phenomena are vital for any type of environmental monitoring, from the practical day-to-day management of rivers, agriculture and weather forecasting, through to longer-term assessment of climate change and glacial retreat. This new edition of *Measuring the Natural Environment* looks at all aspects of past, present and future measurement techniques, describing the operation of the instruments used and the quality and accuracy of the data they produce.

The book describes the techniques and instruments used to measure all of the variables of the natural world: solar and terrestrial radiation, air and ground temperature, humidity, evaporation and transpiration, wind speed and direction, rainfall, snowfall, snow depth, barometric pressure, clouds, lightning, atmospheric chemistry, soil moisture and soil tension, groundwater, river level and flow, water quality, sea level, sea surface temperature, ocean currents and waves, and polar ice. This second edition has been brought completely up to date, and expanded considerably through the addition of six new chapters, and the extension and modification of many of the existing chapters.

Measuring the Natural Environment is the first book to make a thorough enquiry into the origins of environmental data, upon which our scientific understanding and economic planning of the environment directly hang. The book will be important for all those who use or collect such data, whether for pure research or day-to-day management. It will be useful for students and professionals working in a wide range of environmental science: meteorology, climatology, hydrology, water resources, oceanography, civil engineering, agriculture, forestry, glaciology and ecology.

Ian Strangeways is Director of TerraData, a consultancy in meteorological and hydrological instrumentation and data collection. From 1964 until 1989 he was head of the Instrument and Applied Physics sections at the Institute of Hydrology (Natural Environment Research Council).

MEASURING THE NATURAL ENVIRONMENT

Second Edition

IAN STRANGEWAYS

CAMBRIDGE
UNIVERSITY PRESS

PUBLISHED BY THE PRESS SYNDICATE OF THE UNIVERSITY OF CAMBRIDGE
The Pitt Building, Trumpington Street, Cambridge, United Kingdom

CAMBRIDGE UNIVERSITY PRESS
The Edinburgh Building, Cambridge CB2 2RU, UK
40 West 20th Street, New York, NY 10011–4211, USA
477 Williamstown Road, Port Melbourne, VIC 3207, Australia
Ruiz de Alarcón 13, 28014 Madrid, Spain
Dock House, The Waterfront, Cape Town 8001, South Africa

http://www.cambridge.org

First published 2000
Second edition 2003

Printed in the United Kingdom at the University Press, Cambridge

Typeface Times 11/14pt.   *System* LATEX 2$_\varepsilon$ [TB]

A catalogue record for this book is available from the British Library

Library of Congress Cataloguing in Publication data
Strangeways, Ian, 1932–
Measuring the natural environment / Ian Strangeways. – 2nd edn
      p.   cm.
Includes bibliographical references and index.
ISBN 0 521 82205 X – ISBN 0 521 52952 2 (paperback)
1. Earth science – Measurement. 2. Environmental monitoring.   I. Title.
QE33.S79 2003
363.7′063–dc21   2003044038

ISBN 0 521 82205 X hardback
ISBN 0 521 52952 2 paperback

Contents

Acknowledgements

Richard Dawkins, in *The Blind Watchmaker*, says he prefers to include credits at the relevant point in the text, rather than in a separate section. I agree, but although I tried it, I found it did not suit the style of my book, mostly because I have used a fairly impersonal style. This makes it less seamless to include acknowledgements in amongst the technical detail. I have broken with tradition to the small extent, however, that names are in bold and each starts on a new line, as in a reference list.

But why should I need any help? My experience of measuring the natural environment extends from 1964 to 1989 at the Institute of Hydrology (IH), now the Centre for Ecology and Hydrology (CEH), first as head of the Instrument Section and later of Applied Physics, continuing after 1989 as consultant until the present. During this nearly 40-year period, my work has been a mix of new instrument development and the application of existing equipment, new and old, to a variety of projects, many overseas and embracing all of the world's climates. Despite this, experience of every aspect of the subject has not been equal and I felt it advisable to check out some specialised areas and details about which I was not entirely certain. And now in this second edition, I have added new chapters and although the methods and general principles are remarkably similar throughout the whole field of measuring the natural environment, they are different in detail. So further advice had to be obtained and visits made.

I would, therefore, like to acknowledge the advice, help and time of those listed below, who talked me through new topics, helped fill in gaps, corrected errors, checked my text, gave access to equipment for photography, spent time showing me round laboratories or simply reassured me that I was right in the first place:

James Bathurst, *Newcastle University*, commented on, and refined, my summary of the slope-area method of estimating river flow.

John Bell, ex *CEH*, supplied detailed notes describing how soil moisture is measured by the thermogravimetric method.

Ken Blyth, *CEH*, gave me advice on the basic principles of remote-sensing satellite instruments.

Mike Brettle, *Vaisala UK Ltd.*, loaned me radiosonde equipment to photograph (Figs. 16.1, 16.2 and 16.3) and advised on state-of-the-art radiosonde sensors.

Chris Collier, *Salford University*, read the section on weather radar and suggested changes and additions.

Mike Collins, *ex Met. Office*, explained the various definitions and acronyms of lighting detection systems.

J. David Cooper, *CEH*, discussed soil moisture capacitance probes and the time domain reflectrometry method and gave access to instruments for me to photograph Figs. 9.5 (*a*), (*b*), 9.6 and 9.8.

Andy Dixon, *CEH*, guided me through the technical complexities and terms of borehole drilling, and loaned me two photographs (Figs. 9.15 and 9.16).

Jonathan Evans, *CEH*, showed me the latest eddy correlation equipment, now able to measure CO_2 fluxes, and supplied drawings from which I produced Fig. 7.4(*b*).

John Gash, *CEH*, advised on techniques for measuring water vapour flux and gave access to equipment for photographing, shown in Figs. 5.11, 7.2, 7.3 and 7.4(*a*).

Reg Herschy, *CNS, Reading*, read the chapter on Rivers and lakes and suggested changes, in particular concerning river flow measurement.

Wynn Jones, *Met. Office*, supplied information on ocean buoys, provided Figs. 17.3, 17.4 and 17.5 and checked the chapter on ocean measurements (first edition).

Charles Kilburn, *Rutherford/Appleton Laboratory*, showed me a range of equipment at Chilbolton for measuring the upper atmosphere (Figs. 14.7, 16.9, 16.10, 16.11, 16.12). For more details see: http://www.rcru.rl.ac.uk/chilbolton/facilities.htm

Tony Lee, *Met. Office*, clarified the details of lightning radio wave propagation and the basis of the several methods of detecting *sferics*.

Robin Pascal, *Southampton Oceanography Centre*, spent an afternoon answering my questions regarding instrumentation on ships for oceanographic research.

Sarah Price, *My copy editor*, who has suggested many improvements to the text, picked up errors and followed up questions regarding references and copyrights for me.

Jonathan Shanklin, *British Antarctic Survey*, commented on the instrumentation currently in use in Antarctica and helped identify the types of ice in Figs. 18.6–18.10.

John Stewart, *Southampton University*, read, and suggested changes to, the chapter on remote sensing (first edition).

Jon Turton, *Met. Office*, supplied extensive information on the Argo float project, which is an important new ocean observation initiative.

Alan Walker, *Newbury*, demonstrated instruments from his barometer collection at *Halfway Manor*, which I photographed to illustrate Chapter 6. http://www.alanwalker-barometers.com.

1

Basics

The need for measurements

Whether it be for meteorological, hydrological, oceanographic or climatological studies or for any other activity relating to the natural environment, measurements are vital. Knowledge of what has happened in the past and of the present situation, and an understanding of the processes involved, can only be arrived at if measurements are made. Such knowledge is also a prerequisite of any attempt to predict what might happen in the future and subsequently to check whether the predictions are correct. Without data, none of these activities is possible. Measurements are the cornerstone of them all. This book is an investigation into how the natural world is measured.

The things that need to be measured are best described as *variables*. Sometimes the word *parameter* is used but *variable* describes them more succinctly. The most commonly measured variables of the natural environment include: solar and terrestrial radiation, air and ground temperatures, atmospheric humidity, evaporation and transpiration, wind speed and direction, rainfall and snowfall, barometric pressure, soil moisture and soil tension, groundwater, river level and flow, water quality, sea level, sea surface temperature, ocean currents and waves, properties of the upper atmosphere and the concentration of trace gases, clouds and lightning, visibility and the ice of polar regions.

The origins of data

Early instrument development

Measurement of the natural environment did not begin in the scientific sense until around the middle of the seventeenth century when, in 1643, working in Florence,

Measuring the Natural Environment, second edition, Ian Strangeways. Published by Cambridge University Press. © Ian Strangeways 2003.

Torricelli made the first mercury barometer, based on notes left by Galileo at his death. The first thermometer is also attributable to Galileo. Castelli, also in Italy, made the first measurements of rainfall. Sir Christopher Wren in England designed what was probably the first automatic weather station, while Robert Hooke constructed a manual raingauge. In 1846 Thomas Robinson, a clergyman in Armagh, constructed the first instrument for measuring wind speed and in 1853 John Campbell developed the prototype of today's sunshine recorder. Thirteen years later, Thomas Stevenson, the father of Robert Louis Stevenson, designed the now widely used wooden temperature screen. In 1850 George Symons embarked on a lifelong mission to put rainfall measurement on a firm footing, while Captain Robert FitzRoy, who had earlier taken Charles Darwin on the voyage of the *Beagle*, spent much of his later career advancing meteorological measurements at sea. Thus the beginnings of the study of the natural environment started with the development of instruments and the taking of measurements, highlighting the great importance of hard facts in forwarding any science and of the means of obtaining these data. Instrument development continued into and throughout the twentieth century using similar technology until, mid-century, the invention of the transistor presented totally new possibilities. Now, in the twenty-first century, developments progress apace.

The important point about the early designs is not their history, interesting as it is, but that the same designs, albeit perhaps refined, are still in widespread use today. Most of the national weather services (NWSs) of the world rely on them. All the data used by climatologists to study past conditions, and by anyone else for whatever purpose, are derived largely from instruments developed in the Victorian era, and the same instruments look set to continue in widespread use into the foreseeable future. The importance of appreciating their capabilities and limitations is thus of more than passing historical interest. Everyone using data from the past and from the present needs to be aware of where the data come from and of how reliable or unreliable they are likely to be.

Recent advances

In the last forty years, owing to developments in microelectronics and the widespread availability of personal computers (PCs), new instruments have become available that greatly enhance our ability to measure the natural environment. It has been possible to design data loggers with low power consumption and large memory capacity that can operate remotely and unattended, and new sensors have been developed to supply the logging systems with precise measurements. (Sensors may be referred to in some texts as transducers.) In a balanced review of the subject, this new generation of instruments needs to be discussed along with the old, for we are at a time in the measurement of the environment when both types of instrument are equally important, although the change to the new is well under way and accelerating.

Old and new compared

A serious limitation of the old instruments is that, being manual and mechanical, they need operators and this restricts their use to those parts of the world that are inhabited. Most mountainous, desert, polar and forested areas and most of the oceans are, in consequence, almost completely blank on the data map. Thus far, our knowledge of the natural environment comes from a rather limited range of the planet's surface, over a limited range of time.

In contrast the new instruments need the attendance of an operator only once every few months, or even less frequently, and so can be deployed at remoter sites. The same electronic developments have also made it possible to telemeter measurements from remote automatic stations via satellites, making it possible to operate instruments in almost any region of the world, however remote. To this capability has also been added the new technique of remote sensing – its images often being generated by the same satellites as those that relay data.

The old instruments also lack the accuracy of the new. Compare for instance the data from a sunshine recorder giving simple 'sun in or out' information with those from a photodiode giving an exact measure of the intensity of solar energy second by second. Ironically, these improvements are often a hindrance to change, for although changing to a better instrument improves the data, the continuity of old records is lost. This deters many long-established organisations from changing their methods and so the old ways persist. While it is possible to operate a new instrument alongside the old to establish a relationship between them, this is usually only partly successful, the complexity of the natural environment and of an instrument's behaviour meaning that a simple relationship between the two rarely exists.

The new instruments record their measurements directly in computer-compatible form, in solid state memory, allowing their measurements to be transferred to a portable PC or retrieved by removing a memory unit. In the case of telemetry the transmission, reception, processing and storage of the measurements is fully automatic. There is none of the labour-intensive work involved with handwritten records or with reading values from paper charts.

Because automatic instruments do not require the regular attendance of an operator and because their data can be processed with the minimum of manual intervention, staff costs are also reduced.

Both old and new instruments are thus important at this stage in the evolution of environmental monitoring and both are covered in the chapters that follow.

General points concerning all instruments

Definitions of terms

For any instrument, the *range* of values over which it will give measurements must be specified. For example a sensor may measure river level from 0 to 5 m. (In old terms, 5 m would have been called the instrument's full-scale deflection, meaning that the pointer on an analogue meter has been deflected to the end of its scale.)

The *span* is the difference between the upper and lower limits. If the lower end is zero, the span is the same as the range, but if the range of a thermometer is, say, $-10\,°C$ to $+40\,°C$ its span is $50\,°C$.

The *accuracy* is usually expressed in the form of limits of uncertainty, for example a particular instrument might measure temperature to within $\pm0.3\,°C$ of the actual temperature. The term *uncertainty* is sometimes used in place of 'accuracy' and is 'the range of values within which the true value is estimated to lie'. Accuracy may differ across the range covered, such that in the case of a relative humidity (RH) sensor, the accuracy may be $\pm2\%$ RH from 0 to 80% RH while from 80% to 100% RH it may be $\pm3\%$.

The *error* is the difference between an instrument's reading and the true value. It is thus another way of stating accuracy. A systematic error is a permanent bias of the reading in one direction, away from the correct value, as opposed to a random error, which may be plus or minus and might, therefore, cancel out over a series of readings.

Instruments often respond to more than just the variable they are meant to measure. Temperature dependence is, for example, a common problem. In such a case a *temperature coefficient* will be quoted to allow a correction to be made. The coefficient is given as the percentage change in reading per degree Celsius change of temperature. Not only may the sensitivity change with temperature, but so also may the zero point.

The *resolution* of an instrument indicates the smallest change to which it responds; for example, a meter might respond to river level changes of as little as 1 mm. This is also sometimes known as sensitivity or discrimination. But just because an instrument can respond to changes of 1 mm, it does not necessarily mean that it measures them to that accuracy; the instrument might only have an accuracy of ±2.0 mm.

Sometimes an instrument does not respond equally across its full working range, so that its response deviates from a straight line. The extent of the deviation is expressed as its *linearity*, this usually being specified as the maximum deviation (as a percentage of the full scale reading) of any point from a best-fit straight line through the calibration points. The term is used only for instruments that have a close-to-linear response. Some sensors have characteristics that are far from approaching a straight line and may not even have a smooth-curve response. In the case of mechanical instruments, non-linearity may be compensated for in the design of the mechanical links between the sensor and the recording pen; in the case of electrical instruments, a non-linear response may either be compensated electronically, or by recourse to a look-up table or an equation when processing the records in a computer.

Hysteresis is the characteristic of a sensor whereby it responds differently to an increasing reading and to a falling one, the upward curve following a different path to that downward. This might be due to 'stiction' in the mechanical links between sensor and pen; there are equivalent processes in electronic systems. Again the deviation is expressed as the percentage difference (of full scale) between increasing and falling readings.

With the passage of time, a change in instrument performance often occurs, and settings *drift*. This may be due to the ageing of components, which results in a change in sensitivity or a change in the zero point, or both. Stability is the converse of drift and is usually expressed as the change which occurs in a sensor's sensitivity, or zero point, over a year. Because of drift, it is usually necessary to readjust, or *calibrate*, an instrument periodically.

Sensors take time to respond to a change in the variable which they are measuring. Some respond in milliseconds, others take minutes. A *time constant* is given to quantify this. This is the time for a reading to increase to $1 - 1/e$ (about 63%) of its final value, or to fall to $1/e$ (about 37%) of its initial value, following a step change in the variable (Fig. 1.1). It takes about three to four times longer than the time constant to reach 95% of the final reading. Thus a constantly changing variable with a slow-response sensor can result in considerable error. To complicate matters further, the response in one direction can be different to the reverse, owing to hysteresis, as mentioned above. These various terms are illustrated in Fig. 1.2.

The *repeatability* is the degree of agreement between an instrument's readings when presented more than once with the same input signal. Different readings may result each time through a combination of all the various sources of error.

There is also the question of the number of decimal points shown. As mentioned above, just because, for example, a temperature is quoted as being 25.03 °C it does not necessarily mean that it is accurate to hundredths of a degree. The instrument may only be good to ±0.1° or less. Extra decimal points can be introduced into data in many ways, for example during processing to fit a standard format. Decimal points can also be lost in the same way. Users of data should always beware of such possibilities.

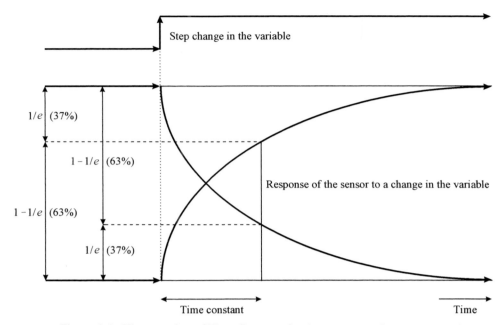

Figure 1.1. The meaning of *Time Constant* for the response of a sensor to a change in the variable.

Choice of site

Where, and how, an instrument is placed in the field (finding a representative site and positioning the instrument on the site) affects how good its readings are, often having as great an influence on overall accuracy as the instrument's own characteristics. Comments are made on this throughout the book. Changes at a site after an instrument has been installed, perhaps slowly over years such as the growth of trees or suddenly such as the erection of buildings, also need consideration.

Maintenance

The maintenance of an instrument, such as its periodic readjustment against a standard, the regular painting of a temperature screen or the levelling of a raingauge, is as important as any other aspect of accuracy and data quality. Neglect of this is widespread and results in unreliable data.

Spatial variability

Most environmental variables vary not just with time but also across space, for example across a local thunderstorm. Faced with this problem, it can be tempting to

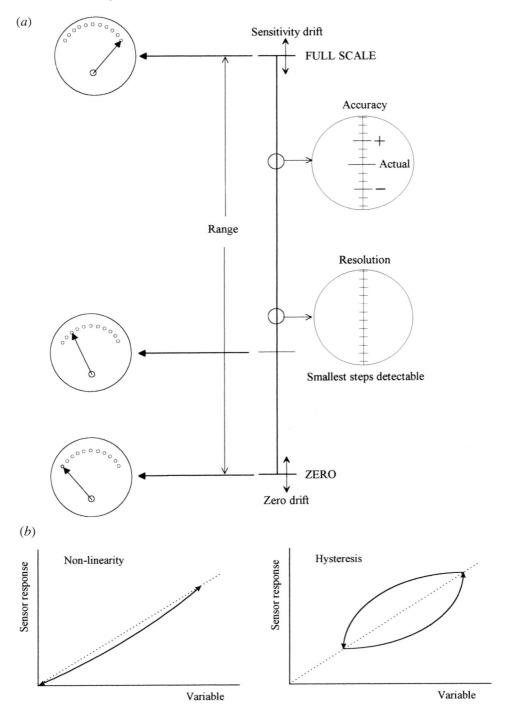

Figure 1.2. Illustration of terms.

conclude that it is pointless to improve instrument performance because instrument error is swamped by spatial variability as well as by the difficulty of selecting the perfect site. But this is a philosophy of defeat. Better, instead, to improve the instrument, choose the site more carefully, take measurements at several points and so get an estimate of variability. Spatial variability cannot be measured unless we have good instruments, for what might be assumed to be spatial variability may instead be instrument error.

Using data of unknown quality

It is difficult to know how good data are if there is no record, along with the data, of what type of sensor was used, how it was exposed or its maintenance schedule (metadata). Often there is no sure way of checking on these matters. In general, our data sets are a mix of measurements collected by a variety of instruments, installed in different ways and operated at different levels of competence, all of which may have changed with time and with the operator. Discontinuities and different levels of error are invariably present.

Plan of the book

The aim of this book is to show how measurements of the natural environment are made and to point out the pitfalls. Each variable, or group of variables, is given a chapter to itself, working through from the early instruments, most of which are still in use, to the latest developments, only a few of which may yet be in use. How the measurements from the new electronic sensors are processed and stored automatically in data loggers or are telemetered to a distant base are the topics of two later chapters. While remote sensing is a specialised topic, already well covered in other publications, a chapter on it is included for completeness, describing how measurements are made remotely, how good or bad they are and how they relate to measurements made *in situ* on the ground.

Mathematical analysis is used sparingly, and where equations are used a line-by-line derivation from first principles is not always given. The SI (Système International d'Unites, NPL 1963, BSI 1969) is used throughout, unless otherwise stated. The method of defining units is, for example, to abbreviate metres per second to m s^{-1} and cubic metres to m^3.

This is not an operator's handbook for the technician, but an outline for the users of data who would like (or need) to know how the data they use were collected and how good or bad they are likely to be. It is also intended for instrument designers who need background information, or for those about to collect field data and install instruments, or for students of meteorology, hydrology and environmental sciences, who should know how measurements are made. Useful additional information on the basic meteorological instruments can be found in three UK Meteorological Office handbooks (1982a, b and 1995) while the World Meteorological Organisation (WMO 1996) periodically issues and updates technical reports on how meteorological instruments should be installed and operated and their principles of operation. These are well worth consulting for details.

Since I started the first edition in the mid 1990s, the wealth of information available on http://www. sites has increased greatly, as has the speed of access to them via broadband. I have included a few web sites here in the text, but there are dozens more. To allow the latest information to be made available, as well as a list of useful web sites and new papers, I have set up my own web site which readers can look at. The URL (address) of the site is *http://www.ianstrangeways.org.uk.*

Prices are occasionally given for equipment or services (such as satellite telemetry charges). Readers will be able to make due allowance for inflation, but the prices are only meant as a very rough guide, and often more to show differences between options rather than absolute values. The majority of the information in this book is fairly timeless and even that liable to change is certainly 5- to 10-year proof.

The order of the chapters was difficult to make entirely satisfactory. However, the sequence is not, as one reviewer said of the first edition, random. The order chosen reflects, as far as is possible, the sequence of the natural process – solar radiation drives the transfer of heat and water vapour into the atmosphere and powers the wind, so these three topics start off the book; evaporation results in precipitation which enters the soil and groundwater, finding its way into the rivers and lakes; there is a natural flow in these processes which is reflected in the chapter order. At this point it is necessary to digress into how the measurements are logged in modern systems and then, optionally, telemetered to a distant base. These two topics would have been out of place at the start of the book before any sensors had been described, but cannot be left until the end because they become increasingly relevant as we progress to the upper atmosphere and the oceans. Returning to the environment, the measurement of more distant, less direct, phenomena including visibility, clouds, lightning and the upper atmosphere follow. Chapter 17 moves from the land to the oceans and their depths, their size and dominance making them vital in understanding both weather and climate. The difficulty of taking measurements in Polar Regions and mountains concludes the earth-base measurements, these parts of the world being only poorly monitored. Remote-sensing, particularly from satellites is a complex matter of increasing importance and only the basics can be included in one chapter, but it is important to look briefly at what can, cannot, is being, and will be, measured from space and how these integrate with earth-based measurements. The penultimate chapter considers the monitoring of atmospheric composition. The last chapter looks forward to where we go next. I cannot see a more logical sequence to the chapters than this, but even so it is necessary to refer the reader backwards and forwards from time-to-time since no order is without its inconsistencies. Many cross-references are included. I am not entirely certain if these are annoying or useful, but they can always be ignored.

I trust those with specialised experience which has taken a lifetime to acquire will excuse what some may see as my naivety of treatment of their subject. I felt them looking over my shoulder frequently as I laboured at the work.

References

British Standards Institute (1969) The use of SI units. Report PD 5686. HMSO, London.
Meteorological Office (1982a) The handbook of meteorological instruments. HMSO, London.
Met. Office (1982b) The observer's handbook. HMSO, London.
Met. Office (1995) The marine observer's handbook. HMSO, London.
National Physical Laboratory (1963) The inclusion of metric values in scientific papers. HMSO, London.
World Meteorological Organisation (1996) Guide to meteorological instruments and methods of observation. WMO No. 8 (generally known as the CIMO Guide).

2

Radiation

The day was ending in a serenity of still and exquisite brilliance. Only the gloom in the west, brooding over the upper reaches, became more sombre every minute, as if angered by the approach of the sun. And at last, in its curved and imperceptible fall, the sun sank low, and from glowing white changed to a dull red without rays and without heat, as if about to go out suddenly, stricken to death by the touch of that gloom brooding over a crowd of men.

Joseph Conrad *Heart of Darkness* (from a boat on the Thames near London).

The variable

Integrated over the whole of its radiation spectrum, the sun emits about 74 million watts of electromagnetic energy per square metre. At the mean distance of the earth from the sun, the energy received from the sun at the outer limits of the earth's atmosphere, at right angles to the solar beam, is about 1353 watts per square metre (W m^{-2}) and is known as the *solar constant*. In fact the energy received is not quite constant but varies over the year by about 3%, because the earth is in an orbit around the sun that is elliptical, not circular. The actual output of the sun itself also varies with time, the most familiar regular rhythm being the 11-year sunspot cycle, although the variations due to this are less than 0.1%. There are other, longer, cycles such as the 22-year double sunspot cycle, and the 80- to 90-year cycles (Burroughs 1994). It is useful to define some terms.

Units and terms

Radiant flux is the amount of electromagnetic energy emitted or received in unit time, usually expressed in watts (1 watt = 1 joule per second).

Measuring the Natural Environment, second edition, Ian Strangeways. Published by Cambridge University Press. © Ian Strangeways 2003.

Radiant flux density is the radiant flux per unit area expressed in watts per square metre ($W\,m^{-2}$), although other units such as $mW\,cm^{-2}$ and $cal\,cm^{-2}\,min^{-1}$ may be used ($1\,W\,m^{-2} = 0.1\,mW\,cm^{-2} = 10^{-3}\,kW\,m^{-2} = 1.433 \times 10^{-3}\,cal\,cm^{-2}\,min^{-1}$). No longer in wide use is a unit called the langley, which is equivalent to $1\,cal\,cm^{-2}$.

Irradiance is the radiant flux density received by a surface; the *emittance* is the flux density radiated.

Insolation is the irradiance integrated over a period of time, typically a day in the case of solar radiation; thus the insolation is the total energy received in the period. On cloudless days the diurnal variation of solar irradiation is roughly sinusoidal. In the UK in summer the maximum intensity of the radiation at solar noon is about $900\,W\,m^{-2}$. With a day length of 16 h, the maximum daily insolation can be calculated to be about $9\,kWh\,m^{-2}$. However, insolation is often expressed in megajoules rather than kilowatt hours, 1 kWh being equivalent to $1000 \times 60 \times 60 = 3.6\,MJ$, giving a UK daily maximum insolation of about $32\,MJ\,m^{-2}$. This is reduced by cloud cover to a summer average of 15 to $25\,MJ\,m^{-2}$, while in winter it varies from 1 to $5\,MJ\,m^{-2}$.

Spectral composition

Figure 2.1 illustrates the spectrum of the Sun's emission, both as received just outside the atmosphere and as received at the Earth's surface after modification by its passage through the atmosphere (see also Fig. 19.1). During its passage, radiation in different parts of the spectrum is absorbed, scattered and reflected. For example, some ultraviolet radiation is absorbed and scattered by ozone and by oxygen while parts of the infrared spectrum are absorbed by water vapour and carbon dioxide. Scattering and diffuse reflection occurs from particles larger than the wavelength of the radiation, such as dust, smoke and aerosols from volcanic activity, while Rayleigh scattering by air molecules and particles smaller than the wavelength gives rise to the blue sky and the blue appearance of the Earth from space (see Chapter 13 on Visibility). Clouds can reflect up to 70% of the energy from their upper surface, the lower clouds being responsible for the loss of considerable amounts of solar energy back to space.

As a consequence of scattering and reflection, the irradiation received at ground level is in the form of both direct and diffuse radiation. In cloudless conditions, the ratio of the diffuse to direct radiation depends on the angle of the Sun and on the amount of aerosols in the atmosphere, the diffuse component varying from 10% in very clear air to 25% in polluted air; typically it is about 15% (Monteith 1975). Just before sunrise, just after sunset and in cloudy conditions, the radiation is of course entirely diffuse.

To differentiate between the radiation from different sources, various categories of radiation have been defined (Meteorological Branch, Dept of Transport, Canada

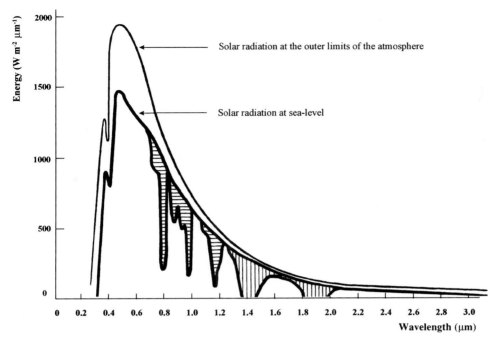

Figure 2.1. Spectra of solar radiation outside the atmosphere and at ground level showing the bands where there is absorption due to water vapour and carbon dioxide in the mid-IR (vertical shading) and due to oxygen and water vapour in the mid- and near-IR (horizontal shading), while oxygen and ozone absorb in the UV at wavelengths shorter than 0.4 μm (not shaded).

1962). The wavelength limits usually quoted are somewhat arbitrary, particularly at the longer infrared wavelengths, and should be seen only as a guide and not as a precise statement.

Solar radiation is the radiation received from the Sun reaching the measuring instrument without a change in wavelength, sometimes also referred to as short-wave radiation (Fig. 2.1). At the surface of the Earth, most of the radiation is in the spectral band from 0.3 to 3.0 μm.

Terrestrial radiation is that emitted by the ground and also by the atmosphere, in particular by water vapour and carbon dioxide – the same gases that also absorb it – reaching the instrument without change of wavelength within the (arbitrary) limits of 3 to 100 μm; it is sometimes referred to as long-wave radiation. Included in this is radiation emitted by clouds in the narrow band from 8 to 13 μm, a region in which little radiation is emitted by water vapour and carbon dioxide. These two general categories are further subdivided as follows.

Direct solar radiation refers to the incoming direct solar energy incident on a surface at right angles to the solar beam. (All other definitions refer to the radiation falling on a *horizontal* surface.)

Diffuse solar radiation is that part of the incoming solar energy incident on a horizontal surface shielded from direct radiation from the sun.

Hemispheric solar radiation (also known as *global radiation*) is the sum of both the direct and the diffuse radiation on a horizontal plane.

Reflected solar radiation is that portion of the total solar energy reflected from the ground (and atmosphere) upwards without a change of wavelength, measured on a downward-facing horizontal surface. The ratio of this to the total incoming solar radiation is the *albedo* of the surface, more correctly known as the *reflection coefficient* of the ground (which must be differentiated from the surface's *reflectivity*, the fraction of the total incoming solar radiation reflected at a specific wavelength).

Net radiation is the difference between all incoming and all outgoing radiation fluxes of all wavelengths (solar and terrestrial) on a horizontal surface.

Total incoming radiation is the sum of all incoming solar and terrestrial radiation.

Total outgoing radiation is the sum of all outgoing solar and terrestrial radiation.

Incoming long-wave radiation is that radiated downwards from the atmosphere and clouds.

Outgoing long-wave radiation is that radiated upwards from the ground and atmosphere.

Photosynthetically active radiation (PAR), covering the band from 0.4 to 0.7 μm, is the spectrum used by plants for photosynthesis.

Ultraviolet (UV) radiation extends from about 0.1 to 0.4 μm and is subdivided into three narrower bands designated A, B and C.

Daylight illumination refers to the visual quality of the radiation in terms of the spectrum to which the human eye responds, which is from 0.38 to 0.78 μm peaking at 0.55 μm.

Sunshine recorders

The Campbell–Stokes recorder

Since the maximum possible level of irradiance at noon for any one place and for any particular date is known, and since the daily variation is approximately sinusoidal, a knowledge of cloudiness will give some indication of daily insolation. A sunshine recorder obviously gives information not only on sunshine duration but also on the extent of any cloud, from which a rough estimate of daily insolation can, therefore, be derived, and which may be relevant to climatic change. For example, the 2001 IPCC reports quote an increase in cloud over the land and sea of about 2% during the last 50–100 years (Folland *et al.* 2002).

The first attempt to record sunshine duration was made in 1853 by John Campbell, who used a glass sphere to burn a trace of the sun's image focused onto the inside of a wooden bowl. This was modified in 1880 by Sir George Stokes, hence its

Figure 2.2. The sun's image focused onto the card of a Campbell–Stokes sunshine recorder burns a track as the sun moves across the sky.

present-day name, the Campbell–Stokes recorder. The modifications by Stokes, and subsequent refinements made by manufacturers, led to the present-day design (Fig. 2.2) in which the sun's image is focused onto a specially manufactured card by a carefully ground glass sphere, which burns a track along the card if the sun is not obscured by cloud (Fig. 2.3). The card is manufactured so that it does not catch fire, but simply chars, and is affected minimally by rain. This simple instrument is still in widespread use worldwide.

If there is uninterrupted sunshine, or none, the trace is easy to read. Rules have been established for interpreting the meaning of the burnt track but the estimate can only be approximate when there are intermittent, short bursts of sunshine or when the sun is hazy. There is also some uncertainty around sunrise and sunset. Further error is introduced through imprecise adjustment of the instrument, notably with regard to the exact positioning of the sphere relative to the card. There is also a tendency to over burn the card when the sun is intermittently visible (Painter 1981) giving an error of up to +100% over an hour and +10% on a year's total (Prior 2002).

For use at different latitudes and in different seasons, a selection of card holders and cards of varying shape and size is needed (Fig. 2.3), there being models covering latitudes 0° to 45° and 45° to 65° (N or S). For latitudes higher than 65°, two recorders must be operated back-to-back, one being extended in latitude by simple modification and the other being both modified and set on an inclined wedge. However, there is a tendency today to use electrical sunshine sensors rather than this complicated dual arrangement.

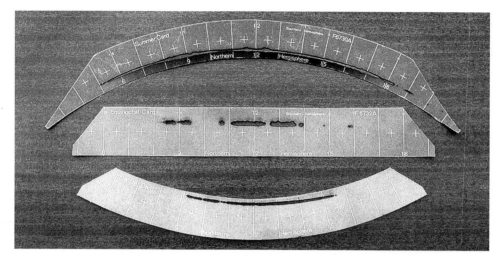

Figure 2.3. Traces burnt on summer, winter and equinoctial cards of a Campbell–Stokes recorder. Upper, 1 May 1997; centre, 4 October 1997; lower, 1 January 1998 (Institute of Hydrology site in southern England).

It is difficult to put a figure on the accuracy of this type of sensor, but it is unlikely to be better than ±5% since the interpretation of the track is to some extent subjective.

Electrical sunshine sensors

Because of the long records of sunshine made by the Campbell–Stokes design, going back to the nineteenth century, there was a need in some quarters to continue sunshine observations into the future, but on an automatic basis. To satisfy this requirement, electrical sunshine sensors were developed.

In the Haenni type, a narrow shutter rotates, several times per second, around a circular array of light detectors, shutting off the direct solar beam momentarily in turn to each detector that has a view of the sun. If the sun is not visible owing to cloud, the passage of the shutter causes little change to the light level detected since the shutter obscures only a small part of the diffuse radiation. If the sun is visible, however, a large change occurs momentarily as the sun becomes shadowed by the shutter. In another, similar, design a slit is used in place of a shutter. Thresholds have to be set to enable the sensor to detect the transition from 'no sun' through 'hazy sun' to 'full sun', and some uncertainty will inevitably be introduced. However, the instrument responds instantaneously to the presence or absence of the direct solar beam, which is an improvement on the Campbell–Stokes instrument, but it requires power to drive the shutter motor and this makes it less suited to remote sites where power is limited. To prevent ice formation, the glass cover can be heated, but this also consumes power. (If the Campbell–Stokes instrument is frost- or snow-covered it will not operate, but there is an operator to clear it; however, the operator will not be there all day, every day.)

The Shiko type of sensor uses two photoresistor cells exposed to the sun in a diffusing glass tube, one being shielded from the narrow band of sky across which the sun passes. In the absence of direct sun the two sensors are of very similar resistance, but they are considerably different when the sun is visible. This imbalance is detected and gives a yes/no signal that can be recorded. It requires no power and responds rapidly (Amano *et al.* 1972).

Meant mostly for amateurs, the Jordan recorder consists of two semicylinders with slits on their flat side, one facing east and the other west, their angle adjustable for latitude. The presence or absence of direct sunlight is recorded on light-sensitive paper of the blueprint type. The records are said to be open to more uncertainty than the Campbell–Stokes recorder because of the difficulty of ensuring that the sensitivity of the paper is consistent, but one would have thought that this could be overcome without much difficulty.

Other types exist, such as the 'Marvin' design, but they are not true sunshine sensors because they cannot differentiate well enough between diffuse and direct radiation. In addition there is the *pseudo-sunshine sensor* based on a pyranometer (see below).

Compatibility between sunshine recorders

Any change from one type of sunshine recorder to another inevitably results in a discontinuity of readings (as it does with any instrument). Differences of up to 20% can occur if a change is made from the Campbell–Stokes to any of the electrical sensors (World Meteorological Organisation 1994, Chapter 1 references). The Campbell–Stokes instrument is the classical and most used sunshine recorder and was selected by the WMO in 1962 as the reference standard against which other types should be compared. By reduction of sunshine totals to this standard the WMO believes that systematic differences can be reduced to $\pm 5\%$.

Radiometers

'Radiometer' is the generic term for all instruments measuring radiation. In the case of solar and terrestrial radiation, they are subdivided into several classes depending on the part of the spectrum they sense, as follows.

Pyranometers measure all incoming solar radiation, direct and diffuse, on a horizontal surface within a field of view of 180°. They are also known as *solarimeters*.

Effective pyranometers measure the same thing as pyranometers except that they include terrestrial long-wave radiation as well as solar. For this reason they are sometimes called *total radiometers*.

Pyrheliometers measure just the direct solar radiation, and do so at right angles to the solar beam, the instrument tracking the Sun's movement. The sensing element is shielded so that it sees only the sun and a small annulus of sky around it.

Net pyrradiometers sense the difference between the total incoming and total outgoing radiation, of both solar and terrestrial origin, that is the net value of the upward and downward fluxes of both long- and short-wave radiation. They are also known as net radiometers and radiation balance meters.

Pyrgeometers respond to long-wave terrestrial radiation only. They can be upward- or downward-looking, depending on whether incoming or outgoing radiation is to be measured. They are also known as infrared radiometers.

Photometers measure that part of the spectrum to which the human eye is sensitive, 0.38 to 0.78 μm.

Photosynthetically active radiation (PAR) sensors measure the spectral band used by plants to photosynthesise, 0.4 to 0.7 μm.

Ultraviolet sensors measure from 0.2 to 0.4 μm. They do not have a special name.

The bimetallic actinograph

An early step towards measuring solar energy, rather than sunshine duration, was the development of the *bimetallic actinograph*, also known as a *pyranograph* because it is a pyranometer which records on a chart. (The term 'actinograph' comes from the word 'actinic'; the property of *actinic* rays, especially the shorter wavelengths of green to ultraviolet, is to produce photochemical effects, although their measurement need not be based on chemical changes.) The instrument was designed by Robitzsch in 1926.

In an actinograph, a blackened bimetallic strip exposed horizontally to the sun beneath a protecting dome bends in proportion to the solar energy incident on it. (A bimetallic strip is made by welding together two bars of dissimilar metals and rolling them into a thin strip. When the temperature changes, the two strips expand by different amounts and the composite strip bends; see also the *bimetallic thermograph*, Chapter 3.) One end of the strip is fixed and the other can move. As it bends the motion is transferred to a pen via levers, recording a trace on a paper chart fixed to a drum that is rotated, usually once a day, by a spring- or battery-operated clock. To prevent air-temperature changes from affecting the record, the fixed end of the strip is supported by a second bimetallic strip, shielded from the sun within the case, which responds only to changes in ambient temperature. In addition a third, shorter, bimetallic strip, at right angles, ensures that the sensing strip remains horizontal so that its angle to the sun does not change. In an alternative design, temperature compensation is achieved by two white bimetallic strips exposed on either side of the black strip, which reflect most of the solar radiation and so bend mainly in proportion to air temperature changes (Latimer 1962).

This type of instrument takes up to five minutes to respond to changes in radiation, in part due to pen 'stiction' but also because of the time taken by the strip to heat or cool. It is not, therefore, a precise instrument, nor can it be used to measure instantaneous levels of solar radiation. However, by measuring the area beneath the curve recorded on the chart throughout the day (with a *planimeter*) a rough estimate of the daily total of energy received can be obtained – to about ±10%. Actinographs are now little used, because of the availability of much superior, and as cheap, electrical methods.

Pyranometers

Thermal solarimeters

There are two kinds of solarimeter – thermal and photoelectric. (The term *solarimeter* is viewed by some as a trade name, not a scientific term, but as it is the more expressive I will use it.) The thermal types, like their mechanical counterparts, sense radiation intensity by measuring the heating effect of the radiation, but they do it electrically. All are similar in principle, having a black surface exposed within a protecting dome, the black surface rising in temperature by an amount proportional to the direct and diffuse radiation it absorbs. By correct choice of black material, absorption can be made equal across the full solar spectrum. The black surface is exposed beneath a glass dome, which, being transparent to all but the far ultraviolet and opaque to any incoming infrared radiation of terrestrial origin, allows just the solar spectral band to reach the sensing element.

The measurement of the rise in temperature of the black surface is relative to the ambient air temperature and this is achieved by measuring the temperature difference either between the black surface and the sensor case, shielded from the radiation (Fig. 2.4), or between the black surface and a similar, but white, surface alongside it (Fig. 2.5). The black surface has traditionally been a coating of lacquer containing carbon black, which produces a matt black finish with wide spectral absorption. Magnesium oxide has been used to produce the white surface because it reflects the solar spectrum very effectively while absorbing wavelengths longer than 3 μm. The latter is important because, although not gaining access from the outside, long-wave infrared radiation does originate within the glass dome itself (since all bodies above absolute zero radiate energy in a frequency band dependent

Figure 2.4. The black (hot) sensing surface beneath the double glass domes of a thermal solarimeter of the shielded type; a white conical cover keeps the case at ambient air temperature (cold junction).

Figure 2.5. The alternate black (hot) and white (cold) sensing surfaces of a 'star' type of solarimeter beneath its single glass dome.

on temperature). By using magnesium oxide as the cold-junction coating, the long-wave radiation originating from within the glass dome is absorbed by both the black and white surfaces alike, thereby (largely) cancelling out. The effect of long-wave radiation that emanates from within the glass dome is minimised in designs which have only a black surface (Fig. 2.4) by using two concentric domes; the inner one reduces convective and radiative heat exchange between the sensor and the outer dome. In some designs the outer dome can be interchanged, allowing materials with different spectral-transmission characteristics to be substituted for the detection of specific bands of radiation. Some designs also ventilate the outer

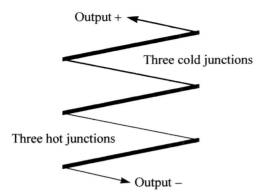

Output +

Three cold junctions

Three hot junctions

Output −

Figure 2.6. A thermopile is a series of junctions between two dissimilar conductors. If one set is maintained at one temperature and the other at a different
temperature, a (very small) voltage is generated that is proportional to the difference in temperature. This is useful where differences in temperature (rather than
absolute temperature) are to be measured, radiation sensors being a good example.

dome by circulating warm air over it to prevent the formation of frost and dew or
the accumulation of snow. The resultant extra heating and consequent long-wave
radiation is taken care of through the use of a double dome or the use of magnesium
oxide as the white surface.

In both solarimeter designs it is the temperature difference which is sensed and
this is most conveniently measured with a thermopile (Fig. 2.6). In both designs, the
hot junctions of the thermopile are in contact with the black surface, while the cold
junctions either make thermal contact with the shielded body of the instrument or
with the white surface exposed alongside the black (Latimer 1962). The thermopiles
are typically manganin–constantan, antimony–bismuth or copper–constantan, the
number of junctions ranging from 10 to 50. Today, thick-film microelectronic techniques can simplify the production of thermopiles. The resultant outputs from such
sensors are in the low-millivolt range and need amplification before logging, recording or displaying. Some recent designs use a circular array of diodes in a bridge
circuit to measure the temperature difference in place of the thermopiles and these
can produce outputs 20 or more times greater.

As the sun moves across the sky, the angle of incidence of the direct beam
varies. The sensor response should be proportional to the cosine of the angle, being
zero when the sun is on the horizon. However, owing to variations in the black
surface, particularly at low sun angles, and also to refraction through the glass
domes, cosine errors of about ±2% can occur. Furthermore, as the azimuth of
the sun (the rotational angle measured eastwards from the north–south direction)
varies, the sensitivity of the sensor may vary, although the error is reduced if the
sensor is rotationally symmetrical. A typical time constant for this type of sensor
is 5–20 s. If all errors are considered together, an overall accuracy of about ±5%

is achievable, but only if care is taken in calibrating, installing and operating the instrument.

Two modifications can be made to solarimeters to extend the range of measurement. A *shadow ring* can be fitted to cast a shadow of the sun on the sensor, thereby giving a measure of just the diffuse radiation. To measure the albedo of the ground, two sensors are used, one facing up and the other down. If their outputs are recorded separately, a subsequent calculation will give the ratio of the two readings and thus the albedo (or the reflection coefficient). If they are connected in opposite polarity and the single composite output is recorded, a measure of the short-wave radiation balance is obtained.

Pyranometers can also be used as pseudo-sunshine sensors 'bright sunshine' being registered when the pseudo-direct irradiance exceeds a value agreed internationally at 120 W m^{-2}, to differentiate between sunshine as opposed to daylight. However, there is some uncertainty as to the correlation between the readings produced by this method and those from a Campbell–Stokes recorder.

By suitable data processing, day length can also be sensed, again by setting a threshold against which the level of irradiance can be compared. These additional functions can be achieved at the same time as using the sensor as a pyranometer.

Photodiode sensors

The light-sensitive diode, photodiode or photovoltaic cell is a useful new sensor capable of measuring solar radiation (Fig. 2.7). It functions quite differently from

Figure 2.7. The light-sensitive diode solarimeter. Beneath the flat, white diffusing cover is a light-sensitive diode, offering a cheaper alternative to the thermal type of solarimeter.

thermal sensors, responding to the individual incoming photons directly and producing free charge carriers – electrons and holes (a 'hole' is the absence of an electron,
effectively a positive charge). It can be in the form of a p–n junction depletion-
layer diode or can have a third layer sandwiched between the p and n layers. There
are also Schottky photodiodes, which use a metal–semiconductor junction. The
open-circuit voltage of photodiode sensors rises rapidly with increasing radiation
levels, soon reaching saturation at around 0.6 volts, but their short-circuit current
is proportional to the irradiation level over a wide range of intensities and it is this
characteristic that is used to measure solar radiation. Because this type of sensor
acts at the atomic level, producing an electrical signal directly, its time constant is
measured in microseconds.

Such diodes have the advantage of being cheaper than sensors of the thermal type.
However, they have the drawback that their spectral response is not flat, nor does it
encompass the full solar spectrum, ranging only from 0.3 to 1.1 μm (Fig. 2.8). Some
photodiode solar sensors are blue-enhanced by filters to compensate for the reduced
sensitivity at these shorter wavelengths (Chappell 1976). The error associated with
the limited spectral range is from ±3% to ±5%, the actual value depending on the
spectral composition of the radiation, which varies with season, time of day and

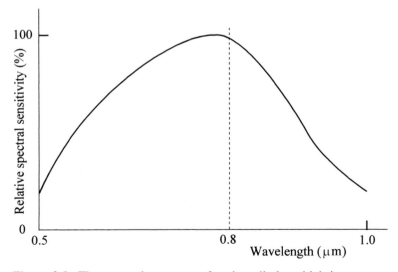

Figure 2.8. The spectral response of a photodiode, which is narrower than that
of thermal radiation sensors, in which the black surfaces absorb virtually all the
incoming radiation irrespective of wavelength. In the case of a basic silicon photodiode, as shown here, the sensitivity ranges from the violet into the infrared,
peaking at around 0.8 μm in the near-infrared. Its performance as a solarimeter
will thus depend partly on the spectral composition of the radiation at any given
time.

cloud cover. Photodiodes have good long-term stability but a temperature coefficient of around 0.1% $°C^{-1}$, for which compensation must be made. Their overall accuracy is somewhat less than that of a well-serviced thermopile pyranometer, but their much lower cost often outweighs this consideration.

PAR sensors

The spectral range of light-sensitive diodes is sufficient to cover all that part of the spectrum associated with photosynthesis, that is from 0.4 to 0.7 μm. With suitable filtering to mimic the response of plants, diodes can be used very effectively as sensors of *photosynthetically active radiation* (PAR). Diodes are also well suited to this task since they measure the incoming quanta of energy directly and this is closely related to the rate of sugar production in plants. The units of measurement used are not watts per unit area but moles per unit area; a mole (abbreviated to '*mol*') is a measure of the amount of a substance, in this case of photons, and in this application it is expressed in the form μmol m^{-2} s^{-1}. (See also pH, Chapter 10.)

Photometers

The spectral range of photodiodes also covers that of the human eye (0.38 to 0.78 μm) and with suitable filters can be made to simulate the eye's response so as to measure natural (daylight) illumination. The unit of illumination is the lux = 1 lumen m^{-2}. 1 lumen = 0.001496 watts, therefore 1 lux = 0.001496 W m^{-2} (or 668.5 lux = 1 W m^{-2}). 1 foot-candle = 10.76 lux.

Ultraviolet sensors

Heterojunction photodiodes, which are based on two different semiconductors having different band gaps, can measure very specific radiation bands. If the materials are, for example, gallium arsenide and germanium the ultraviolet (UV) spectrum can be measured. (The simpler diodes described earlier do not respond to these shorter wavelengths.) Such UV diodes are made covering three bands. The C-band peaks at 0.29 μm; this is the band that causes most erythema (sunburn) and other biomedical processes such as vitamin D production. This is hard UV, and is much influenced by stratospheric ozone. The B-band peaks at 0.31 μm, but is still noticeably affected by the ozone layer. The A-band peaks at 0.34 μm and is little affected by ozone, behaving more like blue light but with increased scattering. The use of the three sensors together can give information about the effects of cloud and aerosols on the transmission and scattering of UV radiation across the full UV spectrum.

UV sensors are similar in design and appearance to sensors using normal photo-diodes; the protective covers, however, are made of quartz, glass not being transparent to the full UV spectrum. Amplifiers are sometimes included within the sensor to boost the signal, which is low since radiation levels are low (being confined to a narrow band). The units used for UV measurement are the same as for other solar radiation bands, that is watts per square metre, a typical maximum sensor coverage being around 50 W m^{-2}. Accuracy ranges from $\pm2\%$ to $\pm7\%$.

Pyrheliometers

Pyrheliometers are specialised instruments for the measurement of direct solar radiation at right angles to the incoming beam. There are several designs, but all are similar in that a thermal sensor is exposed at the bottom of a tube which is kept pointed at the sun. Using circular apertures, only radiation from the sun, and from a small annulus around it, is allowed to fall on the sensing surface, the angle of view to achieve this being 5.7°.

As with thermal solarimeters, it is the rise in temperature produced by the solar radiation falling on a black surface that is used to measure intensity, the black surface being at the bottom of the tube. It is mostly the means of measuring the rise of temperature that differentiates the various designs of sensor. One design is very similar to a normal solarimeter, the temperature of a black disc relative to the body of the tube being measured with a thermopile, the tube being kept at ambient air temperature (by virtue of being white or of polished chromium). It differs from a solarimeter simply in that the receiving disc sees only the sun's disc and is at normal incidence to it.

In another design, the temperatures of two blackened strips of platinum are measured. One strip is exposed to the sun; the other is in shade but electrically heated, the energy required to bring it to the same temperature as the exposed strip being known by means of measurement of the current through it and the voltage across it. This energy is the same as the energy being received from the sun.

For meteorological synoptic (forecasting) stations, readings are made manually at true solar noon and at times when the sun's angle from the zenith is such that the length of the path of the radiation through the atmosphere is 2, 3, 4 and 5 times the thickness of the atmosphere vertically above mean sea level. At these times, the observer must also make judgements regarding the sky's whiteness or blueness, since a variety of aerosols is often present, attenuating and scattering the beam. There are rules for judging the state of haziness of the sky, but the process is partly subjective. Pyrheliometers can also be operated continuously, the sensor being driven by a motor so that it follows the passage of the sun across the sky. Measurements are recorded on a chart or logger, although information on the haziness of the sky will not be available from automatic sites. Most instruments can have a filter placed at the end of the tube in the path of the radiation, allowing specific bands to be measured. Normally the tube is capped with plain glass.

The accuracy of pyrheliometers is similar to that of pyranometers since they are alike in principle. However, as they are always operated at normal incidence to the sun's beam, there is no cosine or azimuth error.

Net pyrradiometers

The measurement of net radiation is particularly important because it gives information on the amount of energy available to heat the atmosphere and the ground, to evaporate water and to power photosynthesis and transpiration.

A *net pyrradiometer* measures short- and long-wave radiation from about 0.3 to 70 μm, the heating of a black surface again being the sensing method. One black surface faces horizontally down, another up; a thermopile measures the difference in their temperatures and thus the energy exchange (Fig. 2.9).

The protecting domes cannot be of glass as this is opaque to the longer-wave radiation. Instead polyethylene, which transmits the full solar and terrestrial spectrum, is used. Such domes are not as strong as glass and so are more vulnerable to damage (for example by birds) and they also age, owing to sunlight and weathering.

Figure 2.9. A net radiometer. Its upper black surface, protected by a dome of polyethylene (which is transparent to both solar and terrestrial radiation), senses the total incoming radiation. An identical, downward-looking, surface senses the total outgoing radiation, the sensor output being the difference between the two, the net energy exchange.

Furthermore, they are less easy to keep clean and there is the problem that water gets into the cavity, both at the seal that holds the domes to the case and also because the material is slightly pervious to water vapour. To protect against this, an interchangeable silica-gel dessicating tube is built into the case; through this the cavity breathes. In an alternative design, thinner polyethylene domes are used and are kept inflated by purging them with a slow flow of dry nitrogen gas, but this is inconvenient owing to the need for gas cylinders and has been largely dropped in favour of smaller, more rigid domes. A heating ring can be fixed around the side of the domes, invisible to the sensing surfaces, to keep them dew- and frost-free. However, one of the difficulties with measuring net radiation is that the domes and the body of the instrument emit infrared radiation themselves, just as in the case of the glass domes of solarimeters, and this can introduce errors.

In an attempt to overcome this problem of radiation from the domes, there have been designs which dispense with them completely, the black surfaces being exposed directly to the radiation (Latimer 1962, Dept of Transport, Canada 1962). But this leaves the surfaces open to the weather and so, to offset the effect of varying cooling by wind, they are force-ventilated by a motor-driven fan. However, not only does this require electrical power for the motor, it also requires corrections to be made. Also during and after rain the sensors are inoperative, since the upper surface becomes wet and cooled. For these reasons, this type of sensor is not much used today, although recently a net sensor without domes and without ventilation became available commercially; it comprises two slightly conical black surfaces. It is not intended for precise work since rain and wind affect its response, tests apparently indicating an error of 0.8% in the radiometer reading for each metre per second increase in wind speed. Even sensors with domes are to some, but a lesser, extent affected by water on the domes, since this affects the passage of radiation and itself radiates in the infrared. There is also some slight dependence on the wind speed over the domes.

Effective pyranometers

Also known as total radiometers, these instruments constitute half of a net radiometer and measure the combined total input of energy across the full radiation spectrum from (nominally) 0.3 to 100 μm. They can be either upward-facing, measuring the total solar input (direct and diffuse) together with the terrestrial radiation from the atmosphere and clouds, or downward-facing, measuring the reflected solar radiation together with the long-wave infrared emissions from the ground. The cold junction of the thermopile is the instrument case, kept at ambient air temperature by a shield. Sensors of this type are not widely used.

Pyrgeometers

Pyrgeometers are sensors that measure just long-wave radiation of terrestrial origin in the spectral band from 3 to 100 μm, and are simply pyranometers modified by changing the glass covers to domes that are opaque to the main part of the solar spectrum but transparent to the longer wavelengths arising from absorption and re-emission. Otherwise they are similar, in that a black surface is warmed through exposure to the radiation, its temperature relative to the case being measured by a thermopile.

By using two pyrgeometers, one facing down and the other up (in a way similar to that in which two pyranometers are used for the measurement of albedo), in combination with an albedo sensor, the same information is obtained as from a net radiometer, since the one pair measures net long-wave energy exchange and the other net short-wave. This combination provides a comprehensive measurement of the radiation environment – incoming and reflected solar radiation, albedo, incoming and outgoing terrestrial radiation and total net exchange of energy.

Exposure of radiation sensors

The Campbell–Stokes sunshine recorder requires a clear view of the sky, with nothing that will cast a shadow on the sensor wherever the sun may be at any time of the day or year, although below about 3° the sun has insufficient energy to burn a trace and so visibility right down to the horizon is not essential. Exposure on the top of a building may actually help in this case, although for most other sensors exposure on a roof is definitely not to be recommended, as will be shown. In the case of electrical radiation sensors, the exposure rules are the same, although in addition there should be no bright reflecting object nearby.

The same criteria must be met by net radiometers; furthermore, because they also measure the outgoing radiation the ground beneath them needs careful attention. The ground should be as representative of the area as possible, whether it be vegetation, soil, rock, snow or water, since its reflective and infrared-radiative properties affect the measurements.

Net sensors are usually mounted at a height of from one to two metres. Figure 2.10 illustrates the cumulative contribution of radiation received from the ground for increasing angles, from which it will be seen that 50% of the total input is received from a cone of semi-angle 45°. In the case of a sensor at 1 m height, this is equivalent to a circle of ground 1 m in radius (3.142 m^2). Raised to 2 m, the area contributing 50% of the input increases by a factor of four. From the graph it can also be seen that 90% of the input originates within a cone of semi-angle about 65°, that is from a circle of 4.4 m radius (60 m^2) if the sensor is deployed at 2 m height. Nothing

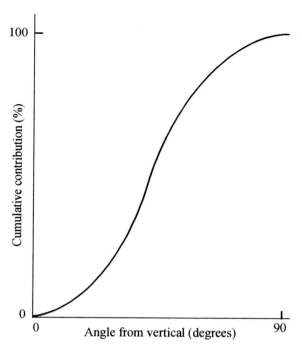

Figure 2.10. The proportion of irradiance received by the downward-looking sensing surface of a net radiometer (or of an albedo sensor) varies with the angle, the proportion of the contribution peaking at 45 degrees; the graph here shows the percentage cumulative contribution for increasing angles from the vertical.

is to be lost, therefore, by exposing the net sensor at two or more metres above the ground and, indeed, as raising the sensor increases the ground encompassed in its view, the measurement becomes more representative the higher the sensor is installed. The ground should, however, be representative up to the maximum distances included, and the mast which supports the sensor should be as unobtrusive as possible.

References

Amano, I., Koseki, K. & Tamiya, K. (1972) Photoelectric Shiko-type sunshine recorder. *Japan Meteorological Agency, Weather Serv. Bull.*, **39**, 77–9.

Burroughs, W. J. (1994) Weather cycles. *Weather*, **49**, 202–8.

Chappell, A., ed. (1976) Optoelectronics: theory and practice. Texas Instruments Ltd. ISBN 0 904 047 19 9.

Folland, C. K., Karl, T. R. and Salinger, M. J. (2002) Observed climate variability and change. *Weather*, **57**, 269–9.

Latimer, J.R. (1962) Laboratory and field studies of the properties of radiation instruments. Meteorological Branch, Dept of Transport, Canada, Circular 3672, TEC 414.

Meteorological Branch, Dept of Transport, Canada (1962) Manual of radiation instruments and observations. Circular 3812 INS 117, Manual 84.

Monteith, J. L. (1975) *Principles of Environmental Physics*. Edward Arnold, London.

Painter, H. E. (1981) The performance of a Campbell–Stokes sunshine recorder compared with a simultaneous record of the normal incidence of irradiation. *Met. Mag.*, **110**, 102–9.

Prior, J. (2002) Sunshine recording. *R Meteorol Soc. Special. Group Observ. Syst. Newsl.* **18**, 15.

3

Temperature

This strange severity of the weather made me very desirous to know what degree
of cold there might be in such an exalted and near situation as Newton. We had,
therefore, on the morning of the 10th (December 1784), written to Mr —, and
entreated him to hang out his thermometer, made by Adams, and to pay some
attention to it morning and evening, expecting wonderful phenomena, in so
elevated a region, at two hundred feet or more above my house. But, behold! On
the 10th, at eleven at night, it was down only to 17 °(F) and the next morning
at 22°, when mine was at 10°! We were so disturbed at this unexpected reverse of
comparative local cold, that we sent one of my glasses up, thinking that of Mr —
must, somehow, be wrongly constructed. But, when the instruments came to be
confronted, they were exactly together; so that for one night at least, the cold at
Newton was 18° (−7.8 °C) less than at Selborne, and, through the whole frost,
10° or 12°.

Gilbert White. *The Natural History of Selborne* (Severe frosts).

The variable

Only about 17% of solar radiation is absorbed directly by the atmosphere as it
passes through it (Lockwood 1974). In fact the atmosphere is heated primarily as
follows: solar radiation heats the ground and the ground's heat is transferred to
the air, firstly by molecular diffusion across the *laminar boundary layer* (a layer
only a millimetre or so thick, which clings to most surfaces); beyond this, in the
turbulent boundary layer, transfer is by turbulence, which is much more effective at
transferring heat than is diffusion. Heat is also, thereafter, transferred by convection,
bubbles of warmer air rising into the cooler air above. This transfer of warmed air
away from the surface is the *sensible heat flux*. While it is difficult to measure the
rate of energy transfer (see The eddy correlation method; Chapter 7), the resultant
changes in air temperature are important and more easily measured.

Measuring the Natural Environment, second edition, Ian Strangeways. Published by
Cambridge University Press. © Ian Strangeways 2003.

A proportion of the heat from the warmed ground is also transferred downwards as the *soil heat flux*, the rate of transfer being influenced by the amount of water in the soil and also by the pore and particle sizes and the presence of vegetative material. Many biological processes are influenced by soil temperature, from the activity of micro-organisms to the germination of seeds and plant growth, and observations of soil temperature are made at depths down to a metre or more.

The importance of the oceans in weather and climatological processes is now firmly established, and in this the sea surface temperature is of particular significance. While the measurement of sea temperature is made with identical thermometers to those used for the air, how the observations are made is dealt with in Chapter 17 on the oceans.

Scales and reference points

The first measurement of temperature dates from the time of Galileo (1564–1642), when primitive thermometers, open to the air at the top and using water and other liquids such as spirit of wine and linseed oil, were made. It was not until 1720 that Fahrenheit used mercury, although the upper end of the tube still remained open. Only later was the end sealed and the air removed, giving the modern form of thermometer.

Various scales have been adopted, initially only one point being fixed, the rest of the scale being equal fractions of the volume. Many different fixed points were used. Fahrenheit later introduced the idea of two fixed points, his original points being the lowest point observed in Danzig in 1709 and body temperature, which for reasons unknown he set at 96°. This scale gave the melting point of ice as 32° and the boiling point of water as 212° (at standard pressure). These two points were later used as the fixed points, when it had been established that they were constant and easily reproduced. In 1730, Reaumur noted that a volume of 1000 units of mercury at ice point became 1080 at the boiling point of water and so marked his scale from 0 to 80.

In 1736 the Swedish astronomer Anders Celsius adopted the convention of 0° for the boiling point of water and of 100° for ice point, thereby creating the centigrade scale. This convention was later inverted, giving the present-day zero for ice point. This reversal is attributed to different people, one being the Swedish botanist Linnaeus, another the Frenchman Pierre Christin. Triggered by an article of mine (Strangeways 1999), there was an exchange of several letters in *Weather* during the year 2000 on this topic (Vol 55, pages 63, 145, 342 and 392) which seems to have put an end to the debate. The centigrade scale is now called the Celsius scale in honour of its inventor. Since 100 degrees Celsius = $212 - 32 = 180$ degrees Fahrenheit, $1\,°F = 5/9\,°C$.

In 1848 the Scottish physicist William Thompson, later Baron Kelvin of Largs, proposed the *absolute* or *thermodynamic* scale of temperature, in which absolute zero corresponds to the temperature at which all molecular motion stops. On the Kelvin scale, one degree Kelvin (known as a kelvin, K) is equal to one degree Celsius and absolute zero (0 K) corresponds to −273.15 on the Celsius scale. The melting point of ice is thus +273.15 K and the boiling point of water is +373.15 K.

For precise calibration purposes, the triple point of water (see below) is used as one of the fixed points, the boiling point of water (equilibrium between the liquid and vapour phases of water) at a standard pressure of 1013.25 hPa being another. The triple point of water, 0.01 °C (or 273.16 K) is produced for calibration purposes in an arrangement of two concentric glass tubes. The outer is sealed and contains liquid water, ice and water vapour in equilibrium. The inner tube, into which the thermometer under test is placed, is open to the atmosphere and contains water in equilibrium with the surrounding outer tube at 0.01 °C.

The scale of temperature used today is the *International Temperature Scale* (ITS-90) agreed in 1990. It is based on a number of equilibrium states, including the triple point of water and the boiling point of water (at standard atmospheric pressure) and also the boiling point of oxygen (−182.962 °C or 90.188 K) at standard atmospheric pressure. There are also other secondary reference points, one of which is the familiar ice point of water (0 °C) but also the triple point of argon (83.805 K) and mercury (234.315 K) and the freezing point of gallium (302.914 K) and of indium (429.748 K). Although called the ITS, it is, in practice, the familiar Celsius scale.

Mechanical thermometers

Liquid-in-glass

Present temperature

Any property that changes with temperature can be used as the basis for a thermometer, the most common in use still being, as in Galileo's time, the thermal expansion of a liquid in a glass tube. As later introduced by Fahrenheit, the liquid is still usually mercury, but below its freezing point of −38.8 °C ethyl alcohol is used instead, which freezes at −115 °C. Although the coefficient of expansion of alcohol is much greater than that of mercury, it is not so effective as a thermometric liquid because its composition changes slowly with time (being organic). It also adheres to the glass, particularly if the temperature changes quickly, and the liquid column tends to break up. Alcohol also responds more slowly than mercury because its thermal conductivity is lower.

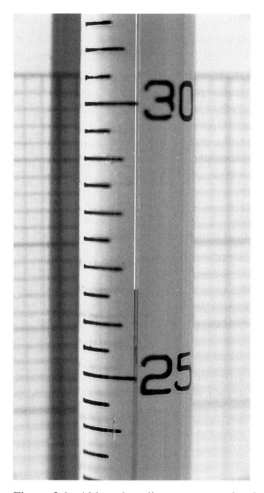

Figure 3.1. Although ordinary mercury-in-glass thermometers are graduated in 0.5° steps, tenths of a degree can be estimated; in the photograph the reading is 26.6°.

Liquid-in-glass thermometers make use of the fact that the liquid expands with temperature more than its glass container and so is forced up a fine-bore stem from a bulb containing the bulk of the mercury. For most meteorological and other environmental applications, the stem of the thermometer is graduated in steps of 0.5 °C or sometimes 0.2 °C, from which the temperature can, with care, be estimated to 0.1 °C (Fig. 3.1). For periodic checking, thermometers are compared with a precise standard, graduated in steps of 0.1 °C.

Meteorological thermometers are manufactured to within certain tolerances. For mercury, from −40 to 0 °C the maximum error is typically between −0.3 °C and

+0.2 °C while from 0 °C upwards it is between −0.2 and +0.05 °C. For spirit thermometers, below −40 °C the maximum error is ±0.06 °C, from −40 to 0 °C and from +25 °C upwards it is ±0.25 °C, while from 0 to +25 °C it is ±0.1 °C. Readings may be in error anywhere within these limits. The figures are typical UK values and may vary slightly from country to country and from manufacturer to manufacturer.

The two main sources of error in mercury-in-glass thermometers are, firstly, changes in the glass of the bulb, which contracts slowly over the years introducing a rise in the zero (and so also in the readings) of about 0.01 °C per year initially, although it slows up later. Secondly, if the liquid column is not read at eye level, a parallax error is introduced which can be up to ±0.2 °C. Small breaks in the mercury column can also form, resulting in a higher reading.

A mercury-in-glass thermometer has a *time constant* (the time to reach 63% of the final value after a step change in the temperature – see Chapter 1) of about one minute for a bulb of 10 mm diameter in an airstream of 5 m s^{-1}. In practice, temperatures tend to be cyclic and uneven and it is difficult to predict theoretically how a thermometer will behave under such conditions, but any temperature fluctuations faster than the time constant will not be registered. When taking a reading of the present temperature it is desirable that the thermometer does not respond too quickly, as it would then be unduly affected by short-term variations. The average over a period of about 5 – 10 min is of the right order, and for this a time constant of no less than about 1 minute is appropriate.

Maximum thermometers

Thermometers able to hold the maximum reading since last read are installed in most temperature screens worldwide. They are similar to the ordinary mercury-in-glass thermometer except that there is a constriction in the mercury column between the bulb and the lowest graduation on the stem (Fig. 3.2). As the temperature falls, the mercury column breaks at the constriction and is not drawn back into the bulb, thereby holding its position. If the temperature rises again and to a higher value, the column is reformed and the reading rises to the new high. It is then held at this new point when the temperature again falls. To prevent gravity from moving the detached mercury column, the thermometer is operated on its side. It is reset by a sharp shake downwards, as with a clinical thermometer. Tolerances of manufacture, errors and resolution are the same as for the ordinary type.

Minimum thermometers

Alongside a maximum thermometer there is usually one for recording the minimum temperature. This differs from the ordinary and maximum types in that it uses clear spirit as the liquid. Into the liquid is introduced a slide indicator, which is free

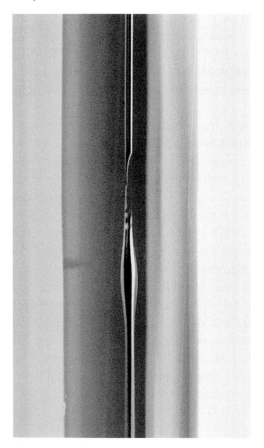

Figure 3.2. The constriction in the bore of a maximum thermometer breaks the mercury column if the temperature drops, preventing the column falling back below the highest reading. It must then be manually reset by shaking it.

to move along the bore but which stops when it gets to the liquid's meniscus. The thermometer is set by inverting it so that the indicator falls and rests on the meniscus; it is then laid on its side. If the temperature rises, the slide stays where it is, the liquid going round it (Fig. 3.3). If it falls, the meniscus pulls the slide back towards the bulb, the slide being unable to break through the meniscus owing to surface tension. The minimum temperature is where the top of the slider rests. This thermometer suffers from the errors mentioned earlier for ordinary spirit-in-glass thermometers.

Mean temperatures

Maximum and minimum readings can be used to derive an approximate daily mean temperature by taking $\frac{1}{2}$(max. + min.). At stations where spot readings are

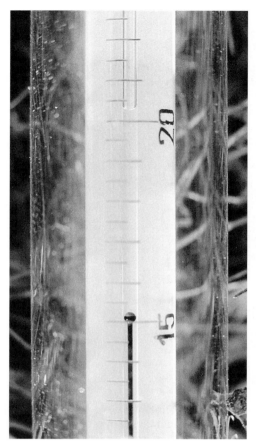

Figure 3.3. A minimum temperature of 15.2 °C indicated by the sliding index of a minimum thermometer, the current temperature of 20.3 °C being shown by the spirit meniscus. Although this thermometer is measuring minimum ground temperature, similar instruments are operated in Stevenson screens (Fig. 4.1).

also taken several times a day, such as every three hours for weather forecasting, a better mean can be derived by using $\frac{1}{8}(03 + 06 + 09 + \cdots + 24)$, where the numbers inside the parentheses refer to the temperature at the time of the spot reading. There are many variations on this, such as $\frac{1}{12}(02 + 04 + \cdots + 24)$ or $\frac{1}{4}(02 + 08 + 14 + 20)$, each giving a different accuracy. The variation in the way the mean is obtained presents difficulties when comparing mean temperatures from different stations or when attempting to derive a global mean, as in climate studies (see Chapter 21). It is also not uncommon for the method used to calculate the mean to be changed during the lifetime of a station, further complicating the situation.

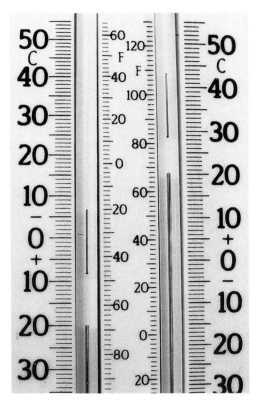

Figure 3.4. The two sides of a Six's thermometer, the lower ends of the index slides indicating a maximum of 28.5° (right) and a minimum of 8.5°, the present temperature, indicated by both ends of the mercury column, being 20°.

Combined maximum and minimum thermometers

It must be stressed that this is not a professional instrument, but one often used by gardeners and horticulturists. Because it is so widely used in such contexts, it warrants a mention. A *Six's* thermometer (Fig. 3.4) consists of a U-shaped tube with a bulb at each end, one being completely filled with spirit, the other only partly filled. The bottom part of the tube is filled with mercury. When the temperature rises, the liquid in the full bulb (left in the figure) expands and forces the mercury towards the partly filled bulb, the reverse occurring when the temperature falls. Both ends of the mercury column indicate the present temperature. In the clear liquid above the mercury, on both arms of the tube, is a movable index, similar to that in a minimum thermometer. As the mercury moves backwards and forwards the indexes are pushed upwards on both sides. They have springs to stop them falling downwards when the mercury moves away (or a magnetic plastic strip behind the tubes), so that they stay at their highest points, which correspond to the maximum and minimum temperatures. The indexes are of steel, and are reset by drawing them down to the mercury with a magnet. This type of thermometer is of lower accuracy than the individual maximum and minimum thermometers because the graduations are not on the tube but on a backing plate to which the tube is fixed and also they are only marked in 1 °C steps. They are not suitable for accurate work.

Bimetallic thermographs

In principle these are like bimetallic actinographs for solar radiation measurement. The sensitivity of a bimetallic strip is greatest when it is straight, although for compactness in a thermograph it is usually coiled into a helix (for measuring solar radiation it is of necessity short and straight). Sensitivity is inversely proportional to the thickness of the strip and proportional to the square of its length and to the difference in the coefficients of expansion of the two metals. New strips suffer an aging process and so are subjected to repeated heating and cooling cycles before use. Sudden large temperature changes, or temperature changes beyond their design limit, can permanently change the sensitivity and zero of bimetallic thermographs or damage them.

One end of the helix is fixed to the instrument's case, in such a way that the zero point can be adjusted by rotation of the helix. The temperature range is adjustable by altering the length of the helix. The other end of the helix is free to rotate, moving an arm holding a pen across a paper chart (Fig. 3.5). A spring- or battery-operated

Figure 3.5. The spiral bimetallic strip of this thermograph (outside the case on the right) indicates a fairly steady 10 degrees on the pen trace of the paper chart.

drum rotates the chart, of which there are several designs to suit different seasons and climates.

The time constant of this type of instrument is about 20 s, compared with the 50–60 s of a mercury-in-glass thermometer. Friction of the bearings is the main source of error and can be minimised by careful initial alignment of the mechanism and subsequent regular maintenance. The effect of the friction of the pen on the chart can be minimised if the bimetallic strip is made fairly rigid, thereby imparting its bending action more positively. A thermograph can, at best, be expected to have an accuracy of ±0.5 °C but, to achieve this, care is necessary in making the initial settings and any necessary periodic adjustments. In practice it is possible that the readings may be out by considerably more than 0.5 °C.

Electrical thermometers

Resistance thermometers

Pure metals

That conductors change in resistance as the temperature changes and can, therefore, be used to measure temperature has been known since the nineteenth century. Various metals and alloys can be used, the choice depending on such factors as the required stability over a long period, the uniformity of product, the degree of change in resistance with temperature and cost. All desirable features do not occur in one metal, but the most commonly used is platinum because of its stability and relative linearity; nickel is often a second choice when cheapness is important. Although all metals have a nearly straight-line temperature response they are not completely linear and for precise work a correction may be necessary.

Sensitivity to temperature change is defined by the material's *temperature coefficient*. For platinum this is 3.9×10^{-3} K^{-1} while for nickel it is 6.8×10^{-3} K^{-1}. Taking the example of a *platinum resistance thermometer* (PRT) with a resistance of 100 ohms at 0 °C (the usual standard today and the most-used metal), the increase in resistance for a 10 degree rise is $100 \times 3.9 \times 10^{-3} \times 10$ (that is, 3.9 ohms) giving a total resistance at +10 °C of 103.9 ohms, and so on, proportionally, for increasing and decreasing temperatures. Grade 'A' PRTs have a manufacturing tolerance of ±0.15 °C from −50 to 0 °C increasing to ±0.25 °C at 100 °C.

To measure the change in resistance, the thermometer can be one arm of a Wheatstone bridge, or a constant current can be passed through it, the voltage across it being measured. In both methods current flows through the sensor during the measurement, thereby heating it. Provided that the current is kept low, that the readings are made quickly and that there is an adequate flow of air past the sensor, the heating effect can be ignored. In the case where a data logger is used, the readings are taken in milliseconds and so heating is negligible.

Although resistance thermometers are now mostly used in automatic instruments, they have also been used at manual stations; the Wheatstone bridge being balanced by the operator. This takes time and so self-heating is more of a concern, but the advantage is that the temperature can be read remotely without having to go to the screen and open it.

When the sensor is some distance from the measuring equipment, be it manual or automatic, it is necessary to compensate for the resistance of the wires between the sensor and the bridge, which would otherwise unbalance the bridge as well as introduce unwanted additional resistance changes due to temperature. Compensation can be achieved by including extra leads in the opposite side of the bridge to balance out those of the sensor's or, alternatively, the current can be fed along one pair of wires while the signal voltage across the sensor is picked off directly by another pair of wires that carry no current (a four-wire system). A further complication of the low resistance of PRTs is that contact resistance within plugs and sockets can become a source of error.

It is possible to use software in a data logger to detect and store maximum and minimum temperatures as well as average or instantaneous values, the one PRT supplying all the readings.

Semiconductor temperature sensors

There is an alternative type of resistance thermometer to the PRT – the *thermistor*. This is a semiconductor heat-sensitive resistor that has a temperature coefficient an order of magnitude greater than those of pure metals. Sensors with both positive and negative slopes can be made and all types have a much higher value of resistance than PRTs. In addition they can be made very small, which is useful in certain applications requiring rapid response times or minimal radiative heating. The higher resistance of the sensors compared with a PRT (typically ranging from 3 to 100 kilohms at $20\,°C$) eases the problem of lead resistance and of plug and socket contact resistance.

The non-linear response of thermistors is a slight disadvantage but it can be corrected by the use of two- or three-element composite thermistors within the probe. Or corrections can be made in the logger circuits or in the computer when processing the readings. Depending on how the linearisation is done and how the corrections are made, the accuracy of a thermistor varies from about ±0.05 to $\pm0.5\,°C$. An aging process occurs and to reduce this thermistors are temperature-cycled by the manufacturer, keeping the long-term drift typically within the range of $\pm0.015\,°C$ per year.

Cheaper thermistors may vary in resistance, one to another, and so are not directly interchangeable without recalibration. (Curve-matched sensors are, however, available at little extra cost, and these can be directly interchanged.) The same problem

Figure 3.6. A pin illustrates the small size of three thermistors and a platinum resistance thermometer. The platinum is not in the form of wire wound on a bobbin, but is deposited on the plastic wafer as a fine track, its resistance trimmed by laser.

occurs with the cheaper PRTs. Figure 3.6 illustrates three thermistors alongside a miniature PRT; the pin gives the scale.

Thermocouples

In 1821, Seebeck discovered that where two different metals touch, a small voltage is generated. If two such junctions are formed and one is kept at a temperature different from the other, the consequent difference in potential between the junctions is proportional to their difference in temperature. This is known as a *thermocouple* and can be used to measure temperature difference. If a number of such junctions are connected in series, the voltage is added and the combination is known as a *thermopile*.

Because thermocouples measure difference in temperature rather than absolute temperature, they are well suited to applications such as radiation sensors (Chapter 2) and the Bowen-ratio method of measuring evaporation (Chapter 7). To measure absolute air temperature, however, it would be necessary to hold one set of junctions at a known temperature (for example 0 °C) and this is not very practical in the field. Thermocouples also produce rather low output voltages and so require higher amplification than the resistive types of temperature sensor.

Thermocapacitors

Intended mostly for radiosondes, these extremely small sensors have a short time constant to allow the necessary rapid response to changing temperature with altitude. They are described in Chapter 16 (Fig. 16.2).

Exposure of thermometers

Air temperature

Stevenson screens

Although solar radiation passes through the atmosphere without much absorption, a thermometer exposed to the same radiation will absorb much of it and heat up considerably, the extent of the heating depending on the colour, texture and size of the thermometer, but the temperature rise can be up to 25 °C above air temperature. Only in the case of very small thermistors, of PRTs made of fine wire, of thermocouples or of the capacitive sensors designed for radiosondes (Chapter 16) will radiative heating be negligible.

So, in measuring air temperature it is necessary to shield whatever type of thermometer is used from both solar and terrestrial radiation, and also from precipitation, while allowing air to pass freely into and through the screen from outside without any change in its temperature. The inside surfaces of the screen should also be as near as possible to the outside air temperature. The first person to realise all of this and to design a suitable screen was the Scottish lighthouse engineer Thomas Stevenson (father of Robert Louis Stevenson), who, in 1866, designed the prototype of the now familiar wooden temperature screen (Figs. 3.7 and 4.1). These wooden structures have double-louvred sides to provide insulation from the outer body, with an opening front to give access to the thermometers. The roof is also double, with an air space, while the base is made of overlapping boards to allow free movement of air.

Standard screens in the UK have internal dimensions of about 1 m width, 0.5 m height and 0.3 m depth, which is large enough to contain a variety of thermometers as well as a bimetallic thermograph and a hygrograph. However, worldwide, there is a broad variety of designs, some bigger than those described above, some so big as to have steps up to them. There is also a half-width model, just big enough to contain the liquid-in-glass thermometers. On board ships, where there is little room, an even smaller screen with single-louvred walls is used, just big enough to hold two thermometers (wet and dry for humidity measurement).

Figure 3.7. The open door of this Stevenson screen exposes its louvred construction and the typical arrangement inside of maximum and minimum (centre) and wet and dry thermometers (see Fig. 4.1 for close-ups) along with a bimetallic thermograph (on the right, see Fig 3.5) and hair hygrograph (on the left).

Little work has been done on the time constants of wooden screens, but the work of Langlo (1949), Bryant (1968) and Painter (1977) points to values of from 4 to 17 min for wind speeds of 10 to 0.5 m s^{-1}, respectively. As regards the effectiveness of the screen at shielding the thermometers from radiation and of ensuring that the air inside the screen is as close as possible in temperature to that outside, Painter found that, in the UK, if the conditions were unfavourable (bright sun and low wind speed) the temperature could be in error by +2.5 °C, while on a cloudless calm night the screen could be colder by up to −0.5 °C. It is important also to recognise that if the screen is not kept clean it will be less effective, and this will cause additional error. Repainting should be done every two years and the screen should be washed on alternate years. Keil (1996) showed that temperatures could be lower by 1 °C after the repainting of an old screen. These factors are likely to be even more of a problem in climates which have many hours of sunshine and high solar radiation intensities.

The opening of the front of a screen to take readings exposes the thermometers to radiation from outside. Readings must, therefore, be taken quickly to avoid error, but not so hurriedly that mistakes are made. Since automatic weather station (AWS) temperature screens are not opened to take readings, this source of error is avoided (see below).

The best site for a screen is at the most exposed position available. The top of a building is not, however, appropriate, since temperatures can change with height quite considerably, especially at night, and also because the building itself will have a great influence on the temperature. Nor, for example, is a hollow in the landscape or a steep slope suitable, as it may well have an untypical microclimate. It is traditional to operate screens over short-cut grass, but obviously if the surrounding country which is to be monitored is desert or rock or snow-covered or an expanse of water or dense forest, the screen should be deployed appropriately. Because quite considerable temperature gradients can occur near to the ground, it is important that screens are mounted at the standard height of 1.25 m. Where this is not possible the height must be known so that corrections can be made. Over a snowpack that changes depth with season, it is usual to move the screen up and down to keep it the same distance above the snow.

Aspirated screens

The Stevenson screen is ventilated naturally, relying on the wind to move air through it. While this is generally good enough, it does introduce some error. To overcome the problem of varying air speed, for precise work, aspirated screens are used in which the air is circulated past the sensors with a fan at a known speed. The commonest type of power-ventilated screen was designed for intermittent, manual operation, that is for taking spot measurements. It was also intended primarily for the measurement of relative humidity using the wet-and-dry-bulb method, where a good passage of air is even more essential, and so it is described in the next chapter, on humidity.

Aspirated screens provide one of the most accurate ways of measuring air temperature and, if higher precision is required, aspiration is sometimes used at automatic stations (if there is sufficient electrical power to drive a fan) and a number of designs are made for this purpose. For more precise results, Stevenson screens are also occasionally artificially ventilated, a fan on top drawing in the air. There are also naturally aspirated screens that turn into the wind to make the best use of any natural ventilation.

The whirling hygrometer provides ventilation manually, by whirling the thermometers around by hand on a frame, with a handle. It has no screen and has to be kept out of direct sunlight and read quickly. It too is dealt with in the next chapter since it is intended essentially for measuring humidity.

Small screens for automatic weather stations

In the early days of AWSs (the 1960s) electrical thermometers were sometimes deployed in ordinary wooden Stevenson screens. While this had the advantage of some compatibility with manual systems, it was no longer actually necessary

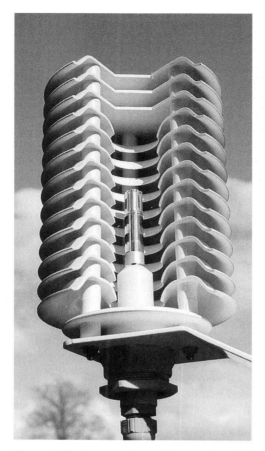

Figure 3.8. Automatic weather station (AWS) temperature screens are much smaller than the large wooden Stevenson screen, although they perform the same function of shielding the sensors from radiation. This shield is cut away to show a cross-section of the louvres. Inside is a small tube containing a miniature resistance thermometer and a humidity sensor (Fig. 4.5).

now that the sensors were smaller, and so the first miniature temperature screens were developed. A typical design is shown in Fig. 3.8. Mostly they are naturally ventilated, although aspirated designs (Fig. 3.9) are made for situations where power permits and where natural ventilation is always very low, for example in a dense forest.

The problem of changing from one type of instrument to a new one, as explained earlier, is that it introduces a step change in the observations, and this occurs if a change is made from Stevenson screens with liquid-in-glass thermometers to AWS screens with electrical resistance thermometers. Huband (1990) investigated the performance of three similar screens of the type shown in Fig. 3.8, comparing

Figure 3.9. An AWS screen may be aspirated by an electric fan to obtain more precise measurements. This is more important where wet and dry bulbs are being measured to derive humidity. In the screen shown here, the small cylinder on the left contains the fan, the larger cylinder on the right containing the wet-bulb water supply.

them with each other and with a Stevenson screen. He found that measurements in the AWS screens agreed with those made in the Stevenson screen to within about $\pm 0.5\,°C$ and for much of the time to within $\pm 0.2\,°C$, although the differences could at times be as much as $\pm 1\,°C$, the latter probably being due to the different time constants of the screens since these larger differences occurred mostly at times of rapid temperature change. Maximum temperatures occurred within 10 minutes of each other but there was more delay in the Stevenson's minimums, probably owing to this screen's greater thermal mass. Earlier, McTaggart-Cowan and McKay (1976) had tested 19 screen designs, including aspirated models, and while they found that the aspirated screens were within the same ± 0.5 and $\pm 0.2\,°C$ range of agreement, the non-aspirated were found to be less accurate, particularly with regard to maximum and minimum temperatures, each one being different.

Ground temperature

The same types of thermometers used for the measurement of air temperature are also used with little modification to measure ground temperature. It is largely how they are exposed that differs.

Surface minimum temperatures

The minimum temperature reached overnight of a surface exposed to the sky is measured with the same type of minimum thermometer that is used for measuring minimum air temperatures in a Stevenson screen (Fig. 3.3). In this application, the thermometer is exposed without any screen, almost horizontally, with as clear and unobstructed a view of the sky as possible.

The minimum temperature of three types of surface is measured: short-cut grass, bare soil and concrete. Their radiative properties being different, they experience different minimums and thus give an indication of what radiative losses are occurring. As with most types of long-term observation, precise working practices have become established, both within national weather services (NWSs) such as the UK Meteorological Office and through the recommendations of the WMO. Measurement of ground-temperature minimums requires that the thermometer is almost horizontal, inclined at around 2°, the bulb just touching the tips of the short-cut grass, bare soil or concrete. The latter is usually a light-grey-coloured slab, 50 mm thick with sides of 1 m by 600 mm, its surface being set flush with the ground. If there is snow, the thermometers in the cases of grass or bare soil are laid on the snow surface, but not quite touching, while in the case of concrete, the slab is swept clear of snow when the thermometer is set. The usual practice is to put the thermometers out just before sunset, but at many sites they are left out permanently.

Because minimum thermometers use ethyl alcohol as the liquid, and suffer from the problem that the alcohol evaporates and recondenses further up the tube, minimum thermometers exposed on the ground often have their top ends covered in a black material that warms in the sun and so helps to prevent any evaporated liquid from condensing in the top of the tube, returning it instead to the main column of liquid.

The same measurements can be made using electrical thermometers, although at remote, unattended stations the sensors get covered if it snows. Logger software detects and stores the minimum readings.

Subsurface temperatures

In addition to the surface minimum temperatures, ground (or 'soil') temperature is also measured at depths of 5, 10, 20, 30, 50 and 100 cm, although not necessarily at all these depths at all sites. There are three designs of mercury-in-glass thermometer for making ground temperature measurements.

Temperatures at 5, 10 and 20 cm depths are measured using a thermometer with a right-angled bend in it (Fig. 3.10) and with graduations of 0.5 °C, allowing estimates to 0.1 °C in the same way as for air temperature. Quality and manufacturing tolerance are also the same as for air thermometers. They have to be installed carefully so as to make good contact with the soil, and with minimum disturbance to it, so that readings are representative. Once installed they remain in place, undisturbed.

For depths of 30 cm and deeper, an ordinary thermometer is suspended in a steel tube of about 32 mm bore, set vertically in the ground. The tube is a permanent fixture, but the thermometer is withdrawn on a chain for reading, being installed in a glass case with rubber 'O' rings on the outside to protect it and also to minimise air convection within the tube. The thermometer bulb is embedded in wax (Fig. 3.11)

Figure 3.10. Soil temperature thermometers for depths of 5, 10 and 20 cm are normal mercury-in-glass instruments but with a right-angle bend in them.

to increase its time constant so that the reading does not change while it is out of the tube being read. This is acceptable because ground temperatures change very slowly.

The second, but less common, method uses thermometers similar to those for depths up to 20 cm, but with a 30° bend instead of 90°. These are used for all depths, right up to one metre (Fig. 3.12) and are left in place once installed. They are very vulnerable to breakage during installation and are difficult to handle. Better contact with the soil and the absence of a steel tube make for a less disturbed environment, although tests with the tube method indicate that in fact errors due to conductivity along the tube, or convection within it, are negligible.

It is usual to measure ground temperatures up to 20 cm depth under bare soil and temperatures at greater depths under short grass, whatever the type of thermometer used. But it does depend on circumstances, and grass would be inappropriate in, say, a desert.

Electrical resistance thermometers can be installed in the ground at any depth and connected to a data logger. They offer by far the simplest method of measuring soil temperatures and can be buried and left undisturbed, with only a thin wire emerging at the surface, thereby minimising any heat conductance from the surface and with the added advantages that automatic logging confers.

Soil heat flux

It is not just the temperature of the ground that is relevant; the flux of heat through the ground is also of importance since it is one indicator of the relative division of net radiation into sensible, latent and soil heat fluxes. Knowing the value of one of these fluxes helps to estimate the others. For measuring the flux of heat through the soil, a *heat-flux plate* is used. This is normally a disc about 10 cm diameter and

Figure 3.11. For measuring soil temperature below 20 cm, a normal thermometer housed in a glass outer tube is lowered on a chain down a permanently installed metal tube. The cap that supports the thermometer chain seals the tube at the top, while two 'O' rings on the glass tube, just contacting the inside of the metal tube, prevent convection. The thermometer is kept at a stable temperature, while temporarily withdrawn for reading, by being permanently sealed in a mass of wax (at the bottom of the photograph).

Figure 3.12. An alternative type of ground thermometer, for use at all depths, is similar to that used down to 20 cm, except that the bend is not 90 but 30 degrees. Such a thermometer is left permanently in place.

a few millimetres thick, of a material that has a similar conductivity to soil. It is buried in the ground and the difference in temperature between the upper and lower surfaces is sensed by a thermopile; some sensors use thermistors.

If the sensor differs in conductivity from the soil, it will disturb the heat flux around it. It is also difficult to install the plate without undue disturbance of the soil. To minimise disturbance, the sensor is sometimes installed by digging a pit and inserting the plate sideways. Even then it is difficult to ensure good thermal contact with the soil and it is also likely that the plate will move slightly with time. And even with careful back-filling, the pit will probably modify conditions and affect the flux (Gilman 1977).

References

Bryant, D. (1968) An investigation into the response of thermometer screens – the effect of wind speed on the lag time. *Met. Mag.* **97**, 183–6, 256.

Gilman, K. (1977) Movement of heat in soils. Institute of Hydrology Report No. 44.

Huband, N. D. S. (1990) Temperature and humidity measurements on automatic weather stations. A comparison of radiation shields. Internal report, Campbell Scientific Ltd.

Keil, M. (1996) Temperature measurements in a Stevenson screen. University of Reading, Dept of Meteorology report.

Langlo, K. (1949) The effects of the solar eclipse of July 1945 on the air temperature and an examination of the lag of the thermometer exposed in a screen. *Met. Ann. Oslo*, **3**, No. 3, 59–74.

Lockwood, G. L. (1974) *World Climatology, An Environmental Approach*. Edward Arnold, London. ISBN 0 7131 5701 1.

McTaggart-Cowan, J. D. & McKay, D. J. (1976) Radiation shields – an intercomparison. Canadian Atmospheric Environment Service unpublished report.

Painter, H. E. (1977) An analysis of the differences between dry-bulb temperatures obtained from an aspirated psychrometer and those from a naturally ventilated large thermometer screen at Kew Observatory. Meteorological Office, UK, unpublished report. Copy available in National Meteorological Library, Bracknell, UK.

Strangeways, I. C. (1999) Back to basics: the 'met enclosure': Part 4 – temperature. *Weather*, **54**, pp. 262–9.

4

Humidity

One morning the view was singularly clear, the distant mountains being projected with the sharpest outline, on a heavy bank of dark blue clouds. Judging from the appearance, and from similar cases in England, I supposed the air was saturated with moisture. The fact, however, turned out quite the contrary. The hygrometer gave a difference of 29.6 degrees (F), between the temperature of the air, and the point at which dew was precipitated. This difference was nearly double that which I had observed on the previous mornings. This unusual degree of atmospheric dryness was accompanied by continual flashes of lightning. Is it not an uncommon case, thus to find a remarkable degree of aerial transparency with such a state of weather?

Charles Darwin *Voyage of the Beagle* (Cape de Verd Islands).

The variable

Just as air, warmed by contact with the ground, is transferred into the atmosphere by processes of diffusion, turbulence and convection, so too is the water vapour produced by evaporation. The ratio in which the net radiative energy is divided between heating the atmosphere, heating the ground and evaporating water is dependent on many factors, such as the amount of water actually available, the nature of the ground and the type of vegetation. Knowing the rate of evaporation of water is useful information in hydrology, meteorology and agriculture, but it is difficult to measure. However, the amount of water vapour in the air, i.e. the air's humidity, is easier to measure and this chapter looks at how it is done; Chapter 7 addresses the more difficult problem of how evaporation rates are measured.

Units and terminology

Hygrometry is the measurement of the water content of solids, liquids and gases; in environmental applications this usually means the water content of the atmosphere.

Measuring the Natural Environment, second edition, Ian Strangeways. Published by Cambridge University Press. © Ian Strangeways 2003.

Water vapour exerts a pressure, the *vapour pressure* (VP) of water, which can be measured in any of the usual units of pressure, such as millibars (or hectopascals now; See Chapter 6). Above a water surface in an air-filled container, the VP rises to a maximum level, the *saturation vapour pressure* (SVP), beyond which it cannot rise any further at that temperature (the rate at which water molecules are leaving the surface of the water being then the same as the rate at which they return due to molecular bombardment). The higher the temperature the higher the SVP: at $0\,°C$ it is 6.11213 hPa over water. There are equivalent SVPs over ice, being 6.11154 $0\,°C$. All these SVPs are given in hygrometric tables.

Cooling a sample of moist air (while keeping its pressure constant) increases its RH, and a temperature will be reached when the RH is 100%. Further cooling then causes dew to form. This is known as the *dew-point* temperature. There is a similar temperature, below $0\,°C$, known as the *frost-point* temperature (or hoar-frost point).

The *relative humidity* (RH) is the ratio (expressed as a percentage) of VP to SVP. Thus 100% indicates saturation. This applies just to the temperature of the sample at that time so, without also knowing the air temperature, the RH is not complete information.

More academic are units such as the *mixing ratio* (the ratio of the mass of water vapour to the dry mass of the air it is associated with); the *specific humidity* or *water content* (the ratio of the mass of water to the mass of moist air); and the *vapour concentration* or *absolute humidity* (the ratio of the mass of water vapour to the volume of the moist air it is associated with), which is in effect the density of the water vapour, hence its additional name of *vapour density*.

As to what this means in terms of the actual amount of water vapour in the atmosphere: At $20\,°C$, 100% RH is equivalent to 23.37 hPa pressure, or an absolute humidity of $17.27\ \mathrm{g\ m^{-3}}$, which at standard pressure is 21.22 ppt (parts-per-thousand) or about 2% of the atmosphere.

Measurement techniques

The first recorded attempts to measure humidity date from around 1500 when Leonardo da Vinci described the effect of humidity on the weight of a ball of wool, while in 1665 Robert Hooke noted that the length of a gutstring changed with the humidity. All methods of humidity measurement, old and new, can be divided into four groups.

The addition or removal of water

Water vapour is added to or removed from the air, the resultant changes being measured. This category includes the *gravimetric*, volumetric and *psychrometric* methods. In the first two, water vapour is removed from the sample by a desiccant and the gain in weight of the desiccant, or the change in volume of the sample, is measured. These methods are laboratory techniques and are not described further. However, the psychrometric method (the *wet-and-dry* method) is one of the most practical and widely used means of measuring humidity, and so is covered fully below.

The adsorption and absorption of water vapour

Water vapour is adsorbed (taken onto) or absorbed (taken into) many materials, to an extent dependent on the level of RH, causing them to change in some physical, chemical or electrical way, the processes being reversible and repeatable. By measuring the change, an indication of RH can be obtained. Several instruments use these properties, in particular the *hair hygrometer*, the *thin-film capacitive* sensor and the *ion-exchange* sensor, all of which are in wide use and so are fully described below.

The condensation of water vapour

One of the most precise ways of measuring humidity is by detecting the temperature at which dew forms. This class of instrument includes the *dew-point* and *dew-cell* hygrometers, but neither are widely used for everyday applications, although the dew-point method provides a good laboratory standard and finds application in the Bowen-ratio method of measuring evaporation.

The radiation absorption method

Measurement of the absorption of infrared radiation passed through air containing water vapour allows rapid changes of humidity to be sensed, but it is little used except for specialised applications such as the eddy correlation measurement of evaporation. It is also widely used in industrial instruments such as Gas Analysers and this is discussed in Chapter 20 on Atmospheric composition.

The psychrometric method

When evaporation takes place from a wet surface into an airstream, the surface cools until it reaches a state of equilibrium, the amount of cooling depending on the relative humidity, the air temperature and the airspeed. By measuring the temperatures of the air and of the wet surface, the VP can be determined. This is the wet-and-dry method and is one of the most commonly used. It is simple, cheap, works from very low humidities to 100% RH, is precise and consumes no power.

Basic principles

A *psychrometer* consists of two thermometers (mercury or electrical), one measuring air temperature and the other, kept wet by a wick dipped into distilled water, measuring the amount of cooling produced by the evaporation of water from it, which is dependent on the relative humidity.

The psychrometric formula $e = e_s - Ap\,(T - T_w)$, relates the vapour pressure e, the air temperature T, the wet-bulb temperature T_w, the atmospheric pressure p, the saturation vapour pressure e_s (at temperature T_w) and the psychrometric coefficient A.

This coefficient has been derived both theoretically (Wylie 1968) and by experiment, but it is the one factor that cannot be entirely precisely quantified. However, over the decades its value has been carefully estimated for ordinary mercury-in-glass thermometers, and with air speeds of 3 m s^{-1}, or more, the coefficient is in the order of 0.667×10^{-3} K^{-1}. For smaller bulbs or greater ventilation rates, the coefficient decreases.

Below freezing point, the wet bulb becomes an ice bulb, although evaporation continues to occur (as regelation), and so the ice bulb is colder than the dry bulb and humidity can still be calculated – using different tables. However, the ice bulb will eventually dry out and will not then give meaningful readings. At an attended site, the bulb can be manually painted with water to make an ice bulb, and for this there are set procedures. This is not possible at automatic unattended stations.

Screens and accuracy

The accuracy of the wet-and-dry method is mostly dependent on the rate of flow of air past the wet bulb and this varies depending on the screen used.

Stevenson screens

The wet-and-dry method is widely used throughout the world in Stevenson screens (Fig. 4.1). Indeed it is probably the most-used method for measuring atmospheric humidity. The ventilation is natural in this type of screen, introducing some uncertainty over the psychrometric constant. Folland (1975) investigated the problem and suggested that in a large screen the mean ventilation rate is around 0.75 m s^{-1}. Others (Bultot & Dupriez 1971) showed that only rarely did the rate reach 1 m s^{-1}. For natural ventilation of this order, the UK Meteorological Office uses a psychrometric coefficient of 0.799×10^{-3} K^{-1} above $0\,°$C and 0.720×10^{-3} K^{-1} below freezing. Other inaccuracies arise if the wrong muslin is used to cover the wet bulb, if the muslin is dirty or contaminated, or if distilled water is not used. How well the wick fits the thermometer is also of importance and there are many rules regarding this, such as are given in *The Handbook of Meteorological Instruments* and *The Observer's Handbook* (Met. Office 1982a, b). If the dry bulb

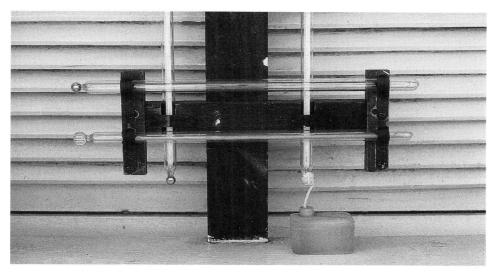

Figure 4.1. Inside most Stevenson screens are four basic thermometers – wet and dry bulbs (the vertical thermometers, right and left) and maximum and minimum thermometers (set horizontally, upper and lower). The wet-bulb wick is kept moist by the distilled water in the small plastic bottle. The difference in temperature between the wet and dry thermometers gives a measure of the relative humidity.

becomes wet, through fog or condensation, it will act as a wet bulb, with consequent errors.

Because of these many factors it is not possible to give a firm figure for the accuracy of humidity measurements in a Stevenson screen, but most manufacturers claim $\pm 1\%$ to $\pm 2\%$ RH, with the warning that this is dependent on air temperature and wet-bulb depression. There must, therefore, be some doubt as to the accuracy and $\pm 1\%$ to $\pm 2\%$ must be viewed as the best obtainable, under optimal conditions and that generally it will be less.

Aspirated screens

For more precise measurements, the air is circulated past the thermometers by a fan in an *aspirated psychrometer* (also known as an aspirated hygrometer or Assman psychrometer). A typical psychrometer consists of a spring- or electric-motor-driven fan that draws air up a central tube, the tube being divided into two paths, one housing the dry thermometer and the other the wet. The wet bulb has a length of wick tied to it, but it does not dip into a water container as in a Stevenson screen, instead being wetted manually just prior to use. The fan is operated until stable readings are reached – after a few minutes. A built-in slide rule is then used to derive the relative humidity, dew point or vapour pressure from the wet-and-dry

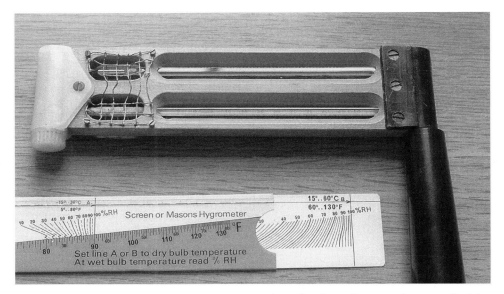

Figure 4.2. The whirling hygrometer is a simple and economic way of taking spot readings of air temperature and humidity; the wet and dry thermometers, mounted in a frame on a handle, are whirled by hand to aspirate them. The lower bulb in the figure is kept moist by the small water container (left), just as in the Stevenson screen. The slide rule enables humidity to be calculated without recourse to tables.

readings. The aspirated psychrometer is an accurate instrument and is often used as a standard against which to compare others.

A more economic way of obtaining aspirated wet-and-dry readings without either a screen or a fan is to mount the wet and dry thermometers on a simple frame. This arrangement is known as a *whirling* or *sling psychrometer* (Fig. 4.2), the whole construction being whirled around on a handle. Whirling at anything more than three revolutions a second produces the necessary wind speed and readings are taken at intervals until stable. The readings must be taken quickly since when movement stops the temperatures change quickly. It is also necessary to prevent the sun from shining on the instrument and if it is raining it must be kept under shelter. Used intelligently, good results can be achieved cheaply.

Because the ventilation is under control in an aspirated or whirling psychrometer, the psychrometric coefficient can be more precisely quantified, being taken as $0.666 \times 10^{-3}\,\mathrm{K^{-1}}$ above $0\,^{\circ}\mathrm{C}$ and $0.594 \times 10^{-3}\,\mathrm{K^{-1}}$ below $0\,^{\circ}\mathrm{C}$, for any ventilation rate above $3.6\,\mathrm{m\,s^{-1}}$. In such conditions, the humidity can probably be measured to within $\pm 1\%$ or $\pm 2\%$ RH. A few Stevenson screens have been equipped from time to time with fans to aspirate them. When this is done, a similar accuracy will be achieved.

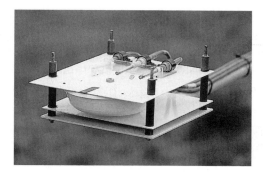

Figure 4.3. The wet-and-dry method is equally applicable to electrical thermo-meters, as in this early design of an AWS screen (top cover removed). The wet bulb (left) is kept moist by a wick that dips into the water container beneath. The dry thermometer (centre) is wired in a bridge circuit with the wet bulb to give a direct measure of the difference in temperature. The third thermometer (right) measures air temperature with a separate bridge circuit.

AWS screens

Wet and dry electrical resistance thermometers can be used in automatic weather station (AWS) screens (Fig. 4.3). While this is a very effective method of measuring RH automatically, the AWS must be visited often enough to replenish the water, or an automatic method used for this; furthermore, if operated below freezing for any length of time, the data will be in error.

Although AWS screens are generally non-ventilated, being smaller than Stevenson screens their natural ventilation is probably better, and thus humidity measurements are likely to be somewhat more accurate than in a large wooden screen. However, while there are carefully prescribed rules specifying exactly how the wet-bulb wick should be fitted to a mercury-in-glass thermometer, there is a much greater variety of resistance thermometers in use, and no rules about fitting the wick, resulting in some uncertainty as to the exact value of the psychromet-ric coefficient. AWS screens are also much less well investigated than Stevenson screens and so the accuracy of humidity measurements made in them by the wet-and-dry method is not as certain. When used under poorly ventilated conditions, such as in a dense forest, or for more precise measurements, AWS screens may be ventilated by fans (Fig. 3.9). These will probably give a similar performance to other aspirated types, that is ± 1–2%.

The hair hygrograph

Many materials change their physical dimensions as the amount of water vapour in the air changes and such changes can be used to measure humidity. Human hair is one such material, and is widely used. (Others are cellophane, wool and gold-beater's skin, the latter being used in early radiosondes.) The hair hygrograph is almost as widely used as the wet-and-dry method, is well developed, well proven, reliable, and has a long and respectable history, despite its primitive concept and mechanisms. It is not, however, very precise.

Basic principles

Human hair, which is one of the most sensitive of the hygroscopic materials, has been used since the eighteenth century, when de Saussure made the first known hair hygrograph. As the RH changes from 0% to 100%, human hair increases in length by about 2.5%, in an approximately logarithmic way, most of the increase being at the lower humidity end. It is not much affected by any other vapours or gases, nor is its length very temperature dependent. The change in length is a function of the RH and not of the absolute amount of water vapour in the air. To use this property, a band of hairs is kept under light tension, the change in length with changes in RH being magnified and linearised by levers and converted into a pen movement, which is recorded on a paper chart (Fig. 4.4).

Accuracy

Spilhaus (1935) found that there is a gradual drift in the length of hair held under light tension, affecting the zero setting. Although most of this can be corrected by periodically saturating the hair with distilled water, there is still a need to adjust the zero on a regular basis. Changes in the hair also affect the reading at the top

Figure 4.4. The change in length with relative humidity of a bundle of strands of human hair, kept under light tension in a hair hygrograph, is transferred to a pen which records a trace on the rotating paper chart. The hair is visible stretching from left to right, held in tension at its centre point.

end of the RH scale; this is checked regularly by covering the instrument with a saturated cloth, which produces about a 95% RH atmosphere, and then making any necessary adjustment. The lower end of the scale is less simple to check since 0% RH is not so easily produced (although placing the instrument in a sealed container with sufficient desiccant such as silica gel will probably come close to achieving 0% if it is left there long enough). However, instead, the instrument can be put in a room having a steady temperature. When its reading is stable, the RH is determined using an aspirated psychrometer.

The time constant of ordinary hair, above freezing point, is in the order of a minute, but at $-10\,^{\circ}$C it has risen to 3 min, at $-20\,^{\circ}$C to over 6 min and by $-30\,^{\circ}$C to about 13 min or longer. If the hair is rolled flat under pressure, the time constant falls to no more than 30 s at $-30\,^{\circ}$C and to only 10 s at $30\,^{\circ}$C, but the price paid is an increased brittleness of the hair and a consequent loss in its strength, causing it to break under tension more easily. The time constant is also dependent on the tension exerted on the hair, on the previous treatment or history of the hair and on the ventilation rate, and it is different for increasing and decreasing humidities (hysteresis).

Manufacturers' claims vary as regards accuracy but a typical specification is $\pm 5\%$ for 20–80% RH. No claims are made for the top and bottom 20%. The hair hygrograph, however, is not intended as a precise instrument, its usefulness being its ability to make a chart recording of the changes in RH, but it does not do this to high accuracy.

Thin-film capacitive sensors

There are several types of *electrical humidity sensor* in which a thin film of material changes its capacitance or resistance in the presence of water vapour by an amount related to the RH. These are the most common type of electrical humidity sensor used today; most AWSs and hand-held instruments use them. Early designs were based on aluminium oxide (Jason 1965) or on thin polymer films (Nelson & Amdur 1965).

Basic principles

The most commonly used type of electrical humidity sensor is made by depositing a thin metal film in two sections on a glass substrate about 0.2 mm thick and about 5 mm square. Onto these two 'plates' is deposited a 1 μm thick amorphous organic polymer layer to act as the capacitor's dielectric. An upper single plate of water-permeable, vacuum-evaporated metal (palladium or gold), about 10^{-2} μm thick, is then formed on top of this (Fig. 4.5). The thickness of this upper plate

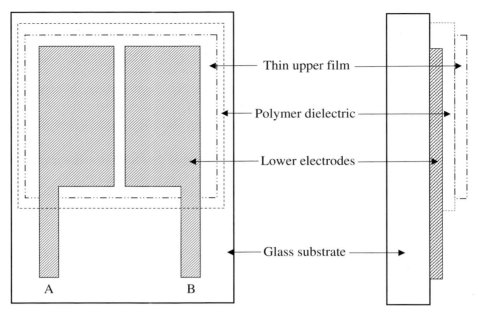

Figure 4.5. Thin-film humidity sensors are made by depositing two thin metal films (A & B) on a 5-mm-square glass substrate. Onto these is then deposited a 1-μm-thick polymer layer acting as the dielectric. A third, extremely thin, water-vapour-permeable metal film is deposited on top of this. The capacitance between the plates A & B changes with changing RH.

is a compromise between the requirements that water vapour should pass through it quickly and that it should be thick enough not to have too high an electrical resistance. (The capacitor is operated in a tuned circuit and a low value of resistance in series with a capacitor leads to a high Q-value for the circuit – a high narrow peak output at the resonant frequency.) Because of the way in which it is constructed, with the lower plate in two sections and just one upper plate, the sensor consists, in effect, of two capacitors in series. It is designed in this way so that electrical contact with the very thin upper plate is avoided (Stormbom 1995).

Water vapour molecules enter the polymer through the upper plate, forming bonds with the polymer molecules and thus increasing the electrical capacitance of the system; the capacitance is almost linearly related to RH (and not to the absolute amount of water vapour). The water molecules form two different types of bond with the polymer, one being very similar to those formed in liquid water, giving the water a dielectric constant of around 80 and thus a conveniently large capacitance change to measure. In the second type of bond with the polymer, the water molecules form dipoles with longer relaxation times. However, these two types of bond can be separated by the use of different frequencies to measure the sensor's capacitance. At radio frequencies, the capacitance changes induced by the liquid-water-like bonds predominate while the non-linear response of the second type only shows up at lower frequencies. Thus by using frequencies above about 1.5 MHz a linear response is obtained. (This is similar to the techniques used to measure the water content of soil; see Chapter 9.) The capacitance measured in this way is also relatively free from temperature dependence and no correction is necessary.

The water vapour absorption process occurs in two stages, the first taking less than a second since there are a fixed number of bonding sites in the polymer. But over a few minutes the number of sites increases owing to swelling in the polymer material and these become occupied more slowly, over minutes, the exact time and amplitude of the second stage of absorption depending on the particular polymer used. According to Salasmaa and Kostamo (1974) this two-stage absorption accounts for the hysteresis that some sensors of this type exhibit (there is no hysteresis in the short-term change). These time constants are dependent on temperature. Below freezing point, the two effects having merged, there is no longer an initial rapid response.

Accuracy

Manufacturers claim that from 0% RH (perhaps 10–20% RH) to about 80% RH the thin-film capacitive sensor is accurate to about ±2% RH, and this is probably the case. From 80% to 100% it is specified as being ±3%. However, accuracy over the top 10%, especially at 100% and especially in cloud, must be open to some question. The long-term stability is also questionable, particularly if 100% is encountered for long periods. Performance at this upper end is expressed by manufacturers' statements such as 'Stability is better than 0.5% RH per year in normal air conditions (0–70% RH)'. While this may be normal for offices and laboratories it is not the case out of doors, where 70% is regularly exceeded. If the sensor is operated in cloud or fog it can experience an RH of 100%+ (supersaturated), while in rain the RH approaches 96% – as it also does daily after sunset in many climates. Another manufacturer expresses it by saying that accuracy is ±3% above 80% if the sensor is kept continuously above 80% RH, which is impracticable out of doors. For use in cloud and in radiosondes, some of the latest sensors have built-in microheaters on the glass substrate to drive off dew or hoar frost, or raindrops (see Fig. 16.3).

Ion-exchange sensors

An alternative type of electrical sensor is based on a conductive ion-exchange surface of chemically treated styrene copolymer (hence its alternative name of polyelectrolyte sensor). The water vapour molecules are adsorbed onto its surface, not absorbed into its thickness, resulting in a faster response time. However, the response time is shorter for increasing humidity, water vapour being released more slowly than it is taken up, with a time constant of about 30 seconds to one minute in the drying phase (depending on the rate of air flow). There is also a hysteresis effect, the difference being $\pm 3\%$ between the upward and downward excursions of RH. As with the capacitive sensors, there is a tendency for the calibration to drift if the sensor is kept at 100% RH for extended periods; furthermore if the sensor is subject to freezing it can be damaged, particularly if damp. Accuracy is in the order of $\pm 5\%$ RH.

The availability of H^+ ions in the copolymer is dependent on the RH, and this regulates the conductivity of the sensor (Musa & Schnable 1965). Interleaving electrodes deposited on the copolymer surface allow its resistance to be measured (Fig. 4.6). The resistance change in response to a change in RH is almost logarithmic, ranging from about 1000 ohms at 100% RH to 10 megohms at 5% RH. As with platinum resistance thermometers (PRTs), the resistance of the ion-exchange sensor

Figure 4.6. This ion-exchange sensor is larger than the capacitive sensor of Fig. 4.5. It changes in electrical resistance (rather than capacitance) with changes in relative humidity.

is measured by making the sensor one arm of a bridge but, as with other sensors that would be polarised by a DC voltage, AC excitation has to be used.

Probably because of the way in which the capacitive sensor has taken over in AWSs, the resistive type is not so widely used. However, development does not stop and the present situation will no doubt change. One area where this type of sensor is used, however, is on the moored buoys operated by the UK Meteorological Office (Fig. 17.3) where it has performed well.

Another form of resistive sensor, no longer much used, was based on a cellulose coating containing small carbon particles in suspension. As the cellulose changed its physical dimensions in response to humidity changes, the contact between the carbon particles changed, varying the resistance.

The dew-point hygrometer

This is a direct and basic method. A mirror is cooled until dew forms on it, the temperature of the mirror then being measured, giving the *dew point* of the air. If the temperature of the air is also known, the VP and RH can be calculated (Wexler 1965). If the dew point is below $0\,°C$, frost forms instead of dew. Cooling is usually brought about by a Peltier device (this, in effect, is the reverse of a thermopile, current passed through it producing heating at one set of junctions and cooling at the other). The dew-point hygrometer is not cheap, owing to its relative complexity. It is, however, a good standard reference technique since it measures dew point directly, all other methods described above relying on the indirect effects of water vapour to achieve the measurement. It is also used to measure very small differences in RH in the Bowen-ratio method of sensing evaporation (Chapter 7).

Dew cells

A substance which has the property of absorbing water vapour from the atmosphere is said to be hygroscopic. Copper sulphate, for example, can incorporate water into its molecular structure, forming *hydrates*–crystals that contain salt and water in fixed and constant proportions. The water is said to be *water of crystallisation*.

Dew cells (also known as *vapour pressure equilibrium hygrometers* or *the heated salt-solution method*) use the fact that the equilibrium vapour pressure (EVP) close to the surface of a salt solution is less than it is over pure water (at the same temperature). This is so for all salts, a fact used to generate atmospheres of known RH for the calibration of humidity sensors (see previous section), but especially so for lithium chloride which has a very low EVP.

If a woven material, such as fibre-glass, is wound on a bobbin and soaked in a saturated solution of lithium chloride, the vapour pressure close to it will be low and water vapour will condense from the air onto the hygroscopic salt. If the material is now heated, say by passing a current through it from a constant voltage source, a temperature will be reached at which its EVP is greater than that of the ambient air. At this point condensation changes to evaporation, accompanied by a phase change from liquid to a solid hydrate crystal. This transition can be detected by sensing the abrupt change in conductance that occurs as the material becomes dry. By measuring the temperature at which the phase change occurs, the ambient VP can be derived (Folland 1975).

There is a lower temperature below which the method will not operate because in order to achieve condensation, the sensor would need to be cooled. However, this does not occur until around $-45\,^{\circ}$C, allowing the instrument to be used over all normal conditions. There are a number of problems of more practical significance. The need to clean and recharge the material with lithium chloride regularly is a practical inconvenience at remotely operated stations. So too is the need to control the ventilation rate so as to prevent ambiguity through effects on the heating process of variable air flow over the sensor. There are also theoretical problems relating to the question of which (of four) hydration states the lithium chloride is at. Down to $-12\,^{\circ}$C the monohydration state applies. Below this temperature, the dihydrate condition occurs, while above 41°C the anhydrous form occurs. Near to these temperatures and the associated changes of state, ambiguity can occur. By adding a small quantity of potassium chloride, the lower limit can be extended to $-30\,^{\circ}$C. But for most of the typical ranges of temperature, the system is reliable.

The impregnated tape can also be heated indirectly, its conductivity being measured by two electrodes in contact with it, spaced some distance apart. This gives improved accuracy, but it is more complex. The sensors can be exposed in normal screens, but aspiration is necessary for best results. Because much depends on the actual construction of the sensor, the WMO does not quote any accuracy for the method.

Radiation-absorption sensors

Water vapour absorbs infrared radiation, and this phenomenon can be used to measure the amount of water vapour present in the air (Moore *et al.* 1976). Water vapour, and other gases, absorb very specific, narrow bands over a wide range of electromagnetic wavelengths, including UV as well as IR. Radiation corresponding to any of these bands will lose energy by an amount depending on the amount of water vapour present, the length of the path and the absorption coefficient at the given wavelength. This can be a precise method, but is rarely used for the every-day measurement of humidity because of its complexity and because it consumes power. But the method does have an important application in the measurement of evaporation by the eddy correlation method (Fig. 7.4) and also the measurement of other fluxes, such as carbon dioxide, as well as the amount of trace gases in the atmosphere (Chapter 20).

Calibration of humidity sensors

The vapour pressure over a saturated salt solution is lower than that over water at any given temperature. This can be used to create atmospheres of known RH for the purpose of calibrating humidity sensors of the thin-film, or similar type.

In theory all that needs to be done is to part-fill a small container with a saturated solution of a salt, such as lithium chloride, place the sensor in the air space above the solution, keep it all at a fixed and known temperature and wait until the humidity stabilises. Equilibrium is reached more quickly if the area of the liquid surface is large compared to the volume of air and if the air is circulated with a fan and the solution is stirred. The temperature must also be kept stable and there should be nothing in the container that might absorb water vapour. Even with such care, it can take several hours for equilibrium to be reached and in the past there was also some uncertainty about the correct RH values for some salts. Tables are available that list the solutions recommended for humidity control (Young 1967) and bottles of saturated solutions can be bought especially for this purpose. At 25 °C, for example, magnesium chloride hexahydrate has a RH of 32.7%, sodium dichromate dihydrate 53.7%, sodium chloride 75.1%, potassium chromate 86.5% and potassium sulphate 97.0%.

References

Bultot, F. & Dupriez, G. L. (1971) Comparaison d'instruments de mesure de l'humidité sous abri. *Arch. Meteotol. Geophys. Bioklimatol, Wien*, **19B**, 53–6.

Folland, C. K. (1975) The use of the lithium chloride hygrometer (dew-cell) to measure dew-point. *Met. Mag.*, **104**, 52–6.

Jason, A. C. (1965) Some properties and limitations of the aluminium oxide hygrometer. In *Humidity and Moisture, Vol. 1, Principles and Methods of Measuring Humidity in Gases*, ed. A. Wexler, pp. 372–90. Chapman & Hall.

Met. Office (1982a) *The Handbook of Meteorological Instruments*. HMSO, London.

Met. Office (1982b) *The Observer's Handbook*. HMSO, London.

Moore, C. J., McNeil, D. D. & Shuttleworth, W. J. (1976) A review of existing eddy-correlation sensors. Institute of Hydrology Report 32, pp. 22–3.

Musa, R. C. & Schnable, G. L. (1965) Polyelectrolyte electrical resistance humidity elements. In *Humidity and Moisture, Vol. 1, Principles and Methods of Measuring Humidity in Gases*, ed. A. Wexler, pp. 346–57. Chapman & Hall.

Nelson, D. E. & Amdur, E. J. (1965) A relative humidity sensor based on the capacity variations of a plastic film condenser. In *Humidity and Moisture, Vol. 1, Principles and Methods of Measuring Humidity in Gases*, ed. A. Wexler, pp. 597–601. Chapman & Hall.

Salasmaa, E. & Kostamo, P. (1974) New thin film humidity sensor. In *Proc. 3rd Symp. on Meteorological Observations and Instrumentation*, pp. 33–38. American Meteorological Society.

Spilhaus, A. F. (1935) *The transient condition of the human hair hygrometer element.* MIT, Cambridge MA, Meteorological Course, Professional Notes No. 8, Part 1.

Stormbom, L. (1995) Recent advances in capacitive humidity sensors. *Vaisala News* **137**, 15–18.

Wexler, A. (1965) Dew-point hygrometry. In *Humidity and Moisture, Vol. 1, Principles and Methods of Measuring Humidity in Gases*, ed. A. Wexler, pp. 125–215. Chapman & Hall.

Wylie, R. G. (1968) Resumé of knowledge of the properties of the psychrometer. CSIRO, Melbourne, Report No. PIR-64.

Young, J. F. (1967) Humidity control in the laboratory using salt solutions – a review. *J. Appl. Chem.*, **17**, September 1967.

5

Wind

It was something formidable and swift, like the sudden smashing of a vial of wrath. It seemed to explode all round the ship with an overpowering concussion and a rush of great waters, as if an immense dam had been blown up to windward. In an instant the men lost touch of each other. This is the disintegrating power of a great wind: it isolates one from one's kind. An earthquake, a landslip, an avalanche, overtakes a man incidentally, as it were – without passion. A furious gale attacks him like a personal enemy, tries to grasp his limbs, fastens upon his mind, seeks to rout his very spirit out of him.

Joseph Conrad Typhoon.

Working on the Cairngorm mountains in Northern Scotland we often experienced gale-force winds while installing experimental equipment (Fig. 5.1). In such places the power of natural forces strikes home. In the winter, on the mountain, there can be a feeling of considerable threat, which I have felt nowhere else. It is not just in a tropical hurricane that the wind's power can be felt.

The variable

Wind is caused by imbalances in the atmosphere due to temperature and pressure differences. The movement of the air is an attempt to attain equilibrium but, owing to solar heating, this is never achieved. Although air movement is three dimensional, the horizontal component is usually by far the greater and this is what is normally meant by the term 'wind'. However, vertical motion also occurs, both at a small scale near to the ground as eddies caused by turbulent flow and convection, and on a large scale as a result of solar heating in the tropics, which powers the general circulation of the atmosphere.

Measuring the Natural Environment, second edition, Ian Strangeways. Published by Cambridge University Press. © Ian Strangeways 2003.

69

Figure 5.1. Working in a 75 mph wind on the Cairngorm plateau at Cairn Lochan presented difficulties when attempting to install delicate equipment. The raingauge being installed is just visible on the ground.

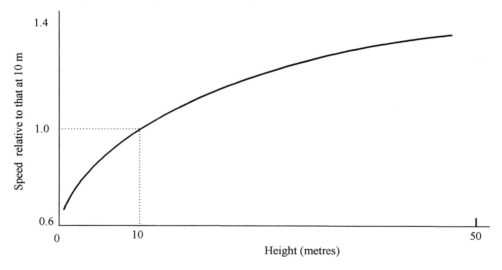

Figure 5.2. Windspeed increases rapidly with height up to about 60 metres, requiring corrections to be made to readings taken at heights other than the standard 10 metres.

For the first 100 m or so above the ground, wind speed increases approximately logarithmically (Fig. 5.2), but with increasing height the influence of the surface has progressively less effect and at an altitude of somewhere between 500 and 2000 m, depending on surface roughness and other factors such as latitude, the speed becomes constant and equal to the *geostrophic wind* (the wind blowing parallel to the isobars). The altitude through which the earth's surface has an influence on the wind is known as the *planetary boundary layer*.

Wind direction is also affected by altitude. At the top of the planetary boundary layer, the direction is the same as that of the geostrophic wind. But descending through the layer, the wind blows at an increasingly oblique angle across the isobars with a component towards the lower pressure region. Plotted from above, the movement of the line of direction marks out a spiral, known as the *Ekman spiral* (Lockwood 1974). Another force acting on moving air (indeed on anything moving) on a rotating object like the Earth is the *Coriolis force*, so named after the French nineteenth century mathematician Gaspard Gustave Coriolis, another name he gave it being the *deflective force*. His interest was not meteorological but industrial, to do with things like water wheels (Persson 2000). The force does not directly concern us in the measurement of wind, but it does affect the wind and how it moves and so at least needs a mention here. It is quite a complex matter and Persson (above) has so far written eight articles on it in the RMS magazine *Weather*, to which readers should refer for an intriguing analysis (Persson 2002). While our main concern here is with measuring wind at the surface, knowledge of what is happening throughout the boundary layer is needed to understand activity at the ground. This is one of the functions of radiosondes (Chapter 16). At the smaller scale, thunderstorms and tornadoes produce strong local winds with significant vertical components, while in sea- and land-breezes, caused by adjacent parts of the earth's surface being at different temperatures, the warmer air rises and is replaced by the cooler. Mountains also affect the flow of air. The same types of instrument are used to measure winds at all these scales.

Units and terminology

The most appropriate unit today for the measurement of wind speed is metres per second (m s^{-1}) although other units from the past continue in use. As a guide for conversion, 1 metre per second is equivalent to: 2.237 miles per hour, 3.600 kilometres per hour, 3.281 feet per second and 1.943 knots (nautical miles per hour; because nautical miles vary with latitude, the internationally agreed nautical mile is 1852 m or 6076.102 feet, thus 1 knot = 1.15 mph). Tradition also provides a way of estimating wind speed without instruments, by using the Beaufort Scale.

Wind direction is specified as the direction from which the wind is blowing and is expressed in degrees clockwise from true north. The resolution used will vary depending on the application of the data. For most purposes the direction is usually expressed in fives or tens of degrees, or as 8, 16 or 32 compass points.

Exposure of wind sensors

Because wind speed increases rapidly with height, and to allow intercomparison of measurements between sites, 10 m is used as the standard reference (Fig. 5.2). Nevertheless, many sensors are exposed at a lower height than this, for reasons of convenience or practicality; where 10 m is not used, it is possible to apply a correction. However, the 10 m rule only applies to an unobstructed site and if sensors are being operated at anything other than a clear site, they may need to be exposed at a greater height than 10 m. There are rules for this, which can be abbreviated thus:

Where obstructions are small and surround the site evenly, the sensor is raised 10 metres higher than the obstructions. If the obstruction's distance away is the same as its height, the sensor is raised to about twice that height. If the obstruction's distance away is 10 times its height, the sensor is raised to one and a half times that height, and at 20 times to one and a quarter the height. If the obstruction's distance away is 30 times its height, no increase in height is needed. When wind sensors are raised above 10 m, an effective height has to be allotted to them, which would give the same readings as those actually observed by the instrument, taking into account the type, size, number and distribution of the obstacles and the actual height of the sensor. Thus although the sensor may be 25 m high, its effective height might be judged to be 15 m. This is a very subjective process and there are no hard-and-fast rules. Whether it is its actual or its effective height that is 15 m, a correction must be made to convert it to 10 m. This applies to land situations; over the sea, the corrections are slightly different. If there is no alternative but to install a wind sensor on top of an isolated building, the procedure is to put it on a mast at least half, and preferably three-quarters, the height of the building. This does not apply to lighthouses!

Wind direction measurement

It is possible to make a rough estimate of wind direction by observing the drift of smoke from chimneys, if they are sufficiently far above the surrounding buildings, but care is needed to avoid deceptive perspective. The movement of clouds, even low ones, should not be used because the direction normally changes with height (see the discussion in the first section of this chapter).

Vane aerodynamics

At its simplest, a wind vane consists of a vertical plate on a cross-arm with a counterbalance, the whole rotating at its balance point on a vertical shaft (Fig. 5.3).

Figure 5.3. A manually read wind vane is a simple device, although the mathematics of its movement are less so. Standing beneath the mast, an observer takes an average reading of direction over 10–20 seconds. This is not as easy as it sounds.

The vane must have a minimum of friction so that it responds to low wind speeds, it must be balanced at the pivot point, and it must be correctly damped. A compromise has to be reached that allows the vane to reposition itself, after a change of wind direction, without excessive delay, but the damping must not be so minimal that the vane overshoots the new position (Wieringa 1967). This characteristic is quantified as the vane's *damping ratio*, which is the ratio of the actual damping of the vane compared with its critical damping – the value of damping that would give the fastest movement to the new direction without any overshoot. In practice it is best for this ratio to lie between 0.2 and 0.7. The ratio is also known as the *damping coefficient*. It is a constant for any vane, being independent of other variables such as wind speed or air density. Another related term is the *damped wavelength*, which is the length of the column of air that passes the vane during the time it takes to move through one damped oscillation. This too is a constant for any vane. A similar term, the *delay distance*, is the length of the column of air that passes while the vane moves half way to the new direction. The *distance constant* (which is like a time constant) is the length of the column of air that passes while a vane responds to 63.2% of a change in direction, a typical value for this being about 4 m.

The damping will depend on such factors as the size of the vane, its inertia and the natural aerodynamic damping that occurs when a plate is moved through a gas. While friction can be introduced to increase damping, it has the disadvantage that its effect reduces with increasing speed, making the vane behave poorly at higher speeds.

The majority of vanes have a single plate. A few have two – either flat, some distance apart and parallel to each other, or as two parallel aerofoils. Another alternative is two flat plates splayed out into a wedge shape from the shaft. According to Wieringa (1967) the wedged-shaped vanes are likely to reduce each other's effectiveness, while the streamlining of the aerofoil plates may reduce their effectiveness at low speeds.

Manual and remote-indicating wind vanes

The simplest type of wind vane is shown in Fig. 5.3. An observer looks at it from beneath and estimates the average direction over about 15 s.

In the pre-microelectronic era, if the direction were to be displayed remotely, a Desynn or Magslip transmitter would be attached to the vane's shaft with a cable connecting it to the appropriate receiver in a nearby office, the dials of the receivers being marked in 5 or 10° steps along with the main compass points. The receivers contained a magnetic rotor that took up the same angle as the transmitter rotor, the transmission of the information requiring just three wires. A comparison of the two types of transmitter is not relevant since today a shaft encoder would more likely be used instead (see the next sub-section). It is possible to make the receiving Magslip or Desynn machine move a pen across a paper chart, and so record the direction.

Automatic wind direction sensors

The great majority of wind direction sensors for automatic logging use a vane, the angular position of the shaft being sensed electrically. Other methods than vanes are used, but these will be dealt with in the section on wind speed, under the headings of pressure, thermal and sonic anemometers.

Sensing shaft angle by potentiometer

The angular position of a shaft can be measured electrically by using a potentiometer or a shaft encoder. Potentiometers can be of the wire-wound or conductive plastic type or they can be constructed as a circular array of magnetic reed switches connected to a resistor network (Fig. 5.4). By using a diode encoder with the reed assembly, a binary signal similar to the Gray code (described below for shaft encoders) can be generated instead of the analogue voltage signal produced by potentiometers.

A problem with wire-wound, and to a lesser extent plastic, potentiometers is that they become worn, particularly in the direction of the prevailing wind, because the vane moves the wiper contact backwards and forwards over the same part of the track continuously. Reed switches avoid this problem while also reducing the torque needed to move the vane; this can be high, reducing the vane's sensitivity at low wind speeds. A typical starting speed quoted by manufacturers, its threshold,

Figure 5.4. This wind direction sensor uses reed switches to convert the shaft rotation into an analogue voltage signal, the magnet arm (left) closing at least one of the 16 reed switches (right), spaced at 22.5 degrees. The reed switches tap off a voltage proportional to direction from a series of fixed resistors. In some designs (Fig. 5.7) a compact, wire-wound, potentiometer is used in place of reed switches.

is about 0.3 m s^{-1} although it can be as high as 1 m s^{-1}. (But threshold information is only meaningful if the angle of attack of the wind is also quoted, and it rarely is.)

The resolution of a wire-wound potentiometer is high, limited only by the closeness of the wire turnings, and for a plastic potentiometer it is theoretically infinite. The relative coarseness of resolution of an array of reed switches is offset by the fact that usually the wind direction is constantly changing through many degrees and so an average over a period of minutes removes much of the resolution problem. In addition, wind direction is rarely needed to high precision. Where high resolution of spot readings is needed, a wire-wound or plastic potentiometer, or a shaft encoder, will be required.

When the wind direction is alternately left and right of north, the output of the potentiometer changes from high to low. If an arithmetic average of several such spot readings is taken, the resultant indication can be south instead of north, although if individual spot readings are recorded the problem does not arise. The simplest solution is to take vectorial means rather than arithmetic means. But an alternative strategy is to extend the angular range of the sensor artificially so that, for example, it works over a range of ±270°, a logic circuit connected to the reed switch assembly switching to a stable position whenever the ends of the range are approached, thereby avoiding any step change.

Digital shaft encoders

An alternative to potentiometers for measuring angular position is the shaft encoder. The most common has an optical disc encoded with a binary pattern, read using infrared light-emitting diodes (LEDs) on one side of the disc and a photodetector on the other, producing a binary code (often, the Gray code; see binary numbers,

Chapter 11). Depending on the angular resolution required of the sensor, the disc may be encoded with as few as four tracks (giving a four-bit word) or as many as nine for high resolution. More information on shaft encoders is given in Chapter 10 under the topic of water level measurement.

Damping ratios

Damping ratios for most of the automatic wind direction sensors vary from 0.2 to 0.4. Values of the damped wavelength typically range from 2.5 to 7 m.

Measuring wind speed

The Beaufort Scale

In 1806 Admiral Sir Francis Beaufort devised his Beaufort Scale for wind speed. The scale subdivides wind into strengths on a scale of 0 to 12, each defined by their observable effects. There are two sets of descriptions, one for land and the other for sea (Met. Office 1995, see the Chapter 1 reference list). Just the land version is summarised in Table 5.1, in abbreviated form. Comparisons between the force-numbers, B, and measurements of wind speed, V (in metres per second at 10 m), have led to the relationship: $V = 0.836 \sqrt{B^3}$.

Cup and propeller anemometers

Sensing wind speed with cups

The cup anemometer was invented in 1846 by Dr Thomas Romney Robinson, a clergyman and astronomer from Armagh (Patterson 1926). The design has changed little since then. When the anemometer is exposed to the wind, the pressure on the open side of the cups is greater than that on their backs, which causes the shaft to rotate (Fig. 5.5). The speed of rotation is nearly linear with respect to wind speed and only for more precise work is any correction required. The response is independent of wind direction and, more or less, of air density. Robinson assumed that the ratio of the speed of the wind to the speed of the cup-centres was a constant, and equal to 3. But this ratio – the anemometer factor – in fact depends on wind speed and instrument dimensions (Patterson 1926), and has subsequently been shown to be yet more complex (Ramachandran 1970). It is not possible to determine theoretically the best design for a cup anemometer, and improvements have come about through wind-tunnel tests, in which each of the cup characteristics is changed individually.

From experiments on a range of anemometers with different cup diameters and shapes and various arm lengths (Acheson 1970, Hyson 1972) it was found that three cups are better than four, because the torque is more constant over a complete

Table 5.1 *The Beaufort Scale*

			Mean wind speed, V	
Name	Description	Force, B	Knots	m s^{-1}
Calm	Smoke rises vertically	0	0	0
Light air	Smoke drifts	1	2	0.8
Light breeze	Wind felt on face	2	5	2.4
Gentle breeze	Light flag extends	3	9	4.3
Moderate breeze	Dust raised	4	13	6.7
Fresh breeze	Small trees sway	5	19	9.3
Strong breeze	Umbrellas hard to use	6	24	12.3
Near gale	Whole trees move	7	30	15.5
Gale	Twigs break off trees	8	37	18.9
Strong gale	Slight structural damage	9	44	22.6
Storm	Trees uprooted	10	52	26.4
Violent storm	Widespread damage	11	60	30.4
Hurricane	No description	12	–	–

revolution and because three cups give more torque weight-for-weight. It was also found that semiconical cups performed better than hemispheres and that beaded edges to the cups rather than a plain finish made the instrument less sensitive to turbulence.

MacCready and Jex (1964) showed that the time constant of a cup anemometer varies inversely with the wind speed, and so it is more meaningful to express the time response of the system in terms of the length of the column of air that passes the anemometer while it responds to 63.2% of a step change in wind speed, this being called the *distance constant* (see Vane aerodynamics earlier in the chapter); it is the product of the time constant and the wind speed. Because the time constant varies with wind speed, the cups speed up more quickly with an increase of wind speed than they slow down with a decrease. The effect of this is that in a variable wind the mean speed measured by the anemometer is higher than it should be.

Schrenk (1929) derived a theoretical relationship expressing the extent of overrun in varying winds and found that the larger the cups, the longer the arms, the slower the changes in wind speed, the higher the mean wind speed and the smaller the mass of the moving parts, the smaller will be the overrun error. As the gustiness of the wind increases, the greater are the errors, so that when the wind fluctuations are large compared with the mean speed, the overrun can be as much as 30% for anemometers with design characteristics that are the exact opposite to those listed above. Nevertheless, Hyson (1972) believed that overrun generally amounts to about only 1% in the real world and that anything in excess of this is probably due to another source of error – the vertical wind. So what are we to believe?

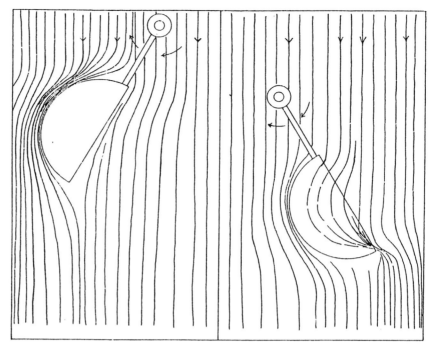

Figure 5.5. The flow of wind around a cup anemometer produces a higher pressure on the open side of the cups than on their backs, giving rise to a torque that rotates the shaft (Patterson 1926).

Sensing wind speed with propellers

The next-most-common sensor of wind speed is the propeller or fan anemometer, described by Dines in 1892. It has three or four blades in the form of helicoids rotating on a horizontal shaft. To keep the propeller facing into the wind, it is usually combined with a wind-direction sensor vane. What has been said about cup-anemometer response applies equally to the propeller, although being more linear it is more precise. There is the extra consideration, however, of the yaw angle, since the propeller will not always face directly into the wind owing to the time it takes to respond to a change of direction. Mazzarella (1954) made measurements that suggested that the ratio of measured to actual speed was $\cos^2 \alpha$, where α is the yaw angle. However, Monna and Driedonks (1979) suggested cos $(1.3\,\alpha)$, although this value may be dependent on wind speed. Again we are left wondering.

In the case of the combined propeller-and-vane construction, the vane performance may be affected by the propeller in a way that is not easy to quantify, and the gyroscopic effect of the propeller rotation may also produce an increase in the vane's damping ratio.

Because propellers are directionally sensitive, they also find an application in situations where it is necessary to measure the three-dimensional nature of air movement, such as in turbulence studies. Such an instrument consists of three propellers, mounted at right angles on a common mast, measuring the along-wind, across-wind and vertical-wind components. Each propeller measures the wind speed in line with its axis. The response to wind at an angle approximates to a cosine law, although with increasing angle the speed of response slows and the distance constant increases. As well as the speed, the direction of rotation of the propellers also has to be sensed.

Measuring cup rotation mechanically

The most usual way of measuring the rotation of a cup anemometer shaft for making manual observations is with a mechanical revolution counter (Fig. 5.6), readings being taken at set times (three-hourly or daily, for example), the difference between the readings giving the amount of wind run over the period (easily converted to average speed). This is obviously not a suitable method for obtaining spot readings of instantaneous speed; for these, hand-held cup anemometers are available, with mechanical or electronic displays, but they are not meant for precise work.

Measuring rotation by generator

For spot readings, the shaft may drive a permanent-magnet AC generator, producing a voltage proportional to speed, the voltage being displayed on a meter some

Figure 5.6. A manually read cup anemometer totals the number of revolutions of the shaft over a period, each turn being proportional to the passage of a known 'run of wind'. Given the elapsed time between readings of the counter, the total run of wind (or the mean wind speed) can be calculated.

Figure 5.7. An electrical anemometer. For automatic measurements, the revolutions of the cups are usually detected by a reed switch closed at each rotation by a magnet, as in this example, although optical sensors are also used. The inverted direction sensor measures shaft angle with a small continuous-rotation potentiometer.

distance away. The AC voltage can also be converted to a DC signal and logged at automatic stations as spot readings. An alternative is to measure the frequency of the AC signal rather than the voltage, giving either a spot reading or (by integration of the cycles) the run of wind. The starting torque of the larger generator anemometers can be high because the magnetic attraction between rotor and stator offers resistance to movement. Generators are used more widely with propeller anemometers than with cups.

Measuring rotation by switch

For the automatic logging of wind speed, cup anemometers that produce a switch contact closure are generally the most useful (Fig. 5.7), the switch being either a magnetic reed switch or an optical equivalent. In most designs, the rotating shaft has a magnet fixed to it that closes a switch once for each revolution. The closures are processed by a data logger, their total over a period being logged to give the run of wind or the average speed, with the possibility also of measuring the time between pulses so as to log instantaneous speeds or peak gusts.

In the pre-microelectronics era, a pulse for each revolution of the cups was too much to handle. To overcome this, a worm-reduction gear was included that caused a magnet to fall past the reed switch for the passage of every tenth of a kilometre (or of a mile) run of wind.

Reed switch operating lifetimes are around 10 million closures, but eventually they fail. An alternative is to use an optical disc through which an LED passes infrared light, which is then detected and produces one pulse per revolution. Alternatively, the disc may be encoded with an optical pattern that produces many pulses per revolution, which can be treated as a frequency or as individual pulses, giving an indication either of instantaneous speed or of total counts over a period. A slight disadvantage of the optoelectronic method is that it takes extra power, while reed switches require only that already taken by the logger, although consumption can be kept low enough to be acceptable.

Cup anemometers vary widely in size and shape, depending on their intended use. Some of the large and heavier designs, meant for general use over long periods, may not start to turn until wind speeds of around 0.5 to 1.0 m s^{-1} are reached, while the more sensitive can respond to speeds down to 0.1 m s^{-1}. Top speeds of between 50 and 75 m s^{-1} are usual. Most manufacturers do not quote a distance constant, but where they do it falls between one and three metres.

Pressure anemometers

Although cup and propeller anemometers dominate the field, other ways of measuring wind have been developed, such as sensing its pressure.

Pressure-tube anemometers

Henry De Pitot, a student of Reaumur in Paris, was superintendent of the Canal du Midi in Languedoc. His greatest claim to fame rests on his invention of the Pitot tube in 1732, which he developed for the measurement of the flow of rivers, rather than air. He was mistaken over the theory of its operation but it was nevertheless a useful invention that helped answer questions about the velocity of flow in rivers at different depths, then an unknown factor (Chapter 10).

The Swiss family of Bernoulli (Johann, Jacob and Daniel) were all mathematicians and somewhat competitive amongst themselves. It was Daniel Bernoulli, in 1738, who analysed pressure–velocity relationships and, although he was unable to derive a general expression, he solved some special cases. The equation

$$p_t = p_s + 0.5\rho u^2$$

is now generally known as the Bernoulli equation. Here the total head P_t is made up of two parts, the static head p_s and the speed head $0.5\rho u^2$, where ρ is the density, and u the speed of the fluid. Although derived for the measurement of the flow of water, the same principle also applies to the flow of air and so to the measurement

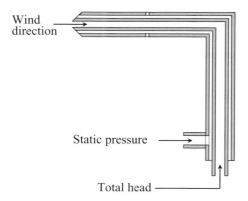

Wind direction

Static pressure

Total head

Figure 5.8. In a Pitot tube sensor, the difference in pressure between the total head and the static head is proportional to the square of the wind speed. The holes sensing the static pressure are in the horizontal tube in a modern design, although in an earlier design by Dines they were in the vertical tube, introducing some slight error.

of wind speed, the difference between the total and static heads being measured using a Pitot tube in the airstream (Fig. 5.8). The tube is kept facing into the wind, giving the total head, while the holes in the wall of the tube give the static head.

Dines (1892) wrote a paper comparing various types of anemometer, including the Pitot tube. His own design, now obsolete, differed slightly from the modern version described above in that the holes that measured the static head were in the vertical tube. This had the effect of causing some suction, thereby reducing the static pressure slightly. Dines describes a method of recording the pressure difference, transmitted through tubes from the sensor head, using a float in a tank of water with pulleys and threads to move a pen over a chart. He concluded that the measurement was very uncritical of alignment with the wind. In fact even with a misalignment of 15° or more, the error in speed indication was less than 1%.

Pitot tubes are not much used today, although it would be possible to measure the pressure difference quite easily with a modern electronic differential pressure sensor. However, a specialised design of pressure anemometer has been developed that measures the pressure on the surface of a tube held vertically in the air-stream through holes spaced at 90° around it, pointing at the four compass points. The tube is divided internally into four chambers behind the holes; differential pressure sensors are connected between the N and S chambers and between the E and W chambers, giving voltage outputs that vary in amplitude and polarity, from which the speed and direction of the wind can be calculated. An accuracy varying from ±0.5 to ±2.5 knots over the range 0 to 150 knots is claimed by the manufacturer. There is the dual advantage that the tube can be heated to prevent ice formation and does not have to be kept pointing into the wind.

Drag-force anemometers

When a rotating sensor is impractical, drag or thrust anemometers can be a useful alternative and several have been developed (Norwood *et al.* 1966). They can also be useful where the sensing of high-frequency wind fluctuations is important or where it is the dynamic pressure that is the primary concern (as opposed to wind speed). They furthermore have an application where three-dimensional wind measurements are required (Doe 1967, Smith 1970), serving much the same purpose as the three-propeller system described earlier, but with a faster response.

The earliest form of anemometer was a plate, kept at right angles to the wind and free to swing about a horizontal axis, the angular deflection due to the pressure of the wind being a function of wind speed. The normal-plate anemometer is a development of this, consisting of a plate held stationary and perpendicular to the wind. Dines (1892) describes such an instrument, comparing it with a Pitot tube, cup and propeller anemometer, and shows how its readings can be recorded using a float in water similar to the one he describes for the Pitot tube.

The drag force on a body is given by $F = \frac{1}{2} A\rho u^2 C_d$ where F is the drag force, A the area of the body, ρ the density of the air, u the air speed and C_d the drag coefficient of the body (which is dimensionless, and typically around 0.7). Drag anemometers are usually in the form of a cylinder for two-dimensional measurements (not being badly affected by vertical components in the wind) or a sphere for three-dimensional wind studies.

In the 1880s Osborne Reynolds developed his theories of fluid flow, forming the basis of present-day aerodynamics (Kuethe & Schetzer 1950). Part of Reynolds' work concerned *bluff bodies* (cylinders and spheres), for which the major cause of fluid drag is *form drag*, whereas for a streamlined object the main cause of drag is skin friction. He found that as the speed, u, of the fluid (in this case air) increased, changes occurred in the nature of the flow, depending on the size of the object (in the case of a cylinder its diameter, d), on the viscosity of the fluid, μ, and on its density, ρ, the ratio $\rho u d/\mu$ being known as the Reynolds number, R_e.

Above a certain critical value of R_e, the flow switches from smooth to turbulent, going through various intermediate stages as the speed increases. Up to an R_e of 3, the flow is entirely smooth; from $R_e = 3$ to 40, vortices form in the shadow of the cylinder, but the flow re-forms downstream; above an R_e of 40 it does not re-form, while above about 70, a train of eddies, known as a *Karman vortex street*, develops (Fig. 5.9). Above $R_e = 200$ the vortices break up into smaller regions of vorticity, the street re-forming downstream. When R_e reaches about 10^5, the stresses become so great that turbulence occurs even in the *laminar boundary layer* adhering to the bluff object, whereupon the drag force suddenly drops. These changes are of significance in the design of a drag anemometer. For example, for a cylinder with a diameter

Figure 5.9. Eddies form what is known as a Karman vortex street downstream of a body exposed to the wind, their frequency giving a measure of wind speed. In the illustration, they are forming in the wake of a cylinder at a Reynolds number of about 70.

Figure 5.10. A drag-force anemometer senses the force of the wind as it passes round a fixed shape, in this case a cylinder, here temporarily shielded from the wind by a cover, allowing the zero setting to be checked. It is part of the cold regions automatic weather station (AWS) discussed in Chapter 18.

of 8 cm the speed at which the drag force suddenly drops is 50 m s^{-1}; a larger cylinder might, therefore, mis-measure higher speeds. While the drag coefficient is not constant for all values of R_e, for an 8-cm cylinder it is relatively so for R_e values of 100 to 10^5, covering most wind speeds.

Instruments of this type usually sense the pressure of the wind on a cylinder or sphere, held stationary, by means of strain gauges fixed to the structure on which the cylinder or sphere is mounted (Fig. 5.10). Other methods have been used in which the drag body moves slightly against a spring, the magnetic coupling between a ferrite core and four coils being used to detect the movement. To prevent oscillation, oil damping may be used. An advantage of drag anemometers is that both speed and direction can be measured by the one instrument. Also, by heating the drag body it is possible to deter the formation of rime ice or the adherence of wet snow, although in one design ice was removed pneumatically (Chapter 18).

However, drag anemometers have many problems – the zero can drift and because of their square-law response, their accuracy varies greatly across the speed range, being poor at low speeds if a wide velocity range has to be measured. They can also be affected by rain.

Vortex anemometers

Another somewhat specialised instrument uses the fact that in the lee of an obstruction, such as a cylinder, Karman vortex streets form (see above, and Fig. 5.9). These vortices are not shed randomly but at a very specific frequency, which Strouhal defined as

$$f = u/Sd,$$

S being the Strouhal number, u the velocity of flow, d the diameter of the cylinder and f the frequency. S is constant, at about 5, throughout all the Karman-vortex-street range of Reynolds numbers. So, for example, at a speed of 50 m s^{-1} and with a cylinder of diameter 1 cm the frequency is 1000 Hz, while at 5 m s^{-1} it is 100 Hz. The spacing of the vortices is about 2.5 times the diameter of the cylinder. An advantage of the method is that the frequency is linearly related to speed and so (theoretically) no calibration is required. Accuracy in operation is simply a matter of ensuring that the vortices are reliably detected. There can be no zero drift or calibration change.

The vortices can be detected by passing an ultrasonic sound wave through their path, the speed and direction variations within the vortices modulating the beam; an alternative method of detection is the use of a *hot wire sensor*. The construction of a vortex anemometer is very much like that of a propeller anemometer; it has a vane to sense the wind direction, and the vortex-producing cylinder, with its down-wind vortex sensor, is mounted on the vane's arm to keep it pointing into the wind.

Thermal anemometers

There are three main types of sensor that use heat to measure wind speed, and, in some cases, direction also. These are heat-transport, heat-pulse and heat-loss anemometers, the latter being the most common although none have wide use; they are all designed for situations in specialised micro-meteorological research where a rapid response is needed. All methods involve heating a small resistance wire by passing a current through it and measuring the effects caused by the wind.

Heat-transport anemometers

A heater wire, powered continuously, produces a stream of heated air which is carried downwind (Taylor 1958, Dyer 1960). Mounted each side of the heater wire, and just a few millimetres away, are heat-sensing detectors, usually in the form of fine platinum wire resistance thermometers. By mounting the detector wires at right angles to the heater wire, the heater being vertical, a wide yaw response is achieved. Wind speed is measured by detecting the difference in temperature between the detector wires resulting from the transport of heat (advection) in the direction of the wind. These instruments have a cosine response over a wide range of angles and can indicate from which direction the wind vector is oriented. However, calibration is difficult at low speeds and, as with all wind sensors using heat, they can be affected by rain and solar heating, although with screening this can be reduced.

Heat-pulse anemometers

The physical design of this sensor is similar to that of the heat-transport anemometer, but the heater wire is pulsed to a high temperature by the passage of current for a few microseconds, introducing a tracer of heated air, the time of flight to the detector wire giving the wind velocity. The method has the advantage of being sensitive to very low wind speeds, although its upper limit is restricted to about 15 m s^{-1}. It has, however, a very fast response, with a short distance constant that is dependent on the distance between the wires, which can be made very small; because it measures the time for the pulse to travel a known distance its output is directly related to wind speed. Wind direction can also be inferred, but the electronics is complex.

Heat-loss anemometers

In the previous two cases, heat transported downwind was detected to give a measure of wind speed. In heat-loss instruments, the amount of heat lost from a heated wire is measured.

In the *hot-wire anemometer*, a short length of platinum resistance wire is supported vertically in the wind and heated by passing a current through it, the rate of heat loss being measured. The loss is made up of two components – that which occurs in still air due to radiation and convection from the wire and that due to the advection of heat when a wind moves past the wire.

The theory of this process was analysed by King (1914), who showed that if the resistance wire was heated to a constant temperature (and thus its resistance was also constant), $I^2 = I_0^2 + Bu^{1/2}$ where I is the total current, I_0 is the current in still air, B is a constant determined by experiment and u is the wind speed. The majority of hot-wire anemometers operate in this constant-temperature mode (Perry & Morrison 1971) because it gives a better response at higher wind speeds than the alternative, simpler, constant-current method, in which the temperature of the wire varies with wind speed (Miyake & Badgley 1967).

If platinum wire with a diameter of 0.002 mm is used, the time constant of the system is very small and the sensor can follow changes of speed down to hundredths of a second. However, this results in a rather delicate instrument and for more rugged designs thicker wires or hot films are used, made by depositing a thin layer of platinum onto an insulating support, such as a cylinder a millimetre or so in diameter.

Heat-loss anemometers are affected by the ambient air temperature unless they are operated well above it. For this reason temperatures as high as 900 °C may sometimes be used, but this can be a problem at low wind speeds because there can be excessive free convection. The simpler sensors do not detect the direction of the wind, although by operating two horizontally deployed wires at right angles, the direction can be calculated (Gjessing *et al.* 1969). The Viking Mars Lander used a multiple hot-film sensor for measuring the speed and direction of Martian winds (Hess *et al.* 1972). Some designs allow three-dimensional measurements of wind speed. Unlike the heat-transport and heat-pulse sensors, the hot-wire anemometer does not have a linear response to wind speed, as the above equation, derived by King, shows.

The great majority of thermal anemometers are used for research into turbulence, only a few finding applications where cups might normally be expected instead. The small number of thermal anemometers available commercially have an accuracy of about ±5% of the actual speed and ±3° to 4° in the direction. However, hand-held hot-wire sensors with a portable meter are more common and these can be particularly useful for measuring low wind speeds in small spaces where portable cup anemometers may not be practicable. A typical specification for this type of instrument ranges from ±3% at the high end of the speed range to ±10% at the lower end.

Sonic anemometers

The speed of sound in moving air is its speed in still air plus the speed of the air. By measuring the difference in time it takes sound to travel with the wind and against it, its speed can therefore be calculated. If the wind is moving at an angle to the sound wave, it is the wind speed component in that direction that is measured, but by keeping the sound path directed in line with the wind (as with a propeller or vortex anemometer) the actual wind speed and direction can be measured. In practice, though, this is usually achieved by using two fixed sound paths at right angles. (The temperature also needs to be measured because the speed of sound is temperature dependent.)

It can be shown that $u = (c^2/2l)\,(t_2 - t_1)$

where u is the speed of the wind, c the speed of sound, l the path length and t_1 and t_2 the times for the sound to travel with and against the wind.

Figure 5.11. This ultrasonic anemometer consists of three pairs of transducers, the ones at the top facing those diagonally opposite at the bottom. Each pair acts alternately as a transmitter and receiver of ultrasonic pulses of sound, the time of flight between each pair giving a measure of the sound in the direction of the line between them and, thus, by vector calculation, of the velocity of the wind in three dimensions.

The sound can be either a continuous wave or a short pulse. In the former, the difference in phase of the two sound waves is measured while in the latter it is the time of flight of the leading edge that is detected. In early designs, continuous waves were used (Kaimal & Businger 1963), but reflections and extraneous noise from outside caused problems, as did the precise detection of the phase difference, especially near zero wind speed. However, the pulse method requires precise detection of the leading edge (Mitsuta 1966) and this too has its problems, although currently it is the most-used method. Three sound paths are often now included, allowing wind speed and direction to be determined in three dimensions (Fig. 5.11).

The technique has a good cosine (i.e. angular) response and can work over a speed range of 0.02 to 40 m s^{-1}. Because the measurements are made over a path length, as distinct from the single-point readings of all the other methods, the result is an average of the speed and direction within the sound path. Rain can cause some loss of data, although the sloping face of the lower sensors (Fig. 5.11) minimises this, and at certain wind directions the physical presence of the sensors produces a wake that affects the readings. Accuracies for speeds averaged over ten seconds are in the region of ±3% to ±5% of full scale, depending on the speed, and of ±2° to ±4° for direction. They have the advantage that there are no moving parts to wear out, but their high cost and the need for power to operate them go against their general use, although this may be beginning to change. However, they are essential tools for research into turbulence, evaporation and other flux measurements (see Chapters 7 and 17).

References

Acheson, T. A. (1970) Response of cup and propeller rotors and wind direction vanes to turbulent wind fields. *Met. Monogr.*, **11, No. 33**, 252–61.

Dines, W. H. (1892) Anemometer comparisons. *Q. J. Roy. Met. Soc.*, **17**, 165–85.

Doe, L. A. E. (1967) A series of three-component thrust anemometers. In *Proceedings of the International Canadian Conference of Micrometeorology, Part 1*, pp. 105–14.

Dyer, A. J. (1960) Heat transport anemometer of high stability. *J. Sci. Inst.*, **37**, 166–9.

Gjessing, D. T., Lanes, T. & Tangerud, A. (1969) A hot wire anemometer for the measurement of the three orthogonal components of wind velocity, and also directly the wind direction, employing no moving parts. *J. Sci. Inst.*, **2**, Series 2, 51–4.

Hess, S. L., Henry, R. M., Kuettner, J., Leovy, C. B. & Ryan, J. A. (1972) Meteorology experiments: the Viking Mars Lander. *Icarus*, **16**, 196–204.

Hyson, P. (1972) Cup anemometer response to fluctuating wind speeds. *J. Appl. Met.*, **11**, 843–8.

Kaimal, J. C. & Businger, J. A. (1963) A continuous wave sonic anemometer thermometer. *J. Appl. Met.*, **2**, 156–64.

King, L. M. (1914) On the convection of heat from small cylinders in a stream of fluid: determination of the convection constants of small platinum wires with application to hot-wire anemometry. *Phil. Trans. A*, **214**, 373–432.

Kuethe, A. M. & Schetzer, J. D. (1950) *Foundations of Aerodynamics*. John Wiley & Sons, London.

Lockwood, J. G. (1974) *World Climatology: An Environmental Approach*, p. 330. Edward Arnold, London.

MacCready, P. B. & Jex, H. R. (1964) Response characteristics and meteorological utilisation of propeller and vane wind sensors. *J. Appl. Met.*, **3**, 182–93.

Mazzarella, D. A. (1954) Wind tunnel tests on seven aerovanes. *Rev. Sci. Instrum.*, **55**, 63–8.

Mitsuta, Y. (1966) Sonic anemometer-thermometer for general use. *J. Met. Soc. Jpn.*, **44**, 12–24.

Miyake, M. & Badgley, F. I. (1967) A constant temperature wind component meter and its performance characteristics. *J. Appl. Met.*, **6**, 186–94.

Monna, W. A. A. & Driedonks, A. G. M. (1979) Experimental data on the dynamic properties of several propeller vanes. *J. Appl. Met.*, **18**, 699–702.

Norwood, M. H., Cariffe, A. E. & Olszewski, V. E. (1966) Drag force solid state anemometer and vane. *J. Appl. Met.*, December 1966.

Patterson, J. (1926) The cup anemometer. In: *Proc. R Soc. Canada, Series III*, **XX**, Meeting of May 1926, pp. 1–54.

Perry, A. E. & Morrison, G. L. (1971) A study of the constant-temperature hot-wire anemometer. *J. Fluid Mech.*, **47**, 577–99.

Persson, A. (2000) Back to basics: Coriolis: Part 1 – What is the Coriolis force? *Weather*, **55**, 165–70.

Persson, A. (2002) The Coriolis force and the subtropical jet stream (Coriolis: Part 8) *Weather*, **57**, 53–9.

Ramachandran, S. (1970) A theoretical study of cup and vane anemometers, Part II. *Q. J. Roy. Met. Soc.*, **96**, 115–23.

Schrenk, O. (1929) Errors due to inertia with cup anemometers in fluctuating winds. *N.Z. Tech. Phys.*, **10**, 57–77.

Smith, S. D. (1970) Thrust anemometer measurements of wind turbulence, Reynolds' stress and drag coefficient over the sea. *J. Geophys. Res.*, **75**, 6758–70.

Taylor, R. J. (1958) A linear unidirectional anemometer of rapid response. *J. Sci. Inst.*, **35**, 47–52.

Wieringa, J. (1967) Evaluation and design of wind vanes. *J. Appl. Met.*, **6**, 1112–14.

6

Barometric pressure

It stood very low – incredibly low, so low that Captain MacWhirr grunted. The match went out, and hurriedly he extracted another, with thick, stiff fingers. There was no mistake. It was the lowest reading he had ever seen in his life. Perhaps something had gone wrong with the thing! There was an aneroid glass screwed above the couch. He turned that way, struck another match and discovered the white face of the other instrument looking at him from the bulkhead, meaningly, not to be gainsaid, as though the wisdom of men were made unerring by the indifference of matter. There was no room for doubt now. The worst was to come, then – and if the books were right this worst would be very bad.

Joseph Conrad *Typhoon.*

The variable and its history

The Torricelli experiment did not come about by accident, but had its origins early in seventeenth century Italy in a question first asked 2300 years earlier by Aristotle as to whether a vacuum could be made or could exist naturally. Aristotle's view was that it could not, because he believed that there would be no dimensions in a vacuum – no up, down, north, south, east or west, and that light could not pass through it. This was still a common view even in the seventeenth century. There was also uncertainty over whether air had weight or exerted pressure on objects submerged in it. The same doubts existed concerning water.

A group comprising Giovanni Baliani, Gasparo Berti, Isaac Beckman, René Descartes, Athanasius Kircher, Rafael Magiotti, Emmanuel Maignan, Jean Rey, Vincenzio Viviani, Nicolo Zucchi, Michelangelo Ricci and of course Evangelista Torricelli and Galileo Galilei were debating these matters earnestly in the early 1600s in Italy; the development of the barometer was not done alone by Torricelli working alone.

Measuring the Natural Environment, second edition, Ian Strangeways. Published by Cambridge University Press. © Ian Strangeways 2003.

A crucial step was taken around 1641 in Rome by Berti, who did an experiment that proved for the first time the testable existence of a vacuum. The idea might not have been Berti's own but possibly Magiotti's or even Galileo's. The test used equipment that resembled what would now be called a water barometer; full details of the experiment are well documented by Middleton (1994). The equipment was *not* a barometer, it was designed to make and to test the reality of a vacuum, not to measure barometric pressure. Middleton (1994) suggests that an experiment of comparable significance today would receive a prestigious honour, yet it was not much discussed and few came to hear of it. The reason for this becomes clear later. Torricelli probably did not see the experiment performed and only heard of it later in a letter from Magiotti.

In 1641 Torricelli went to work with Galileo in Florence, but Galileo was 77 and had just three months to live. Torricelli then took over from him as mathematician in the court of Tuscany, only to die himself six years later aged 39. Among Galileo's papers Torricelli found reference to mining engineers finding it impossible to raise water by single-stage suction-pump to a height of more than about 10 m. From this complex background emerged the Torricelli experiment. It is thought that he did not actually do it himself but explained to Viviani, a close friend, exactly what to do, what would happen and why. After Viviani had done the tests, Torricelli sent his famous letter to Ricci on 11 June 1644.

As regards the aim of the experiment, Torricelli says it was '... not simply to produce a vacuum, but to make an instrument which might show the changes of the air, now heavier and coarser, now lighter and more subtle.' As to the means, Torricelli says 'When these (glass tubes) were filled with quicksilver, their mouths stopped with the finger, and turned upside-down in a vase which had some quicksilver in it, they were seen to empty themselves, and nothing took the place of the quicksilver (in the tube), the neck (height of the mercury column) always remained full to a height of an ell and a quarter and a finger more.' As regards the cause of this, Torricelli explains that 'It has been believed until now (not everyone did) that it was something inside the tube, either from the vacuum, or from that extremely rarefied stuff (aether or ether), but I assert that it (the cause) is external.... We live submerged at the bottom of an ocean of elementary air which is known by incontestable experiment to have weight.... On the surface of the liquid in the basin presses the height of 50 miles of air; yet what a marvel it is, if the quicksilver... rises there to the point at which it is in balance with the weight of the external air that is pressing on it!'

Textbooks usually give 1643 as the date of the experiment, although it is probable that Torricelli was simply planning it then. His letter of June 1644 looks, to Middleton (1994), like a report of an exciting result just obtained, not reported a year later, making 1644 the most probable year of the experiment. Like the earlier

Ricci experiment, the work was kept very quiet; secrecy surrounded it. Why? All those involved were associates of Galileo and remembered all too well what had happened to him at the hands of the Catholic Church in 1633 in his Trial by Inquisition over his view, based on his astronomical discoveries (made with the newly invented telescope), that the Earth went round the Sun, as Copernicus had said – the Heliocentric Theory. Torricelli and his colleagues felt that their punishment would be less lenient than that of the great man himself (which was mere threat of torture on the rack, not carried out, and house arrest for the rest of his life, which was). The idea of a vacuum was seen as heretical by the church because of religious dogma and hostility to the atomist theories of Democritus and Lucretius. The Church was omnipotent; safer to let the whole thing drop, and so no mention was made of it again until after Torricelli's death.

Shortly after, Pascal and Perrier found that the pressure at the top of a mountain was less than at its base and it was also soon realised that the barometer reading varied with the weather. Robert Boyle was studying in Italy at the time, heard of the Torricelli experiment and on return to England formulated his theories describing the relationship between the pressure, temperature and volume of gases – Boyle's laws.

Fifty years prior to the Torricelli experiment, however, a simple device known as the *Dutch weather glass* or *thunder glass* was in production (Fig. 6.1) in which a liquid, usually coloured water, moved up and down the spout in response to changes in barometric pressure. Its existence suggests that the idea of barometric pressure, and its relation to the weather, was already astir before Torricelli. Bolle (1982) cites a document from 1619 in Ghent that describes an instrument invented by Gijsbrecht de Donckere, an engineer in Holland, that 'showed the several moods of the weather', and it may well have been in production some time before the Ghent paper (Crane 1981). Middleton (1994) makes no mention of any of this; perhaps it is just too much of a 'toy' for him. And so it is. The weather glass is what is now called a *sympiesometer* – a barometer in which the column of liquid has air or other gas above it instead of a vacuum; such a device must compensate for the expansion of the gas above the liquid as it warms. Although the glass appears to have been used to predict the weather, it is most improbable that any of the principles were understood. It was simply that, by chance, it had been noted that there was a vague association between the weather and the level of the liquid, no more. It was left to Torricelli to understand the principles. This is like the situation with evolution, the idea of which was around some time before Charles Darwin, who, along with Alfred Russell Wallace, explained the process of evolution through the mechanism of natural selection, putting the idea on a sound scientific footing. So it is difficult to pinpoint just when and where the idea of barometers originated; this is not an unusual situation in the history of science. But without doubt Torricelli was the

Figure 6.1. A reproduction eighteenth century Dutch Weather Glass. The coloured water in the spout moves up and down in response to changes in barometric pressure. Due to the expansion and contraction of the air in the container, the water also moves up and down in response to temperature changes, and corrections would need to be made if it was an accurate instrument. However, it is only an historical curiosity and a decorative ornament.

inventor of the barometer. He also pioneered the understanding of the principles involved and the concepts of atmospheric weight and pressure. ('Barometer' is from the Greek root *baros* meaning weight.)

Barometric pressure changes with the weather because the number of air molecules directly overhead varies, changing the weight of the column, the main cause of this being changes of air temperature, and thus its density. Changes can occur to differing amounts at different heights in the same column, such that at one height the air cools while at another it warms, the integrated effect over the full depth of the atmosphere controlling the total weight at the surface. Thus at one height a barometer may indicate a rise in pressure while at another it may show a fall. As a general rule, barometric pressure falls in response to rising temperatures aloft, as the air density falls, and rises when cooling occurs (Burton 1993). Changes of temperature can occur, for example, through solar heating near the ground, or when a cold or warm front passes. Changes in pressure are also influenced by the more complex processes of *convergence* and *divergence*, in which air molecules are added to or extracted from the air column, usually at altitudes of from 8000 to 12 000 m (Young 1994). All of these processes are interlinked.

Units

The SI unit of force is the *newton* (symbol N) which, by definition, is 10^5 dynes, which is equivalent to 0.102 kg.

The SI unit of pressure is the newton per square metre (N m^{-2}), which has been given the name *pascal* (symbol Pa). So 1 Pa = 1 N m^{-2}.

However, in meteorology the *millibar* is used as the unit of pressure (1/1000 bar). The international symbol for this is *mbar*, but some meteorologists use *mb*.

By definition, 1 bar = 10^5 N m^{-2}, or 10^5 Pa.

So 1000 mb = 10^5 Pa, or 1000 hPa (hectopascals), making hectopascals interchangeable with millibars. Hectopascals are rapidly being adopted as the preferred unit in meteorology.

But 1 Standard Atmosphere (atm) = 1.0132 bar (1013.2 mb, or 14.7 lb in^{-2}). So 1 bar = 0.9869 atm.

Expressed as the equivalent height of a mercury column:

> 1 bar = 750.06 mm Hg (29.53 in), and so,
> 1 atm = 760.00 mm Hg (29.92 in), both at 0 °C.

The equivalent of 1 atm for a column of water is 10.33 m (33.9 ft) at +4 °C.

Modern mercury barometers

Despite the church abhorring a vacuum, mercury barometers became commonplace in the second half of the seventeenth century, although mostly for domestic use and decoration rather than for scientific study. These early barometers are now very collectable antiques (Banfield 1985). Initially they were used more as altimeters, but their meteorological significance soon became appreciated.

Mercury barometers are still amongst the most accurate instruments for measuring barometric pressure, although aneroid barometers (see below) are now becoming their equal in precision, while their greater ruggedness and portability make them much easier to use. Carefully maintained mercury instruments, however, are still the primary standards.

Soon after the Torricelli experiment was performed, there were many improvements made to the basic inverted tube – to expand the scale, correct errors, increase accuracy, improve portability and use different liquids in place of mercury. The diversity and ingenuity of designs evolved in the following centuries is impressive, as Middleton illustrates beautifully in his book on barometer history (1994), but here we will look at just the two barometers that came to dominate meteorology and set the standards in precision for scientific routine use.

The Kew Pattern barometer

The Kew Pattern (Fig. 6.2) has been used widely throughout the world for synoptic and climatological applications since about 1855 when it was designed by John Welsh at Kew Observatory, specifically for use on ships. The firing of guns and the motion of ships had long posed problems, probably first addressed by Robert Hooke around 1667 and later by Admiral Robert Fitzroy in the 1860s while Magellan in the eighteenth century appears to have had a barometer modified to cope with the hazards of ships. The Kew pattern differs from the original seventeenth century design in small, but important ways. The cistern is sealed to protect it from dust by a leather washer, with small holes to allow air pressure changes to diffuse through. The internal diameter of the cistern is precisely 5 cm while the glass tube emerging from it has a bore of 1.6 mm, widening part-way-up to form an air trap, narrowing again before opening at the top to 8 mm, giving the meniscus a broad surface for accurate reading. The mercury is distilled and filtered to remove impurities that would affect its density and to guarantee the formation of a meniscus that does not stick to the tube.

A tube, with a slotted window, surrounds the mercury column, engraved in millimetres or inches of mercury, in millibars, or hectopascals, a vernier scale allowing readings to be made to 0.1 hPa (Fig. 6.3). Because of the precise bores of tube used, only one setting – that of the vernier – is required, the vernier taking account of the change of mercury level in the cistern.

Two models are usually available from a manufacturer, one for sea level to 450 m (1500 ft), spanning the range 1060 to 870 hPa, another covering the range 1060 to 780 hPa for higher altitudes (up to 1070 m or 3500 ft).

The Fortin barometer

The second most commonly used mercury barometer remains unchanged in concept since it was designed by Nicholas Fortin around 1810 (Fig. 6.2). Unlike the Kew

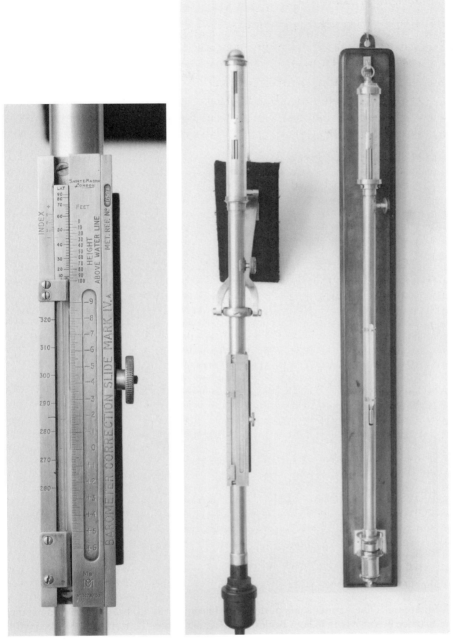

Figure 6.2. On the right is a barometer by Fortin and to its left (in the centre), one designed at Kew. Above the cistern of the Kew model is a 'Gold Slide' (seen in close-up at the left of the figure). The slide was designed for use aboard ships for the correction of gravity, temperature and height above the water level, during the ship's travels around the globe. The slide centres on a thermometer and a movable scale, marked in height above water level (on the ship), the height being set against latitude. The required correction is read from opposite the thermometer mercury column. Although it makes approximations, the resulting correction is acceptable.

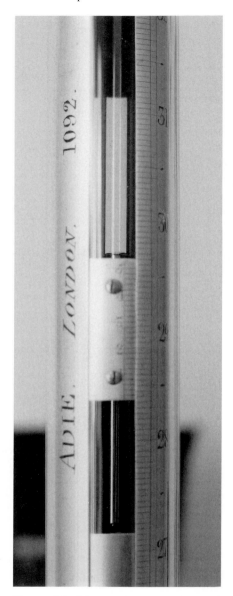

Figure 6.3. The vernier scale of this Kew barometer is adjusted by the knob some distance below it (Fig 6.2). A London instrument maker, Patrick Adie, worked with John Welsh at Kew Observatory, where they developed the barometer based on recommendations made at the International Conference on Marine Meteorology, in Brussels in 1853.

pattern it requires two adjustments, a vernier scale at the top of the mercury column, just as for the Kew instrument, the second being at the bottom in the cistern. This is because the tube bores are not made to precise dimensions. To allow for this, the mercury level in the cistern has to be adjusted to a fixed level. To facilitate this, the mercury is made visible by using a glass cistern, in which is fixed a downward-pointing needle, originally made of ivory (Fig. 6.4). The base of the cistern is made of flexible leather supported on a plate that can be moved up and down with a screw, thereby adjusting the level of the mercury until it contacts the point of the needle – the *fiducial point*. This ensures a precise zero reference, against which the level of the mercury in the main tube is measured (Fig. 6.5). In this design, the vernier does not need to allow for the level changes in the cistern, since they are kept constant at the fiducial point. Despite the general view that Fortin invented these various design features, this was not the case, but what he did do was to bring them all together into one excellent instrument. The design has been in wide use throughout the nearly 200 years since Fortin first made the prototype. As with the Kew pattern, two models are usually available, with the different ranges of altitude cover.

It is rather more portable than the Kew design, although the transport of any mercury barometer is not easy. Many early design modifications were aimed at overcoming this problem, especially for use on ships (Barlow 1994), but it was not until the invention of the aneroid barometer that the problem was satisfactorily resolved.

Accuracy and calibration

Mercury barometers give the correct reading at $0\,°C$, but because the density of mercury (and so its weight) changes with temperature, a correction must be made if readings are made at other temperatures, and for this barometers are supplied with a thermometer. Barometers should not be exposed to direct sunlight (or other heat) because this causes temperature gradients along the length of the mercury column which introduce errors. If gravity is different from the standard $9.80665\ \mathrm{m\ s^{-2}}$, a correction must be applied, because the value of g also affects the 'weight' of the mercury. If small amounts of air or other gas gain entry into the vacuum above mercury, the readings will be in error to an extent depending on the level of the contamination and on its nature. The World Meteorological Organisation (1996) specifies that the reporting resolution for mercury barometers should be 0.1 hPa, while the maximum permitted error, at 1000 hPa, is ±0.3 hPa.

The wheel barometer

Although not designed for scientific use, the *wheel barometer* is worth commenting on here simply because of its widespread popularity as a domestic instrument

Figure 6.4. The mercury cistern of a Fortin barometer has a screw at its base that allows the level of the mercury to be moved up or down via a flexible base until it just touches the *fiducial point*. This corrects for level changes in the cistern as the mercury in the column moves up and down, providing a fixed reference point against which the meniscus in the main tube is measured.

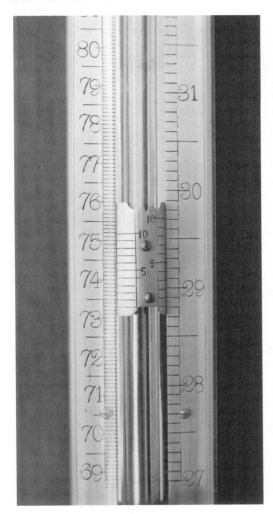

Figure 6.5. The level of the mercury meniscus of a Fortin barometer is measured with a vernier scale, just as it is in the Kew design, but because of the initial setting to the fiducial point, the Fortin vernier does not need to compensate for changes in the mercury level in the cistern, as is required in the Kew.

(Fig. 6.6). The prototype is ascribed to Robert Hooke, in which the movement of the mercury in the cistern is sensed by a float, a wire and counterbalance turning a wheel to which is attached a pointer that moves round a dial (Fig. 6.6, right panel). It became known as the Banjo barometer and was one of the most popular and cheapest domestic barometers (Banfield 1976, McConnell 1994). While it remains an interesting and common antique, it lost favour when the aneroid barometer was invented. These have now largely replaced mercury barometers, even in professional

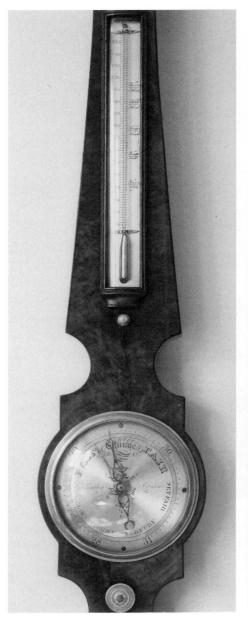
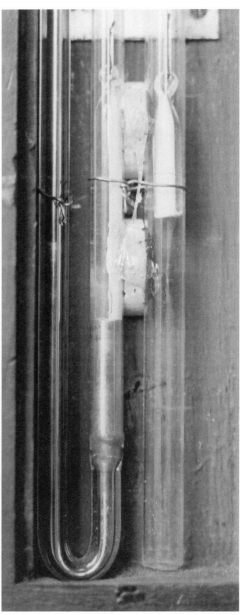

Figure 6.6. In a wheel barometer, a pointer (left) is rotated by a glass float in an upturned cistern (centre-right) which senses the changes of mercury level. Balanced by a counterweight, the float moves a cord over a wheel to which the pointer is attached. This design removes the need for the careful manual adjustment of a vernier scale, but it is not an accurate instrument.

applications, having been refined to the point where they can be just (almost?) as accurate as the Kew and Fortin instruments.

Aneroid barometers

The idea of the aneroid barometer originated around 1700 when Gottfried Wilhelm Leibniz described in a letter to Johann Bernouilli a 'little closed bellows which would be compressed and dilate by itself as the weight of the air increased or diminished'. The technology of the time, however, was unable to make it. Then in 1797 Nicolas Jacques Conté proposed a similar instrument in the shape of a pocket watch, made of a solid brass container with a thin flexible steel top. The top was supported internally by springs and the container evacuated of air, the deflections due to changes in air pressure being magnified and displayed by a pointer. Although it was actually fabricated, the available technology was still insufficient and it was very temperature sensitive. Forty-six years later, however, a practical instrument was finally made, in 1843, by Lucien Vidie. He also coined the name *aneroid*, which means 'without liquid', and took out patents in 1844–46 in France, the UK and the US. However, a pressure gauge was also patented in 1846 by Schinz, a German railway engineer, based on a curved tube with an elliptical cross section which changed its curvature when subjected to internal pressure. These gauges were in use as steam-pressure gauges on trains in Germany by 1848. Bourdon patented a very similar instrument three months later, although it is most probable that he invented it quite independently. He patented nearly every possible form that its construction could take – curved, coiled in a helix and twisted. Both the Vidie and Bourdon gauges were shown at the Great Exhibition of 1851 in London and both were awarded a Council Medal. Subsequently legal squabbles started, Vidie taking legal action against Bourdon. Vidie's design, however, was so similar to Conté's that he lost his case as well as losing two subsequent appeals. A third appeal succeeded, and Vidie was awarded damages, but by then the patent had expired and renewal was refused. Bourdon profited well from his designs because he saw its wider use in industry, but Vidie, being a poor business man, failed to make money, others reaping considerable financial rewards instead.

The non-recording aneroid barometer

Despite this acrimony, Vidie's capsule is the one used in today's barometers, and as with mercury barometers many models have been made for domestic use (Fig. 6.7).

The Vidie aneroid sensing element is a circular, thin, corrugated metal capsule about 5 cm in diameter and 3 mm deep, with the air removed from it. Changes in

Figure 6.7. Made in about 1880, this aneroid barometer is of the type known as an 'open dial', since the mechanism inside is visible. It has two curved thermometers, the one with the centigrade scale (left) is filled with spirit, the Fahrenheit (right) with mercury. This is traditional, but the reason is not clear. Inside the case, a large spring (top) supports the capsule. As the spring moves up and down in response to pressure changes the movement is magnified by the straight rod (right), its movement being converted, via the *arbor chain* and hair spring (centre), to a rotation of the pointer.

pressure compress or relax the capsule, causing it to contract or expand. One side is fixed to a base, the other moving as the pressure changes, a spring preventing the capsule collapsing under the pressure of the atmosphere. Depending on design, the spring may be outside or inside the capsule, or alternatively the capsules can be made of tempered steel, doubling-up as the spring (Middleton 1994). The movement of the free side is magnified mechanically, moving a pointer round a dial. The pressure ranges covered are similar to that for a mercury barometer.

To compensate for small dimensional changes in the capsule and in the lever mechanisms due to temperature fluctuations, a small amount of air may be left in the capsule, its expansion as the temperature increases acting in opposition to the unwanted temperature movements. An alternative method is to make one of the levers in the mechanism a bi-metal strip, which bends and so compensates as the temperature changes. As the balancing of atmospheric pressure is against a spring, not against the weight of mercury, no correction is required for variations in g. Using modern materials and precision engineering, aneroid barometers can now be made as accurate as mercury barometers and, because of their greater convenience of operation and portability, have largely replaced the Fortin and Kew pattern, even in professional applications.

But perhaps they are not quite so perfect. They do require regular checks because of slow changes in the elasticity of the capsules and of the supporting spring, and because of the possibility of air leakage into the capsules (which is nonetheless rare), which require annual checks to be made with a standard instrument, and this is preferably a mercury barometer. The World Meteorological Organisation (1996) specify that errors at any one point on the scale should not be greater than ± 0.3 hPa and that the barometer should retain an accuracy of ± 0.2 hPa over a period of months. Readings should be taken to the nearest 0.1 hPa. For greater precision, individual calibrations can be carried out, and a card showing the required corrections, unique to that instrument, supplied with each barometer. These corrections may vary across the full range and such instruments still need periodic rechecking. So perhaps they are not quite the equal of a good mercury barometer. They are like convenience food.

Barographs

Barograph is the name given to a barometer that records the pressure changes on a paper chart. Before the aneroid capsule was available, more than 100 barographs were designed which, by one means or another, recorded the change in level of the mercury column, ranging from derivations of the wheel barometer, to some acting as a balance, others using flotation, photographic methods and electrical techniques (for details see Middleton 1994). With the arrival of the aneroid capsule, barographs became simpler to design. They are so familiar as to hardly warrant detailed description, but some comments are necessary.

A single aneroid capsule is sufficient to move the pointer of a non-recording instrument, but a barograph needs several in series, up to 14 but typically 8, to provide enough power to overcome the friction of the pen against the paper chart (Fig. 6.8). Excessive mechanical magnification from a single capsule causes the pen to move in steps rather than smoothly. A spring is normally used to prevent the collapse of the capsules, but in an unusual design it is replaced by a weight hanging below the capsule stack, which is suspended from the top (Breeze 1999). Accuracy is again about ± 0.3 hPa, its exact value depending on overall friction within the system, on the quality of maintenance and on

Figure 6.8. In this eight-capsule aneroid barograph, the top lever magnifies the vertical movement of the capsules, further magnified by the arm to which the pen is fixed. The springs that prevent collapse of the capsules under barometric pressure are inside the capsules.

ensuring routine calibration against a standard. Readings are made to the nearest 0.1 hPa.

Electronic barometers

Aneroid-capsule type

Manual measurement

Although now being replaced by electronic sensors, aneroid barometers combining manual and electrical methods to measure the movement of aneroid capsules have been used in the recent past. A stack of about three capsules is arranged to move an electrical contact, in line with a second contact fixed to a micrometer screw. The full range of atmospheric pressure change produces a contact movement of about 1–1.5 mm, its position being determined by turning the micrometer screw until the two contacts just touch; the turning of the micrometer is sensed electrically, producing a digital signal. But it needs an operator. The accuracy of such methods is around ±0.3 hPa, although the discrimination is higher, at 0.02 hPa.

Capacitive measurement

For automatic measurement of the aneroid capsule, its physical displacement can be sensed by capacitive plates installed within the vacuum cavity. In one such design, three separate identical units are used, their outputs being averaged to give a more accurate mean reading. Errors arise largely from the aneroid capsule itself, rather than from the capacitive sensing element, and so accuracy is akin to that of other precision aneroid methods.

Flexing diaphragm

Sensed by straingauges

The aneroid capsule is of course a flexing diaphragm, but with the arrival of microelectronic fabrication techniques came the ability to make silicon diaphragms that flex under pressure (see also 'Sensing water depth by pressure', sub-section 'By electronic pressure sensor', Chapter 10). Onto these thin wafers, using microelectronic etching techniques, are formed straingauge bridges that give an analogue voltage output proportional to the deflection of the diaphragm, which is amplified and either displayed visually or automatically logged. The accuracy for most pressure sensors of this type is, at best, about ±0.1% of the range covered. Thus if the range is 950 to 1050 hPa, 0.1% is ±0.1 hPa. But this is the best achievable, and a more realistic accuracy, and one typically quoted, is between ±0.3 and 1.0 hPa – comparable with aneroid sensors.

Sensed capacitively

Instead of straingauges, the deflection of the silicon diaphragm is now frequently measured capacitively. The flexing diaphragm and a fixed plate within the reference vacuum cavity act as a small capacitance, the value of which varies as the diaphragm flexes and the distance between the plates changes (see also 'Sensing water depth by pressure', sub-section 'By electronic pressure sensor' Chapter 10, and Radiosondes, Chapter 16). This capacitor forms part of an oscillator circuit, the frequency of which is related to the degree of flexing and thus to the pressure. Quartz is sometimes used as the diaphragm since it has the possible advantage over traditional diaphragm materials of better elasticity, thereby improving repeatability and hysteresis performance. Ceramic materials are also used, with little difference in the accuracy achieved, a value of ±0.3 hPa again being typical although some claim that ±0.1 hPa is possible. It would seem, however, that ±0.3 hPa is about the practical limit to the accuracy of all but mercury barometers, and the vibrating cylinder sensor, which will now be discussed.

Vibrating cylinder

A quite different technique is used in this design of pressure sensor. Originally developed for use in aircraft and rockets as altimeters and for sensing pressures associated with engines and fuels, the method has been modified to measure barometric pressure. Figure 6.9 shows a cross-section of the sensor, the space between the two concentric cylinders being a near vacuum, the inside of the thin-walled inner cylinder (of magnetic alloy) being open to the atmosphere. Three drive coils half-way along the sensor cause the inner cylinder to vibrate in a hoop mode (that is, not longitudinally). Changes in pressure change the hoop stresses and thus the natural frequency of vibration of the cylinder; these changes being detected by electromagnetic pick-up coils. Their signal is also fed back to the drive coils to maintain the resonance of the cylinder. But the frequency is also influenced by the density of the air, although correction for this is easily made by measuring the temperature of the sensor. The sensors are calibrated by measuring the frequency of the cylinder

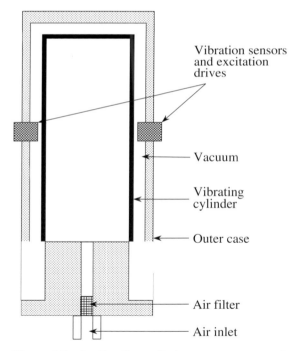

Figure 6.9. A *vibrating cylinder* pressure sensor is about 10 cm long by 6 cm diameter. The inner (dark) cylinder is thin-walled, open to atmosphere inside and surrounded by a near vacuum outside contained within the thicker cylinder. Spaced at 120 degrees through the walls of the outer cylinder are three coils that both induce vibration magnetically in the thin steel cylinder and also detect its frequency of oscillation. The frequency is proportional to the pressure.

at up to ten different pressures and temperatures, producing a unique calibration table for each sensor. The method is sufficiently accurate to measure barometric pressure to 0.1 hPa with a long-term drift of typically 0.05 hPa, making the sensor suitable as a standard.

Installation of barometers

Meteorological instruments usually have to be in direct contact with what they are measuring; barometers, however, are able to sense pressure equally well inside a building. This is convenient, since it protects them from temperature gradients. Pressures measured indoors are correct on a windless day, although the opening and closing of doors causes pressure surges. Wind blowing through windows and doors and across openings such as chimneys presents problems and can induce large errors (many small openings are preferable to a few large ones). In turbulent conditions fluctuations of up to 3 hPa can be introduced (World Meteorological Organisation 1996).

For precise work, a *static pressure head* has to be used to reduce the effects of wind. The head consists of a vertical tube that is kept facing into the wind by a vane, holes positioned around the cylinder relative to the wind-flow ensuring that the pressure inside the tube is independent of wind effects. The barometer is then housed in a container connected to the pressure head by a tube. Such precautions may also be needed if the barometer is housed in an air-conditioned room. Static heads are described in detail by Miksad (1976) and the United States Weather Bureau (1963). (See also 'Aircraft measurements', sub-section 'Civil airline commercial flights', sub-sub-section 'Pressure', Chapter 16.)

At AWS sites, the barometric pressure sensor will usually be housed in a hut or in the enclosure containing the logging or telemetry equipment. It might, therefore, be exposed to greater variations in temperature than ones housed in a building. However, electronic pressure sensors are much smaller and can be housed in a heat-insulated box. Wind might still be a problem, depending on the details of the housing.

Pressure at aeronautical stations

Because barometric pressure varies, aircraft altimeters have to be set to read the correct height when the aircraft is on the ground at the airfield. The readings are referred to in Q code, as QFE and QNH. QFE is defined as the pressure at an elevation corresponding to the elevation of the airfield (at that particular moment). QNH is defined as the pressure at which an altimeter should be set so that it will indicate the official elevation of the airfield when the aircraft is on the ground at the airfield.

Pressure is computed in tenths of a hectopascal but is rounded down to the nearest whole hectopascal. Aneroid barometers are used at airfields but they need to be regularly checked against either a second aneroid barometer or a mercury barometer.

The storm glass

To make any treatment of barometers complete, some mention must be made of the curiosity known as the *storm glass* – a glass tube around 25 cm long and 3 cm in diameter containing a solution of camphor, alcohol, potassium nitrate, ammonium chloride and water, the top being bunged with sealing wax (Banfield 1976, 1985, Bolton & Rae 1992) (Fig. 6.10). It probably originated in the early eighteenth century, and there is some suggestion that it was an accidental offshoot of alchemy. The visual state of the liquid changes with time, fine camphor crystals forming and sinking to the bottom of the tube, while changing in appearance. At other times 'little stars' or 'woolly snowflakes' appear, which float to the top. Or the liquid can become opaque, while at other times perfectly clear, with an assortment of variations between. It was supposed, by some, that the appearance of the solution could be used to predict the weather.

Bolton and Rae (1992) record that Admiral Robert FitzRoy was interested in storm glasses during his voyages. He also called them *camphor glasses* and believed that their action was due in some way to 'electrical tension' connected with the weather. FitzRoy corresponded with Michael Faraday concerning this explanation, but a letter from Faraday makes it quite clear that he could not see how the state of the mixture was affected by anything other than temperature. FitzRoy describes in *The Weather Book* (1863) how Daniel De Foe gave an account of the weather glass in his book *The Great Storm* in which he recorded the changes in appearance of a camphor glass as the storm passed in 1703. (De Foe is best known for his novel Robinson Crusoe, published in 1719.) Sutton (1965) expresses surprise at FitzRoy's faith in what Sutton describes as nothing more than a toy. So to be blunt about it – the storm glass is not a barometer; it is, at very best, a crude, unpredictable, qualitative thermometer. It is of no use whatsoever to meteorology and, as Sutton remarks, the device has no prognostic value. But it is a fascinating and fanciful thing, more akin to fortune telling by tea leaf patterns than to a scientific instrument. No surprise, therefore, that it is beneath Middleton (1994) to even mention it, as indeed he also failed to mention the Dutch weather glass for similar reasons no doubt (even though the Dutch weather glass at least has a more rational relationship to barometric pressure). Bolton and Rae (1992) are more charitable, suggesting that the storm glass might have had some use at sea to provide an indication of temperature, more easily 'read' by sailors unskilled in the reading of graduated mercury thermometers, but this seems unlikely. As far as I know, no one has ever

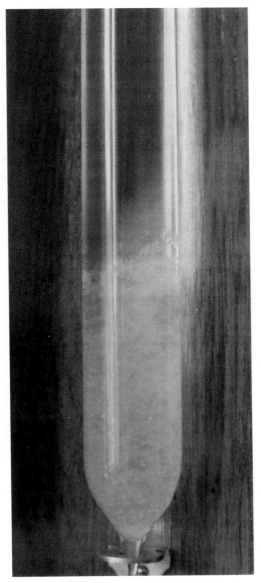

Figure 6.10. This storm glass, showing one of its many (but quite meaningless) states of crystallisation, is part of a FitzRoy barometer, manufactured by Joseph Davis & Co. of the Royal Polytechnic Institution in London sometime between 1870 and 1885. Such a glass is of no use to meteorology, does not measure pressure and is no more than a toy.

'calibrated' such a device as a thermometer and got consistent results. A storm glass is on display in the Met. Office entrance hall, or it was when I was last there when it was in Bracknell.

References

Banfield, E. (1976) *Antique Barometers*. Baros Books, Trowbridge, UK.
Banfield, E. (1985) *Barometers: Stick or Cistern Tube*. Baros Books, Trowbridge, UK.
Barlow, D. (1994) From wind stars to weather forecasts: the last voyage of Admiral Robert FitzRoy. *Weather*, **49**, 123–128.
Bolle, B. (1982) *Barometers* [English translation of the original (published 1978)]. Argus Books Ltd., Watford.
Bolton, H. C. & Rae, I. D. (1992) The Admiral's storm glass: coffee-table renaissance of an old weather instrument. *Weather*, **47**, 89–97.
Breeze, E. W. (1999) An unusual barograph. *Weather*, **52**, 239–241.
Burton, B. J. (1993) Atmospheric pressure, and temperature aloft. *Weather*, **48**, 141–147.
Crane, A. J. (1981) Dutch Weather Glass. *Weather*, **36**, 338–339.
FitzRoy, R. (1863) *The Weather Book*. Longman, Green, Longman, Robins & Green, London.
McConnell, A. (1994) *Barometers*. Shire Publications, Princes Risborough, UK.
Middleton, W. E. K. (1994) *The History of the Barometer*. Baros Books, Trowbridge, UK.
Miksad, R. (1976) An omni-directional static pressure probe. *J. Met. Soc. Jpn.*, **15**, November.
Sutton, G. (1965) Admiral FitzRoy and the storm glass. *Weather*, **20**, 270–271.
United States Weather Bureau (1963) *Manual of Barometry (WBAN)*, first edition. US Government Printing Office, Washington DC.
World Meteorological Organisation (1996) *Guide to Meteorological Instruments and Methods of Observation*, sixth edition, No. 8. WMO, Geneva.
Young, M. V. (1994) Back to basics: depressions and anticyclones: part 1. *Weather*, **49**, 306–311.

7

Evaporation

In the evening it rained very heavily, and although the thermometer stood at
65 °(F), I felt very cold. As soon as the rain ceased, it was curious to observe the
extraordinary evaporation which commenced over the whole extent of the forest.
To the height of 100 feet the hills were buried in a dense white vapour, which
rose like columns of smoke from the most thickly wooded parts, especially the
valleys. I observed this phenomenon on several occasions. I suppose it is owing
to the large surface of foliage, previously heated by the sun's rays.

Charles Darwin *Voyage of the Beagle* (Rio de Janeiro).

The variable

The rate of evaporation from the ground's surface and from plants is controlled by
the relative humidity and temperature of the air, the amount of net radiation, the wind
speed at the surface, the amount of water available and the nature of the surface (for
example its roughness). The type of vegetation and the state of the plant's stomata are
also involved since transpiration is generally lumped in together with evaporation.
To allow for this, the ungainly word *evapotranspiration* has been coined; many do
not like it, but it comes in useful and acts as a reminder. Open water presents another
situation, as do ice and snow. The net incoming solar energy is apportioned to the
three fluxes – sensible, latent and soil heat, and to photosynthesis – according to
the infinite variety and combination of circumstances. Here we are considering the
latent heat flux.

Units and terms

Because evaporation varies depending on the nature of the surface, various 'standard rates' have been defined. For example Penman (1948, 1956), the first to put

Measuring the Natural Environment, second edition, Ian Strangeways. Published by
Cambridge University Press. © Ian Strangeways 2003.

evaporation on a sound scientific footing, looked initially at the simple situation of evaporation from a large area of open water, which he defined as *potential evaporation*, that is the amount of water evaporated per unit area per unit time (hours or days) from a water surface. Although it takes no account of energy exchanges within the water or of turbulent transfer above, it bears some useful relation to reality. Gangopadhyaya *et al.* (1966) then proposed the concept of *potential evapotranspiration*, which is the amount of evaporation from a large area of vegetation, with unlimited water supply to the roots. But this assumed that the amount of evaporation was not greatly influenced by the type of vegetation (based on the experimental evidence of that time). When it subsequently became known that the type of vegetation had a considerable effect, Tanner (1967) proposed the idea of *reference crop evapotranspiration* and there are now several variations on these different classes of evaporation. However, it is Penman's original, classical method that is discussed later. Although his original concept of potential evaporation referred to open water, Penman refined this to include land surfaces, more relevant in practice than open water, and made his measurements over open grassland at the Rothamstead Research Station in the UK.

It is much more difficult to sense the rate of loss of water from the surface through evaporation than it is to measure the gain of water through precipitation. Nevertheless, the rate of evaporation is expressed in the same units as precipitation, being the equivalent depth of liquid water lost into the atmosphere as water vapour, expressed in millimetres lost over an hour or a day. By using the same units, water balances are easily calculated.

Evaporation can be measured either as the loss of liquid water from the surface or as the gain of water vapour by the atmosphere, but few of the methods involve a direct measurement, most inferring the amount indirectly.

Measuring the loss of liquid water

Measuring water loss by the use of a natural catchment

The water-balance equation $E = P - (V_r + V_s + V_l)/A$ relates the various elements in a catchment balance, E being the evapotranspiration, P the precipitation, V_r the runoff in rivers, both above and below the surface, V_s the volume of water stored in the system, V_l the loss due to leakage and A the area of the catchment.

By measuring the precipitation, the runoff in rivers and the amount of water stored in the ground (the soil moisture), the losses (evapotranspiration and leakage) can be calculated (Rodda *et al.* 1976). The measurement of rainfall, of soil moisture and of river flow (see Chapters 8, 9, 10) can be done to varying degrees of precision. Provided, then, that the catchment is a simple one, with bedrock beneath that is known not to leak and with no other ways out for the water (so that V_1 is zero) the method is workable, although expensive.

However, the error will be large unless the variables are measured to high accuracy, the evapotranspiration E being the small difference between the three much larger values P, V_r, V_s. Any error in one of these will exaggerate or minimise the apparent evaporation, perhaps by many times its real value. Also the equation only tells us what the evaporation is from the whole catchment, lumping together evaporation from the ground and plants and transpiration. Nor does it explain any of the details of the evaporation process. It can, however, answer such questions as 'Do forests evaporate more water than grassland?' (Law 1956).

Measuring evaporation through soil moisture changes

Provided there is little drainage from the soil (either on the surface or below), the measurement of changes in soil moisture content can be attributed entirely to evapotranspiration, with allowances for rainfall. Accuracy is improved if measurements are also made of soil tension (see 'Measuring soil tension', Chapter 9), from which the vertical motion of the soil moisture can be established, helping to quantify any downward movement to the water table that does not contribute to evaporation. However, it is difficult to select a representative piece of land to monitor, since evaporation and soil moisture can vary a great deal spatially. Furthermore, the method does not take account of the evaporation of rain directly from the surface of plants, especially large ones, this water never getting to the soil. The method is, therefore, limited in use.

Measuring water loss using a lysimeter

A lysimeter is a container that is isolated from the surrounding soil but filled with soil similar to that outside it and planted with the same vegetation that surrounds it. By suitable design, the contents of the lysimeter can be made to behave in a similar way to their surroundings, provided that the temperature of the soil is kept the same and the drainage is similar (Fig. 7.1). To produce the correct drainage it may be necessary to install some means of suction at the base of the container, to ensure that the soil moisture tension is the same as in the freely draining ground outside the lysimeter. 'The bigger the better' is generally the case for lysimeters, since then edge effects are reduced and internal and surface differences are smoothed out. However, small lysimeters can still give very useful results.

There are two main types of lysimeter – those that measure the weight of the container and its contents and those that do not. John Dalton experimented with the measurement of evaporation in 1795 at a site near Manchester, using what we

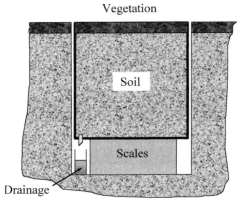

Figure 7.1. A lysimeter is a container filled with soil, replicating as closely as possible that outside, and with the same vegetation. It is weighed either with a simple flexible container filled with a fluid with a manometer tube to the surface or with a mechanical or electronic balance. Changes in weight indicate both precipitation and evaporation. Excess water is drained at the base and measured.

would now call a non-weighing lysimeter (Biswas 1970). It took the form of a container 25.4 cm in diameter and 1 m deep, with a tube at the bottom to collect drainage and one at the top to collect surface runoff. It was filled with sand and gravel at the bottom, with soil on top, left bare for a year and then covered with grass for the second year. A record was kept of the rainfall and the quantity of water collected from the two tubes for two years. Dalton concluded that the annual evaporation was 25 inches (63 cm) and that evaporation increased with rainfall (but not proportionally) and was the same for soil and grass. Simple lysimeters of a similar design can still offer useful results today.

In the case of a weighing lysimeter, a mechanical balance is the best way of measuring the weight because it is the very small changes in weight (due to rainfall and evaporation), not the absolute total dead weight of soil and container, that are wanted. Load cells can be used, but they weigh the total mass and so must be able to resolve small changes in weight in excess of the very large overall weight. It is also possible to use a hydraulic method of weighing, in which the lysimeter stands on a flexible container filled with oil connected by tubing to a manometer above the ground where it can be read. The change in weight, with due allowance for rain (which is measured separately) and for any drainage (which is stored and measured below the lysimeter), is the amount lost by evaporation. A large mechanical balance, perhaps supporting several tonnes of container, is expensive, and so too is the structure in the ground that houses them. Weighing lysimeters are not, therefore, for general use, but they are one of the best methods of measuring evaporation.

Observing strict terminology, it is more correct to call lysimeters that are sealed and have no drainage evapotranspirometers and only to call those that drain lysimeters.

Measuring transpiration and evaporation from plants

It is sometimes the case that the largest contributor to water loss is transpiration. This can be measured directly by cutting through the stem of a plant, or the trunk of a tree, and immersing it in a tank of water. By measuring the fall of water level in the tank, plant uptake is sensed directly, any form of water-level-sensing technique being suitable (Chapter 10). But it is necessary to keep a watch on the health of the plant or tree and to be certain that it remains representative of the surrounding vegetation, particularly as regards stomatal resistance since the degree of openness of the leaves' stomata controls the rate of transpiration. This is difficult to achieve, however, because the plant in the tank has an unrestricted water supply and there may well be differences in transpiration from plants growing naturally in the ground, as their water supply varies with time. The method can only be used for a short time, as in the tank the plant cannot grow naturally for long.

An alternative is to allow the plant to remain in the ground and to measure the *sap flow*, a tracer of heat or chemical being injected into the trunk. In one such design, intended for larger trees, two thermocouple needle-probes are inserted into the sapwood, one above the other a few centimetres apart, the upper one being heated a few degrees. The difference in temperature between the two probes, resulting from heat being carried away by sap flow, gives an indication of the volumetric flow rate. As flow rate can vary round the circumference of the tree, some designs have up to eight pairs of differential temperature sensors (Steinberg *et al.* 1989, Smith & Allen 1996). Designs are also made for stems as small as 2–5 mm.

Measurements can also be made of the amount of rainwater lost by evaporation directly from the plant's surface. This is most usually done only for large vegetation such as forests; the rainfall above a forest canopy and the amount that reaches the forest floor (the *throughfall*) are measured, the difference between the two measurements being the amount lost from the trees by surface evaporation – the *interception loss*. Account must also be taken of that part of the rain that finds its way to the ground down the tree trunks – the *stem flow*.

The above-canopy rainfall is measured by installing a raingauge funnel on a mast or tower, level with the top of the forest, a tube feeding the water to a tipping-bucket raingauge at ground level (Chapter 8). The throughfall, arriving at ground level, is measured in one of two ways, depending on the type of forest. In dense vegetation, large collectors of plastic sheeting are suspended below the trees on frames and also attached to the tree trunks; these collectors direct the collected throughfall and stemflow to a large tipping bucket (Chapter 8). For sparser forests, it is possible to use a cheaper method in which simple funnels and bottles are positioned on a random grid that is changed from time to time to avoid any bias. Stem flow can be measured in a similar simple way.

These techniques illustrate well the difficulty of measuring evaporation and why extremely roundabout and expensive ways may have to be used.

Measuring the flux of water vapour

While the previous methods estimate evaporation by measuring indirectly the net loss of liquid water to the atmosphere, direct measurement of the actual flux of water vapour requires the sensing of the rate of flow of the vapour from the ground into the atmosphere. There are several ways of doing this, but the two most used and most precise are the energy-balance and the eddy correlation methods.

The Bowen ratio (energy-balance) method

Water vapour and heat diffuse from where they are more concentrated, near to the evaporating surface, to where they are less concentrated, in the atmosphere above. The Bowen ratio technique estimates the ratio between the sensible heat flux and the latent heat flux, the Bowen ratio, by taking measurements of temperature and humidity at different levels, thus giving their vertical gradients. Over forests, large towers are needed and complex instruments have to be moved up and down (to avoid instrument bias) to measure the small temperature and humidity gradients within and above forests. Simpler systems can be used, however, over short vegetation, because the gradients are larger (since the flow is less turbulent) and so there is less mixing. In the system illustrated (Fig. 7.2), two tubes, which can be separated vertically by a height of between 0.5 and 3 m, draw air samples into a dew-point sensor (housed in the logger box at the base of the tripod), the samples being switched alternately every few minutes from one intake to the other, so giving measurements of the dew point, and thus the humidity, at the two heights; from these the gradient is calculated. By using one dew-point instrument and switching the inputs, any sensor bias is avoided. The temperature at each height is also measured, by thermocouples (Fig. 7.3; Chapter 3). The readings are typically averaged over half to one hour while, over the same period, measurements are also made of net radiation and soil heat flux, their difference giving the energy available to power the sensible and latent heat fluxes. The Bowen ratio then allows this energy to be partitioned between the two fluxes to give a measure of evaporation (Lloyd 1997).

Figure 7.2. The Bowen ratio apparatus for measuring evaporation. On the end of each of the two horizontal arms is a thermocouple (a close-up is shown in Fig. 7.3), which senses the temperature gradient between the two heights. Beneath each arm are two air intakes (the button-like object, top arm); a fan draws air alternately from each level to a dew-point sensor in the logger box at the base of the mast, giving a measure of the humidity gradient. Out of view is a net radiometer and, below ground, a soil-heat-flux plate. An electronics unit in the box combines these six measurements to give an estimate of evaporation.

The eddy correlation method

Irregularities at the Earth's surface interact with the wind to produce eddies which carry away water vapour and heat (and trace gases such as carbon dioxide and methane) through the process of turbulent diffusion. Although at each height above the ground the mean wind direction is parallel to the surface, within turbulent eddies the movement is in three dimensions, changing rapidly with time and carrying water vapour and heat away from, or towards, the ground. This produces rapid, but small, fluctuations in the water vapour concentration and temperature in these eddies.

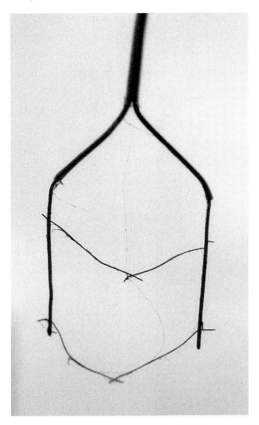

Figure 7.3. The thermocouple of a Bowen ratio instrument. It is sufficiently small to be unaffected by radiative heating and so does not need a screen, allowing unimpeded measurements of temperature to be made with a rapid response time. Two (single) junctions are operated in parallel, in case one fails.

Through this process, an upward flux of evaporated water and warmed air occurs, since, generally, an upward movement of air correlates with a higher than average humidity and temperature – and conversely. (The same logic applies to many trace gases, such as CO_2 – see below.) It is, therefore, possible to measure the actual evaporation by taking simultaneous, rapid, instantaneous readings of the vertical component of the wind, and of humidity, the total evaporative loss being found by integrating all the rapid readings over a period of 10–60 min and comparing this integrated value with a longer-term humidity average. By sensing the rapid temperature changes, as well as those in humidity, the sensible heat flux can also be measured directly in the same way, giving extra information against which the evaporative flux can be checked.

In practice these measurements are difficult to make owing to the high speed at which the readings have to be taken and the small values of the changes involved. The readings must be taken quickly in order to include all the higher frequencies in the turbulent motion, while still being able to sense much slower changes without undue drift in sensor response (Moore *et al.* 1976). The size and frequency of the eddies depend on the nature of the surface, the height of the measurement and the wind speed, but sensors must be able to take 10–20 readings a second, while also being able to see changes occurring over tens of minutes. It is also necessary for the instrument to be able to do the calculations in real time at least as fast as the readings are taken, although with modern electronics this is not difficult; 25 years ago it was less simple.

Suitable sensors to meet these requirements have already been mentioned in earlier chapters. The vertical fluctuations of wind speed are generally measured using a sonic anemometer (Figs. 7.4(*a*) and 5.11) although the use of pressure and thermal wind sensors was, in the earlier stages of design in the 1980s, considered a possibility, as were three-dimensional propeller anemometers (Chapter 5). All have been used at one time or another in eddy correlation instrument design, although the sonic method is without doubt the best today. A cup anemometer (Fig. 7.4(*a*)) is used to measure the average horizontal wind speed.

Infrared radiation passing through air is absorbed by the water vapour it contains, so, by detecting the amount of infrared energy absorbed, a measure of humidity is obtained. (Such techniques are commonly used in industrial infrared Gas Analysers (IRGA) for a variety of trace gases. See also 'Greenhouse gases' in Chapter 20.) The response of such a sensor is instantaneous and the electronics is fast enough to sense the rapid changes of humidity in the turbulent air. The radiation bands used for water vapour are those in the mid-infrared (see later). An infrared source, a light chopper and a detector constitute the sensor. The purpose of the chopper is to produce an AC signal rather than a DC one, since drift is then less of a problem and the signal-to-noise ratio is also improved, allowing background noise to be filtered out. If absolute humidity was the requirement, the instrument would need to compensate for drift in the strength of the light source, in the sensitivity of the detector and in the electronics and for dust in the air, making it a more complex and expensive instrument. But it is changes in humidity that are required, in the relatively short term, and techniques such as comparing the rapidly fluctuating instantaneous readings with their running average achieve this with greater simplicity. It is also possible to operate a capacitive, or wet-and-dry, humidity sensor, deployed in a miniature screen (Fig. 7.4(*a*)), to give an accurate measure of humidity over the longer term, against which the averaged readings from the infrared instrument can be referenced.

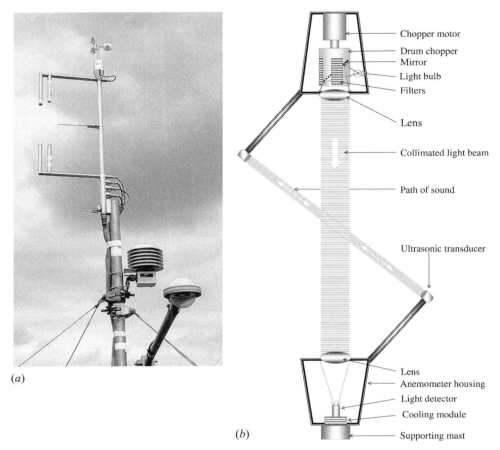

Figure 7.4. (a) The basic sensors of the eddy correlation method of measuring evaporation are the two vertical sets of tubes, seen at the left of the photograph, which measure the rapid fluctuations in vertical wind speed and humidity; the small horizontal tube measures fluctuations in temperature (it is fine enough not to be heated by radiation). In addition, the cup anemometer measures the long-term horizontal wind speed while, in the screen, sensors measure the average temperature and humidity. On the right, a net radiometer senses the available energy and, below ground, a soil-heat-flux plate measures the third energy flux, which together with the other measurements gives a direct measurement of evaporation. A recent development of this instrument is shown diagrammatically (b) which combines a three-dimensional sonic anemometer, of the type illustrated in Fig. 5.11, with the water vapour (and carbon dioxide) IR sensor. This ensures that the same volume of air is measured by both the anemometer and the IR sensor.

The rapid changes in temperature are more easily and cheaply measured than those in humidity; thermocouples, small PRTs or thermistors are the most effective devices for this. All three can follow rapid fluctuations in temperature, fulfilling the same role for sensible-heat-flux measurement as the humidity sensor does for latent heat flux. A slower-responding temperature sensor, such as a larger PRT, is operated in the miniature screen to give a long-term reference.

As a check on the performance of the whole instrument and its calculations, a net radiometer and a soil-heat-flux plate can be operated close by, allowing the overall energy balance to be checked – the available energy measured by the net radiometer should be the same as the total of the sensible heat flux and latent heat flux, measured by the eddy correlation instrument, and the energy measured by the soil-heat-flux plate.

The accuracy of such an instrument is from $\pm15\%$ to $\pm20\%$ of the actual evaporation for hourly values and over longer periods $\pm10\%$.

The instrument illustrated in Fig. 7.4(a) is an early design by the CEH in Wallingford, who have recently developed a new model (Fig 7.4(b)) that puts the IRGA at the centre of a three-dimensional ultrasonic wind sensor of the type illustrated in Fig. 5.11, thereby measuring exactly the same air that is sensed by the IR humidity sensor, rather than being slightly displaced to one side as in the earlier models (Fig 7.4(a)). This new design also measures the rapid variations in CO_2 concentration in the same way as the water vapour flux is measured. This is of considerable current interest because by measuring the CO_2 flux over different surfaces, such as forests or the sea, it is possible to determine whether the plants or water are a net sink or source of the gas (Gash & Nobre 1997, Grace 1995, Moncrieff 1997) – given a long-enough run of measurements. But it is much more difficult to measure the CO_2 flux than that of water vapour because the concentration of the gas is less (about 0.036% compared with up to 2% for water vapour at $20\,°C$ and 100% RH). However, since, like water vapour, CO_2 also absorbs infrared radiation, its measurement is simply a matter of selecting a different frequency band to detect the CO_2. A more complex sensor system was used in the past because of the much lower concentration of the gas. Before going on to describe the new instruments, it is interesting to look at how the CO_2 flux was measured previously.

Until recently the only equipment available to measure the low concentrations of CO_2 in the atmosphere were IRGAs designed for industrial or laboratory applications, too big for mounting in the open close to a wind sensor. To overcome this problem, air had to be drawn down a tube, from close to the sonic anemometer, into the sensor unit housed a few metres away (in a similar way to the air in the Bowen ratio system, Fig. 7.2). This is known as the *closed path* method. But as it takes time for the air to reach the sensor, a correction has to be made to ensure that the wind and CO_2 measurements are in synchronism. Although it might be thought that fluctuations in the CO_2 level would be lost as the air flowed along several metres of tubing, this is not the case for the lower frequencies, although the higher frequencies are smoothed out through turbulent mixing in the tube. This is less serious than it might appear, since it is at the lower frequencies that most of the transfer occurs, and in addition some correction can be made for the lost higher frequencies. There is also a slight advantage in the smoothing effect of the tube in that the small temperature fluctuations in the air are also smoothed out, this being useful since temperature fluctuations, through their effect on air density, affect both CO_2 and water vapour fluxes. The same, closed path, method can equally well be used to measure the fluctuations of water vapour, and some eddy correlation instruments use this construction instead of the more traditional open path design illustrated in Fig. 7.4(*a*).

Let us return now to the new instrument that measures both water vapour and CO_2 fluxes together in one instrument, using the open path technique (Evans & McNeil 2003). This has been realised through the design of a compact, but sensitive, IRGA that fits into the head of the sonic anemometer, the details being illustrated in Fig. 7.4(*b*). The IR source is a miniature filament lamp, its light being chopped at around 350 Hz by a rotating drum. However, this drum also incorporates four light filters, the modulated beam in four separate narrow IR bands being collimated by a sapphire lens and directed downwards, through the air, to the detector in the lower section of the anemometer. Here a similar lens focuses the beam onto a lead selenide IR detector, the latter being kept at a controlled $12\,°C$ to ensure acceptable signal-to-noise ratio with an adequate time constant. The output of the sensor is amplified, digitised and demodulated, giving four separate signals corresponding to the four filters.

Two of the signals represent IR signal levels received at 2.65 and 2.25 μm, the former being absorbed by water vapour in the air, the latter not absorbed by any atmospheric gases, acting as a reference to allow compensation for slight changes in light output and contamination of the external optical surfaces. The rapidly fluctuating difference between the signals thus gives a measure of the variations of water vapour in the path of the beam. The two other signals represent IR levels received at 4.25 and 3.95 μm, the former being absorbed by CO_2 in the air, the latter acting as the reference, thereby giving an indication of the fluctuating levels of carbon dioxide. These measurements, combined with the vertical windspeed measured by the sonic anemometer, give the fluxes of both water vapour and CO_2.

Whereas in earlier designs, a PRT or thermistor was used to detect the fluctuations in temperature, this can now be done using the same sonic pulses that measure the wind speed, since the speed of sound is related to temperature.

The system is calibrated in a sealed chamber into which known amounts of water vapour and CO_2 can be introduced, precise IRGA laboratory instruments measuring their concentrations. Readings are checked at five levels of concentration of both gases (covering the natural levels found in the environment) at three temperatures.

The high cost of these systems, the need for experienced operators and for expert interpretation of the results restrict the use of the eddy correlation method to specialised applications (Gash *et al.* 1991, Gash & Nobre 1997). It must be of some concern that measurements from instruments of this type, in inexperienced hands, could be misinterpreted. However, specialised instruments of today can become the operational techniques of tomorrow, as experience is gained and designs are refined; sonic anemometers are a good example of such a transition.

Estimating evaporation

Ways have long been sought of estimating evaporation by making measurements of other variables and deducing the evaporation from them. From the strict scientific viewpoint these methods may not be ideal, but they can be much more economical than the precise methods discussed above, and they have a wide and useful place in the everyday world.

The Penman method

Penman's method of estimating potential evapotranspiration requires the measurement of energy input, air temperature, humidity and wind speed. In one version of the calculation, the energy input information is obtained from the direct measurement of net radiation, in another the net radiation is estimated from solar radiation or sunshine hours (from a Campbell-Stokes recorder). The direct measurement of net radiation gives much better results since the measurement is more accurate than those derived indirectly from sunshine records.

The Penman equation takes the form

$$E = [\Delta/(\Delta + \gamma)]H_T + [\gamma/(\Delta + \gamma)]E_A,$$

where H_T is the *heat budget* and E_A is the *aerodynamic term*. Δ and γ are *efficiency factors* controlling the relative effectiveness of the energy supply and ventilation (Penman 1948), Δ being the slope of the temperature versus saturation vapour pressure curve and γ the psychrometric constant (Chapter 3). The evaporation is thus calculated in two parts, the heat budget and the aerodynamic term, both expressed in equivalent millimetres of water evaporated in a given time.

The heat budget is calculated as R_N/L, that is the measured or estimated net energy R_N (with an assumed proportion taken up by the sensible and soil heat flux) received over the period of the calculation, usually an hour or a day – depending on the interval for which the net data are available – divided by L, the latent heat of vaporisation of water. Ritjema (1965) showed that, providing the net radiation is measured over the vegetation concerned, the method provides an estimate of *reference crop evaporation* rather than potential evaporation.

Figure 7.5. To estimate evaporation by the Penman method, an AWS has sensors for measuring net radiation (left lower arm), temperature and humidity (in the screen, right lower arm) and wind speed (right upper arm). To provide additional climatological background, there are also sensors for solar radiation (top of the mast) and wind direction (left upper arm) and, out of view, a raingauge. Soil temperatures may also be measured, while for meteorological use there will be a barometric sensor. The antenna is for telemetering the measurements via Meteosat (Chapter 12).

The aerodynamic term is expressed as $E_A = 0.35\,(1 + u/100)\,(e_a - e_d)P_c$, where u is the wind run in metres per day at a height of two metres and e_a and e_d are the saturation vapour pressure and the actual vapour pressure, to which a correction, P_c, has to be applied for altitude: $P_c = 1 + h/20000$, h being the altitude in metres (McCulloch 1965).

The Penman equation as shown above is the simplest physically based model that can be used to estimate evaporation. Over the years, other workers have refined the method, notably Monteith (1965), while Thom and Oliver (1977) modified the original formula to include extra terms with a physical basis that could be changed empirically, to give an estimate of evaporation from different crops and for use over different vegetation including forests.

The instrumentation required for the estimation of evaporation by the Penman method, which needs the measurement of net radiation, temperature, humidity and

Figure 7.6. A 'met. enclosure'. In the 1950s, before AWSs were generally available, Penman developed his technique of estimating evaporation using conventional met. enclosure data. Such sites are still the mainstay of most of the National Weather Services (NWSs) for synoptic data collection (i.e. for weather forecasting) and they are also still widely used for many research projects. The site shown here is in Java, at the University of Bogor, and in addition to a wide selection of raingauge types and an AWS, a conventional Stevenson screen and anemometer (right) can be seen. Out of shot is a selection of radiation-measuring instruments and a wind-direction vane.

wind speed, has already been described in earlier chapters. However, one of the motivations for developing the first automatic weather stations (AWSs) in the 1960s and 1970s (Strangeways 1972, Strangeways & Smith 1985) was to measure this complement of Penman variables. Rainfall was also included to give a measure of the precipitation input. In addition, solar radiation, wind direction and barometric pressure may be included to form a complete climatological station (Fig. 7.5) and to these can be added soil temperature, soil-heat-flux plates and any other sensors that may be needed for the particular application. AWSs are multipurpose instruments that are increasingly being used instead of what is commonly called the met. enclosure, a fenced-off area that houses a collection of old-style instruments; see Fig. 7.6 (Strangeways 1995). It was data from such conventional enclosures, however, that Penman used when developing his methods in the 1940s and 1950s.

Other methods of estimating evaporation, which make more empirical assumptions than the Penman method, have been developed; these rely on the measurement of fewer variables and thus on simpler instrumentation.

Radiation methods

Penman (1956) commented that, in Europe at least, the heat-budget part of his equation was usually at least four to five times greater than the aerodynamic part. A fair approximation of evaporation might thus be obtained by omitting the second term but making some notional allowance for it (Priestley & Taylor 1972). This approach means that it is then not necessary to measure humidity, just radiation. However, Shuttleworth (1979) and Shuttleworth and Calder (1979) showed that the method did not work correctly for forests, although by changing one of the factors in the equation (we will not pursue this here) the method might give an estimation of transpiration. It is only a second best to the original Penman method, but of the various simplifications probably the best.

Humidity methods

Despite Penman's observation that the radiation term is by far the larger of the two, it is nevertheless possible to relate just the second term to evaporation. This involves the vapour pressure deficit $e_a - e_d$ and the wind speed: the dryer the air and the windier the day the greater the evaporation. The method was advocated by John Dalton (1802), who also did experiments with lysimeters (see earlier) and with evaporation pans (see later). But where such humidity and wind measurements exist, it is better to use the full Penman method, using the estimated net radiation (or the measured net radiation if possible). Humidity methods, therefore, are not particularly practical or recommended.

Temperature methods

There are many empirical formulae that try to relate evaporation to temperature alone, all based on the assumption that the two terms in the Penman equation probably have some loose correlation with temperature. Since the first term of the equation is usually by far the greater, it is the relation between temperature and radiation that is important; but there is a lag between the annual temperature cycle relative to radiation and this is a problem unless allowance is made for it. The Blaney–Criddle equation (1950) is one of the better known of this type and was designed to give daily estimates of evaporation averaged over a month. It is expressed as:

$$E = c_u d_l (0.46T + 8),$$

where c_u is a *consumptive factor*, best determined for the particular site concerned, d_l is the fraction of daylight hours occurring in the month and T is the average temperature (in °C). The constants reflect the fact that the originals were derived using °F, as is often the case in equations of this type. However, the easier availability of net radiation data today, from AWSs, means that this approach is not often used.

These alternative, simpler methods can of course help predict roughly what the evaporation might be in the future using available past data from old-style instruments; it is often the case that engineers have to make decisions, and design structures, with nothing more than crude past data to work on. Such estimation methods may then have a vital place, but they do not when instruments are to be set up to measure evaporation in the future. They are worth discussing, however, as they put the Penman method in its rightful place at the top of the estimation methods.

Evaporation pans

While strictly speaking these devices are just further methods of estimating evaporation, they will be treated here in more detail because of their extensive use. Evaporation pans are undoubtedly the most common means in use today of estimating or measuring evaporation and so need special consideration. Like raingauges, but to a lesser extent, evaporation pans have a history going back at least to the eighteenth century. In Southport, near to Liverpool, Dobson (1777) operated a small pan of 12 inches (30 cm) diameter by 6 inches deep, with a raingauge of the same diameter nearby. Based on a recent estimate, using the Penman method, it seems that his measurements of evaporation were too high. John Dalton later (probably around 1785) used Dobson's method at Kendal in the Lake District, for 82 days (May to June inclusive) and measured 5.41 inches (13.7 cm) of evaporation.

Gangopadhyaya *et al.* (1966) described 27 different designs of pan and suggested that the list was probably far from complete. In its multiplicity, it is probably only exceeded by raingauge diversity. This alone is a drawback since it prevents meaningful intercomparisons between sites, but the main problem is whether pans give a measurement that can be related to potential evaporation, or indeed to any type of evaporation.

The energy exchange of pans will differ from that of the ground and the vegetation around them. In deep pans, such as the British Meteorological Office's standard tank (Fig. 7.7(*a*)), which is 1.83 m square by 610 mm deep and is installed with the top 75 mm above the ground, the stored energy will tend to be greater than that of the vegetation so that the surface temperature will be lower in the day, when most evaporation takes place, and higher at night. Russia has two designs of pan, one being 618 mm in diameter by 600 mm deep, set in the ground, with the top 75 mm above ground; the second is very large, with a diameter of 5 m and a depth of 2 m.

If a pan is raised above the ground and is shallow, as is the Class A pan (Fig. 7.8), there will be radiation exchange through the exposed sides, and this will cause a response different from that of the ground and plants. Performance will also be affected by the colour of the pan, a darker surface absorbing more solar radiation, and the colour of the water (for example if algae are allowed to grow in it). But pans exposed above the ground are less expensive and are easier to install, less debris will blow into them and less rain will splash into them, while leaks are more easily detected and corrected.

(a)

(b)

Figure 7.7. UK evaporation pans are sunk into the ground (a). Changes in water level are measured manually, using the hook gauge in the far right-hand corner of the pan (b).

Figure 7.8. The Class A evaporation pan, chosen by the WMO as the world standard, is smaller than the UK model and normally sits above the ground on a frame (left). A pan-level anemometer has been installed to provide additional information to enhance the readings. To the right is a similar pan sunken into the ground to compare performance. Photograph courtesy of the Centre for Ecology and Hydrology, Wallingford.

As with all instruments, exposure is an important factor, the conditions needed for an evaporation pan being similar to those for raingauges. A raingauge must also be operated nearby, to allow corrections to be made for rainfall falling into the pan. Birds and larger animals might drink from pans, and must be stopped by fences or by gratings over the top. Any fence around the pan should be open so as not to obstruct the wind. However, the WMO warn that gratings can reduce evaporation by up to 10% because of the shielding from solar radiation that they cause. If evaporation from a lake is to be measured, readings from a pan floating in the lake will more closely approximate to the lake's own evaporation than measurements from a pan on land, but operational difficulties are considerable.

Like many of the older instruments that are still widely used, much work has been done in comparing the relative performance of the different pans and, as a result, the Class A pan was selected by WMO as the world standard.

The principle of operating a pan is similar for all types. Usually once a day the level of the water in the pan is measured with a hook gauge in a small stilling well in the tank (Figs. 7.7(*b*) & 7.8). The hook (a pointed rod) is attached to a graduated scale with a vernier allowing reading to 0.1 mm, the hook and scale being turned up and down by a rack and pinion movement. It is moved up until the (upturned) point just cuts the surface of the water (this is easier to detect than lowering the point until it meets the water). The change in reading since the previous occasion, with due allowance for precipitation, gives the amount of evaporation that has occurred from the pan. The water in the pan is kept between 5 and 7.5 cm from the top of the tank, either by adding or removing water from time to time. This is done immediately after taking a reading, followed by a second reading to establish the new depth. An alternative is to use a fixed level, the amount of water that has to be added or removed being measured in a graduated cylinder (with one-hundredth the area of the pan).

Extra measurements may be made to help interpret the pan measurements more meaningfully, including wind speed at pan level (Fig 7.8) and the maximum and minimum temperature of the water – measured either by thermometers floating on the water or lying on the bottom. A further enhancement would be the addition of an automatic water level recorder in place of the manual hook gauge. These would usually be external units in which a float turns a potentiometer (Chapter 10) through a rack and pinion mechanism, giving an analogue output for logging. I have seen many evaporation pans around the world, but never one with such an attachment.

Atmometers

Atmometers are not widely used, but they need to be mentioned, their attraction being low cost and simplicity. They operate by measuring the loss of water from a wetted surface (Gangopadhyaya *et al.* 1966). Although there are several designs, the Piche Evaporimeter is probably the most common, consisting of a graduated tube, closed at the top and with a filter paper disc fixed to the bottom. As the water evaporates from the paper, the level falls in the tube and is read periodically. In one design the level is recorded on a chart. The energy for evaporation is derived from the radiation falling on the disc and from the heat conducted through the water supply from the exposed surfaces of the container. So how atmometers are sited greatly affects how they perform, and there is no agreed satisfactory way of exposing them. They can also be affected by dust on the evaporating surfaces and by the surroundings. Although a well-maintained instrument can sometimes have a passable correlation with potential evaporation, a different relationship is to be expected for each situation and any idea of universality is an illusion.

What might be described as a miniature evaporation pan, although in concept it is an atmometer, is the *recording evaporation balance*. It consists of an open water container with an area of 250 cm^2 (17.84 cm diameter) and a depth of about 3 cm; this is placed on a mechanical spring balance, which records the change in weight on a paper chart as the water evaporates. It is not a very common instrument and will suffer from the same problems as other atmometers. No doubt it has its advocates and used intelligently may give some rough indication of evaporation.

References

Biswas, A. K. (1970) *History of Hydrology.* North-Holland, Amsterdam and London.

Blaney, H. F. & Criddle, E. T. (1950) *Determining Water Requirements in Irrigated Areas from Climatological and Irrigation Data.* [TP-96]. USDA, Washington DC, 48p.

Dalton, J. (1802) Experiments and observations to determine whether the quantity of rain and dew is equal to the quantity of water carried off by the rivers and raised by evaporation; with an enquiry into the origin of springs. *Mem., Lit. Philos. Soc. Manchester,* **5**, part 2, 346–72.

Dobson, D. (1777) Observations on the annual evaporation at Liverpool in Lancashire; and on evaporation considered as a test of the dryness of the atmosphere. *Philos. Trans. R. Soc. Lond.,* **67**, 244–59.

Evans, J. G. & McNeil D. D. (2003) An integral low-power open-path instrument for the measurement of surface carbon flux. (In press.)

Gangopadhyaya, M., Hurbeck, G. E., Nordenson, T. J., Omar, M. H. & Uryvayev, V. A. (1966) *Measurement and Estimation of Evaporation and Evapotranspiration.* WMO, Geneva, Technical note No. 83.

Gash, J. H. C. & Nobre, C. A. (1997) Climatic effects of Amazonian deforestation: Some results from ABRACOS. *Bull. Am. Meteorol. Soc.,* **78**, 823–30. (ABRACOS is the acronym for the Anglo-Brazilian Amazonian Climate Observational Study.)

Gash, J. H. C., Wallace, J. S., Lloyd, C. R. & Dolman, A. J. (1991) Measurements of evaporation from fallow Sahelian savannah at the start of the dry season. *Q. J. R. Met. Soc.,* **117**, 749–60.

Grace, J. (1995) Carbon dioxide uptake by an undisturbed tropical rain forest in Southwest Amazonia, 1992–1996. *Science,* **270**, 778–80.

Law, F. (1956) The effect of afforestation upon the yield of water catchment areas. *J. B. Waterworks Assoc.*, **38**, 489–94.

Lloyd, C. R. (1997) An intercomparison of surface flux measurements during HAPEX Sahel. *J. Hydrol.*, **188–9**, 385–99.

McCulloch, J. S. G. (1965) Tables for the rapid computation of the Penman estimate of evaporation. *J. E. Afr. Agric. Forest.*, **84**, 286.

Moncrieff, J. B. (1997) A system to measure fluxes of momentum, sensible heat, water vapour and carbon dioxide. *J. Hydrol. (HAPEX-Sahel special issue)* **188–9**, 589–611.

Monteith, J. L. (1965) Evaporation and the environment. *Symp. Soc. Exp. Biol.*, **19**, 205p.

Moore, C. J., McNeil, D. D. & Shuttleworth, W. J. (1976) A review of existing eddy correlation sensors. *Inst. Hydrol. Rep.* No. 32.

Penman, H. L. (1948) Natural evaporation from open water, bare soil and grass. *Proc. R. Soc. Lond.*, **A193**, 120p.

Penman, H. L. (1956) Evaporation: an introductory survey. *Netherlands J. Agric. Sci.*, **1**, 9–29.

Priestley, C. H. B. & Taylor, R. J. (1972) On the assessment of surface heat flux and evaporation using large scale parameters. *Mon. Weather Rev.*, **100**, 81.

Ritjema, P. E. (**1965**) An analysis of actual evaporation. Agricultural Research Report 659, Pudoc, Wageningen, 107p.

Rodda, J. C., Downing, R. A. & Law, F. M. (1976) *Systematic Hydrology*. Butterworths, London.

Shuttleworth, W. J. (1979) Evaporation. *Inst. Hydrolo. Rep.* 56.

Shuttleworth, W. J. & Calder, I. R. (1979) Has the Priestley–Taylor equation any relevance to forest evaporation? *J. Appl. Meteorol.*, **18**, 639–46.

Smith, M. D. & Allen, S. J. (1996) Measurement of sap flow in plant stems. *J. Exp. Bot.*, **47**, 1833–44.

Steinberg, S. L., van Bavel, C. H. M. & McFarland, M. J. (1989) A gauge to measure mass flow in stems and trunks of woody plants. *J. Am. Soc. Hortic. Sci.*, **11 (4)**, 466–72.

Strangeways, I. C. (1972) Automatic weather stations for network operation. *Weather*, **27**, 403–8.

Strangeways, I. C. (1995) Back to basics: the met. enclosure: Part 1 – its background. *Weather*, **50**, 182–8.

Strangeways, I. C. & Smith, S. W. (1985) Development and use of automatic weather stations. *Weather*, **40**, 277–85.

Tanner, C. E. (1967) Measurement of evapotranspiration in irrigation of agricultural lands. *Agronomy*, **II**, 534.

Thom, A. S. & Oliver, H. R. (1977) On Penman's equation for estimating regional evaporation. *Q. J. R. Met. Soc.*, **103**, 345p.

8

Precipitation

After wandering about for some hours, I returned to the landing-place; but,
before reaching it, I was overtaken by a tropical storm. I tried to find shelter under
a tree which was so thick, that it would never have been penetrated by common
English rain: but here, in a couple of minutes, a little torrent flowed down the
trunk. It is to this violence of the rain we must attribute the verdure at the bottom
of the thickest woods: if the showers were like those of a colder climate, the
greater part would be absorbed or evaporated before it reached the ground.

Charles Darwin *Voyage of the Beagle* (in Brazil)

The variable

A history of raingauging

The first written reference to rainfall measurement was made by Kautilya in
India in his book Arthasastra in the fourth century BC (Shamasastry 1915). The
next reference comes from the first century AD in The Mishnah, which records
400 years of Jewish cultural and religious activities in Palestine (Danby 1933). But
neither the Indian nor Palestinian measurements continued for long. They were just
isolated events, doomed to be ignored and discontinued. There were to be no more
quantitative hydrological or meteorological measurements for another 1000 years –
a period in which scholars believed, or were forced to believe, that one turned to the
sacred scriptures for answers to questions such as 'Where do springs arise from?'. It
was from China, around the year 1247 that the next known reference to quantitative
rainfall measurement comes, and during the fifteenth century the practice of
measuring rainfall was introduced into Korea, probably from China (Fig. 8.1).

As far as we know, the first raingauge to be operated in Europe was made in
Italy by Benedetto Castelli, a Benedictine monk and student of Galileo. In 1639

Figure 8.1. Reproduction of a raingauge design used widely in Korea from the fifteenth century right up to the twentieth. It was probably introduced into Korea from China. This replica is at the Science Museum in London and was kindly made available for photographing by Jane Insley.

Castelli measured rain, just the once, using a graduated glass cylinder about 12 cm in diameter and 23 cm deep. He does not seem to have considered doing this a second time or on a regular basis. In the 1660s Sir Christopher Wren made the first known British gauge and later designed a second, which was probably the first ever that used a tipping bucket (Grew 1681). The first continuous record of rainfall was made by Richard Townley, in Lancashire (Townley 1694), from 1677 to 1703; the gauge consisted of a 12-inch (30-cm) funnel on the roof connected by a lead pipe into his house, the collected water being measured in a graduated cylinder. Interest in measuring rainfall increased rapidly in the eighteenth century, worldwide, and so only the key points can be recorded here.

Dobson (1777) was amongst the first to expose a raingauge to today's standards, on a large, open, grassy patch, overlooking Liverpool. It was 12 inches (30 cm) in diameter and was part of a larger experiment concerned with evaporation. Most gauges, until 1770, had been exposed on roofs. Exposure on roofs has its problems, as William Heberden (1769) was the first to suspect. To investigate this, he operated two identical gauges, one on a chimney of his house and one in the garden, with a third gauge on a 150-foot (45-m) tower of Westminster Abbey. Readings were taken every month for a year and they showed that the gauge on the chimney caught only 80% of that in the garden while the gauge on the tower caught only just over 50% (Reynolds 1965). He could not explain why this was, and it was 100 years before Jevons (1861) demonstrated that the reduction in catch with height was due to wind effects.

At about the same time George Symons, a young assistant in the newly formed Meteorological Department of the Board of Trade (the first UK Met. Office), became interested in rainfall and its measurement, fired by the drought years of 1854–8. It became a lifetime commitment and personal crusade. He amassed rainfall data from many sources and experimented with the various gauges then in use. In 1860 he published English Rainfall, which contained 168 annual totals, mostly from southern England but also from Guernsey and the Isle of Man (Symons 1864, 1866). He resigned his post at the Board of Trade in 1863 to devote all his time to the task of improving rainfall measurement and data collection, much to the debt of meteorology and financial difficulties for himself. By the time he died in 1900, he was receiving records from 3500 sites. Undoubtedly Symons was one of the most significant figures there has been in rainfall measurement and data collection.

Units and definitions

Precipitation includes rain, drizzle, snow and hail, but not condensation in the form of dew, fog, hoar frost or rime even though they can produce trace readings in a gauge of up to 0.2 mm. The total precipitation is the sum of all the liquid collected (including the water produced from melted solid precipitation), expressed as the depth it would cover on a flat surface assuming no losses due to evaporation, runoff or percolation into the ground. While inches have been used in the past to measure precipitation, and still are in some countries, millimetres are the more usual unit today (0.01 in is about equivalent to 0.25 mm). The measurement of snowfall is treated separately later.

Factors affecting all conventional raingauges

The rain collector

Its diameter

In the 1860s, Colonel M. F. Ward did experiments in Wiltshire to compare the performance of rain collectors of different diameters, from 1 in to 2 ft (2.54–60 cm) as well as square funnels. Catch was found to be independent of size, certainly above 4 in. This was confirmed by tests by the Revd. Griffith, also in England, in the 1860s, during which he operated 42 types of gauge (Kurtyka 1953, Reynolds 1965). Although one set of tests in the US suggested that small gauges caught more than larger ones, it is probably safe to say that the area of the funnel does not matter provided it is bigger than about 10 cm in diameter.

The use to which the gauge is to be put in part decides the best size of the collector. For a daily manual reading, too little water makes for inaccuracy, while too much needs a large container to hold it and makes measurement inconvenient. The UK Meteorological Office at one time preferred an 8-in gauge but later changed to Symons' 5 in, which is what is used today for manual, daily observations. (My opinion is that a 5-in gauge catches too little water for convenient or accurate measurement.) If the gauge is of the chart-recording type, there must be sufficient water to operate the float mechanism and move the pen across the paper chart positively, while if it is a tipping bucket gauge there is a lower limit (in bucket size) below which it is inadvisable to go, since a very small bucket tends to have higher errors. Because the water measured by recording gauges is not usually stored but allowed to run away, there is no collection problem and so the bigger the collector, within reason, the better.

Constructional materials and evaporative loss

Traditionally, copper has been used to make raingauge funnels (in the UK), although today anodised aluminium, stainless steel, galvanised iron, fibreglass, bronze or plastic are all used (although not for the UK standard 5-in gauge, which retains its copper construction, for the present). What is important about the material is how well it allows the water falling in it to run off and be collected or measured. Materials do not all behave in the same way with respect to water retention and flow.

Water first wets the surfaces of the funnel and the tubes leading to the bottle or tipping bucket and only after this does it start to flow. When the rain stops, the water left adhering to the surfaces does not drain into the bottle or bucket but evaporates. In the case of a 5-in UK manual gauge, the amount lost is about 0.2 mm per rain event. There can also be some loss from within the storage bottle, although this can be kept low by minimising ventilation, and by housing the bottle below ground to keep it cool. Gratings within the funnel, acting as filters to keep debris out, hold water which is lost by evaporation. Any debris lying in the funnel will also hold water, which will thus be lost. If rain falls as intermittent light showers, loss by evaporation can amount to a significant proportion of the total catch, since it is cumulative; in the case of occasional heavy rain, however, the proportion will be much less.

Instructions given with regard to the UK 5-in gauge is that it should not be polished (although it should be kept clean) – a matt, oxidised surface encourages runoff better than a shiny clean one. Essery and Wilcock (1991) observed a *settling-in period* of a year after new unweathered gauges had been installed, demonstrating that shiny new copper tends to suffer more from evaporative losses. At one time gauges were painted, but this is not advisable (on the inside of the funnel) because paints can cause the water to form large drops that stick, and can also absorb some water if they become porous.

Depth of the collector

One of the important factors in Symons' design is the depth of the collector, which is reflected in the UK Meteorological Office 5-in gauge (see later), which has vertical sides 4.5 in (11.4 cm) deep. This combination of diameter and depth prevents virtually any outsplash of large drops or the rebound of hailstones. Many gauges are to this pattern, but many others have less deep sides above the funnel and this probably results in loss of catch through outsplash. The problem with deep sides, or at least the resulting tall cylindrical shape of the gauge, is that this increases wind-induced errors (see below) and it also increases evaporative loss. At the time of writing, there is a new British Standard being discussed that addresses these matters.

Levelling

It is not uncommon to see a raingauge leaning over at an angle. It may never have been installed correctly, or the ground may have settled after installation, or it may have been knocked by a lawnmower. Whatever the cause, the gauge will read wrongly. In windless conditions a tilted gauge will catch less rain, while if tilted into the wind it will catch more, and if away from the wind less, than it should. An error of about 1% occurs for each 1° of tilt (Kurtyka 1953).

Precision of collector dimensions

If the diameter of the collector is not as specified, its area will be wrong and the wrong amount of rain will be caught. For example, if the area is supposed to be 200.00 cm^2 (requiring a diameter of 15.95 cm) but the diameter is out by 0.5 mm, the catch will be in error by 1.25%. Distortion of the rim of the funnel, by knocks and dropping, will also cause errors.

Siting of gauges

Siting a raingauge near to a large object, such as a tree, is unwise. Nevertheless the number of gauges that are so exposed is surprisingly high. I have seen many, even at supposedly good sites. Perhaps the trees grow and are not noticed; perhaps there is nowhere else for the gauge. A raingauge should be installed no nearer to an object than four times the height of the latter. This applies not only to trees and buildings but also to other instruments such as temperature screens. Some sheltering may be beneficial, however, provided it is kept to the distance specified, because it lessens the adverse effects of wind.

In a forested area, one solution is to site the gauge in a clearing, provided the clearing is big enough to meet the above criteria of distance. If this is not possible, gauges are sometimes installed on a mast just above the forest canopy. If it is necessary to measure how much rain reaches the ground beneath the trees, the situation is more complex (see Chapter 7 on Evaporation).

The effects of wind on raingauge catch

The effects of wind on raingauge performance can cause considerable errors (Strangeways 1996). The raingauge acts as an obstruction to the wind, causing the wind to speed up over the top of the collector and around its sides, while eddies form within the funnel. Speeding up over the top causes some of the drops that should have been caught to pass over the funnel instead of falling into it, while eddies can lift drops out of the funnel that have successfully been caught (Fig. 8.2). These effects increase with wind speed and, since this increases with height, so too do the losses. In 1881 Symons demonstrated, through many experiments, that wind was indeed the cause of a reduced catch with height and showed that, if a 12 in

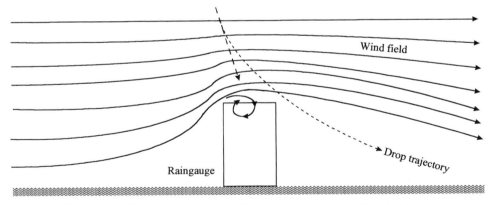

Figure 8.2. The airflow over a cylindrical raingauge is distorted by the presence of the gauge; it speeds up over the top, carrying away drops that should have been caught. Eddies within the collector can also lift out drops and carry them away. However, recent work I have carried out (see Fig. 8.7) suggests that this is an oversimplification, applicable only to wind tunnel conditions. In the real world the wind flows in a very turbulent way, resulting in frequent changes in speed and direction, including the vertical, and this may mean that gauge behaviour in relation to wind is probably more complex than had previously been suggested.

(30 cm) high gauge is taken as the reference, a gauge at only 2 in (5 cm) above the ground caught an extra 5% while gauges at 5 and 20 ft (1.5 and 6 m) lost 5% and 10% respectively (Kurtyka 1953). These were averages and will vary from one rain event to another. In 1893 Abbé showed (see Kurtyka 1953) that the effects of wind depend on the texture of the rain, being more pronounced in drizzle than for larger drops. The problem is even greater for snow with its light flakes and larger area.

Wind shields

An early attempt to combat the problem was made in the US by Nipher (1878), who surrounded the gauge with an inverted trumpet-shaped shield (Fig. 8.3) that deflected the wind downwards. Such a shield can, however, become filled with snow, overcapping the gauge, and so, in 1910, Billwiller (in Germany) cut the

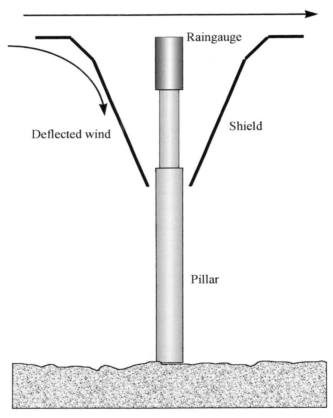

Figure 8.3. The Nipher shield reduces the effect of wind on the catch of a gauge by deflecting the wind downwards, but it tends also to collect snow. The Billwiller screen is similar but has the bottom and the flat collar removed to inhibit snow accumulation.

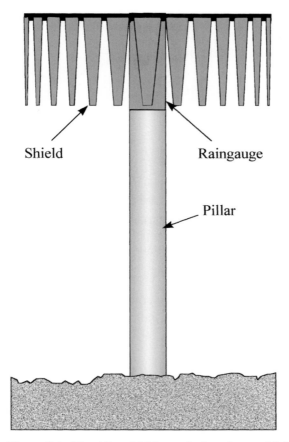

Figure 8.4. The Alter shield was designed to avoid the problem of snow accumulation, its individual leaves swinging freely, preventing the build-up of snow. In strong winds the leaves on the windward side swing in to form a shape similar to the Nipher shield, deflecting the wind downwards. The Tretyakov shield has the Nipher profile but is made of individual (fixed) leaves similar to the swinging leaves of an Alter shield.

bottom out of the Nipher shield to allow snow to pass through. He also removed the outer horizontal collar, although this was later shown to be an important part of the Nipher design. Also, in the US, Alter (1937) designed a shield consisting of loosely hanging strips of metal (Fig. 8.4) which had the property of helping to prevent snow accumulation. It is, however, large and somewhat expensive.

The turf wall

In the UK, Huddleston (1933) designed his *turf wall* (Fig. 8.5), as part of a series of experiments from 1926 to 1933 near Penrith, the object being to find the best

Figure 8.5. The turf wall, seen here at the Institute of Hydrology in Wallingford, was invented by Huddlestone in 1933, its purpose being to shield the gauge from wind. In this particular case it has a Mark 2 5-in manual gauge at its centre. But, like the Nipher shield, it can collect snow.

raingauge exposure for mountainous and windy conditions. He also tested fences and sunken gauges but concluded that the turf wall was best and then went on to determine the optimal design for it.

The pit gauge

Stevenson (1842), in the UK, was probably the first to do anything to combat the wind problem, and his design, with modifications, is still the best. He buried the raingauge so that its rim was just above ground level and surrounded it with a mat of bristles to prevent insplash. In this way the gauge did not interfere with the flow of the wind at all.

Later, in Germany, Koschmieder (1934) designed a modern pit gauge (Fig. 8.6) and this has been shown to be the best way of exposing a raingauge to minimise wind effects (Rodda 1967a, b). By keeping the gauge out of the wind flow, all wind effects are avoided. The grating around it minimises eddies within the pit. It has its problems, however, since it can become snow- or sand-filled (as can the turf wall) or flooded, while at unattended sites it can become overgrown with vegetation. Nevertheless a pit gauge is the best way of exposing any raingauge, as set out in several WMO reports (Sevruk & Hamon 1984). It was currently the topic of a European Standards Working Group on raingauges (CEN/TC 318, WG 5, of which I was Convenor).

Aerodynamic gauges

Another possible way of combating the effects of wind on raingauge catch is to design the profile of the gauge so as to present as little interference to the air flow as

Figure 8.6. The best technique for exposing any raingauge, in order to prevent wind effects, is to install it in a pit surrounded by a grating. The construction seen here conforms to WMO specifications. But like the turf wall it is liable to fill with snow, sand and vegetation, if left unattended. And it is expensive.

possible. Most gauges are cylindrical and these cause the maximum of interference to the passage of the wind. Robinson and Rodda (1969) tested several raingauge designs in a wind tunnel, observing the flow of air with smoke trails. Field tests were also carried out, comparing the catch of the gauges, which included a funnel-shaped gauge of 5 in (12.7 cm) diameter with a half-angle of 45°, the funnel gauge catching most of the rain except when it was heavy, when outsplash was suspected of causing losses.

Adopting a mathematical modelling approach, Folland (1988) analysed the flow of air over a cylinder and proposed a first-guess new design, which he called the flat *champagne-glass* raingauge, a combination of a cone with a diameter of about 25 cm and a half angle of 35°. Wiesinger (1996) also reports the development of an aerodynamically shaped gauge in Austria which includes the facility to measure snowfall.

While Folland was undertaking his theoretical work, I (Strangeways 1984) was also experimenting, independently, using the findings of Rodda and Robinson. The designs that emerged from these developments were similar to those of Folland's. Field tests on two such designs (Hughes *et al.* 1993) confirmed that an improvement in catch was obtained with cone-shaped collectors, but that they did not give results as good as those from pit gauges, nor could the increased catch be unquestionably related to improved aerodynamics. Recently, I have carried out further and more detailed field tests over a three-year period in which gauges conforming to Folland's suggested design were tested, alongside collectors of modified and different profiles

Figure 8.7. By changing raingauge profiles, wind flow can be modified. This figure shows just two of several new designs on test at a field site which I have operated since May 2000. A five-in manual gauge is out of shot. The collected water is weighed manually to 0.1 g to detect small differences, this being essential for the accurate intercomparison of performance. However, not only is aerodynamics of concern, but evaporative loss too can produce considerable error and shallow gauges can suffer from outsplash. It is necessary, therefore, to arrive at an optimum design that balances all of these factors.

(Fig. 8.7). This work is causing a revaluation of the processes involved in wind flow over, around and within raingauges in a natural turbulent situation, as compared with the simpler situation in a wind tunnel.

Correction procedures, field intercomparison and wind-tunnel tests

According to Sevruk and Klemm (1989) there are over 50 different types of raingauge in common use worldwide, all differing in size, shape, material, colour and the height at which they are exposed above the ground (from 0.2 to 2.0 m). The number is greater if the newer automatic gauges are included. Because each gauge has its own characteristic errors, accurate intercomparison of rainfall data from around the world is not possible. To deal with this problem, the WMO organised field intercomparisons of the major national gauges of the world, with pit gauges as the reference, at 60 sites in 22 countries (Sevruk & Hamon 1984). However, such intercomparisons lump many rain ~~ ther over a period and can only give a rough correction ~~ size are the two key factors in how ~~ these are also available it becomes ~~ n.

~~ collectors have shown that it is ~~ e orifice rim: the smaller the rim,

~~ ical simulation techniques of the ~~ nerical procedure. First the air ~~ ectories of different drop sizes ~~ is a formula giving the wind- ~~ infall and drop size for each ~~ dvantage that the corrections ~~ need for field or wind-tunnel ~~ it is necessary to know the ~~ ds to the complexity of the ~~ although it can sometimes

The pu ~~) is usually to estimate
areal av ~~ n. Over anything other
than a v ~~ to obtain a reasonable
average. ~~ gh density of gauges,
even the ~~ one gauge for every
25 km^2 (~~ diameter of 12.7 cm (5 in) the area is
126.7 cm^2 ~~). This means that the area from which rain is collected is around 1 in 2×10^9 of the area over which rainfall is to be estimated. In most parts of the world the raingauge density is much smaller than this and over the seas almost non-existent. While some argue that in a situation like this it is not worth measuring rain very carefully, the contrary is the case, for without accurate point measurements, the spatial variability of rainfall cannot be established because it will be swamped by raingauge errors. Also, a rainguage with a systematic error

(such as undercatch through wind effects) will always give a bias to areal estimates. It is all the more important, therefore, to try to get the point readings as accurate as possible, through careful choice of site, careful positioning of the gauge at the chosen site and use of the best raingauge possible.

Having done everything possible to get good point measurements, how can these be converted into areal estimates? Imagine an area completely covered in square collecting funnels. The area's average rainfall for any period would be the average reading of all the individual gauges. But in practice most of these gauges will be missing. So how is it possible to estimate the average from just a few gauges?

It would be possible to take the average reading of them all. But there might be several gauges clustered together, and they might be in a region that is not typical of the whole area. A technique often used is to work in percentages rather than actual millimetres of rain, each raingauge reading being converted into a percentage of the gauge's average annual catch, determined over a period. This percentage is then converted back to rainfall amounts using the catchment annual average. While this produces more uniform readings, the method confers on each raingauge the same importance, even though some may be clustered together, or perhaps be in an unrepresentative part of the area. One approach to this problem is to give each of the gauges a weighting that takes account of the relative importance of its contribution to the average (Jones 1983). A widely used technique for doing this is the polygon method (Thiessen 1911) in which the area is divided into polygons such that the polygon around each gauge is that part of the area nearer to that gauge than to any other. The weighting for each gauge is then the ratio of the area of its polygon to the total area. But the gauge surrounded by the largest polygon may not be at a representative site.

An alternative to the polygon is to derive the gauge's weighting by dissecting the area into triangles. Briefly the procedure is to construct a mesh over a map of the area and for each mesh point to search for a triangle of raingauges that surround it, but limited to a certain distance. If no such triangle can be found the nearest three gauges are selected. How the mesh is constructed, the weights determined and the maximum distance is selected, is complex (Jones 1983). Other methods calculate the weightings of each gauge based on criteria such as the angles each subtends at the major and minor axes of the area.

Where a network of raingauges is to be installed from scratch, the opportunity exists to choose the sites. How can this best be done? Gauges could be placed on a regular grid, but this might mean that some are in very unsuitable places such as on a hilltop. In practice it is more realistic to site gauges where there is easy access, or where an observer lives (in the case of manual gauges). But a choice made in this way will probably be biased, the remoter, higher and least accessible all being omitted. The area could first be saturated with gauges, to get somewhere near the ideal of covering the whole area with gauges, irrespective of the difficulty of getting to them. Then, when there has been a chance to study their relative catches, the number of gauges could be reduced, keeping a watch on how this reduction affects the accuracy of the mean areal rainfall. Sometimes gauges turn out to be key ones, being very representative of a large area. Another technique is to identify topographic domains (Rodda 1962), in which factors such as the slope, aspect and elevation extend over a clearly definable distance, a gauge being placed at random within the domains; but this is not realistic in flat areas or where the landscape is rugged. There is no ideal solution, and the best method for any one situation needs to be selected with an informed knowledge of the options. A combination of two or more of the above methods may be appropriate.

Manual gauges

As Sevruk and Klemm (1989) have pointed out, there are over 50 different types of daily-read manual gauge in general use worldwide, but as most gauges fall into clearly defined types it is only necessary to describe typical examples of each, and the UK case will be used for this.

The classic five-inch manual gauge

The UK Meteorological Office Mark 2 gauge demonstrates the general features of most manual gauges and is based on Symons' 5-in model (12.70 cm diameter, with an area of 126.7 cm^2), which has now been standard for all of the twentieth century in the UK (Fig. 8.8) and continues into the twenty-first century.

A copper funnel with an accurately turned bevelled brass rim and deep sides fits onto a copper base, set into the ground, with a splayed-out base to give it more stability. For interchangeability, the cylindrical parts are made of drawn copper tube (not of rolled, soldered sheet). Inside the case is a removable copper container and inside that is a collecting bottle. The gauge is installed so that its rim is 12 in (30 cm) above the ground, which should be of short-cut grass or, failing this, gravel. A hard smooth surface should not be used, to avoid insplash. The bottle holds the equivalent of 75 mm (3 in) of rain. The removable copper container provides extra capacity in case the bottle overflows in heavy rain or the gauge is left too long, together holding 140 mm (5.5 in). A cheaper model, known as the Snowdon pattern, has straight sides and so is less stable. A still cheaper version is made in galvanised steel. Another cheaper, but larger, gauge is also available with a funnel aperture of 20.32 cm (8 in) and a total measuring capacity of 180 mm of rain (7 in). In the US, 8 in is most commonly used. From my own experience, this is probably a better size.

Figure 8.8. The 5-in, Mark 2 raingauge preferred by the UK Meteorological Office. Rainwater collected by the funnel is stored in a bottle within the case, which extends below ground, and is measured daily in a graduated cylinder of the type shown. The base of the measuring cylinder tapers to allow small volumes of water to be measured more precisely.

The gauges are read daily using a graduated glass cylinder into which the collected rain is poured. The cylinders are either flat-based or tapered, the latter decreasing in diameter from 1 mm of rain down, for greater accuracy when small amounts of water are collected (Fig. 8.8). They are marked in 0.1-mm steps with an additional mark at 0.05 mm, figuring being at 0.1 mm, 0.5 mm and every 1 mm between 1 mm and 10 mm. The maximum error for the measure is ±0.02 mm up to 2 mm and ±0.05 mm above 2 mm. In reading the amount of water in the cylinder, the procedure is to note the reading closest to the bottom of the water meniscus to the nearest 0.1 mm. Weighing the water gives a more accurate measure, if needed.

When gauges cannot be read daily, a larger container is necessary so that weekly or monthly amounts of rain can be stored. In the UK two such models are used, both having a funnel diameter of 12.70 cm, one holding 680 mm of rain (26.8 in) and a larger model holding 1270 mm (50 in), for areas of high rainfall or where the gauge may have to be left for up to two months. It is known as the Octapent gauge because it is a merging of a 5-in funnel with the base for an 8-in funnel. It has a splayed base similar to the Mark 2 gauge. These two gauges have a precisely fitting removable inner container to hold the water, the water first being measured roughly in the container by means of a dipstick to get a rough estimate of the amount; thereafter it is measured more precisely using a graduated cylinder. Provision is made to insert a flexible tube that will collapse under the pressure of freezing. A modified version of the Snowdon gauge, known as the Bradford gauge, is deeper and can hold 680 mm of rain; it was designed for use in similar circumstances to the Octapent.

Other types of manual raingauge

There are more than 150 000 manually read gauges in use throughout the world. Sevruk and Klemm (1989) presented an analysis of their distribution and type and showed that the most widely used (in 1989) was the German Hellmann gauge, with 30 080 in use in 30 countries. The Chinese gauge was second with 19 676 used in three countries, and the English Mark 2 and Snowdon third, with 17 856 operated in 29 countries, all three types totalling 67 612 and accounting for about half the world's raingauges. The Hellmann gauge has a funnel of 200 cm^2 or 15.96 cm diameter (6.28 in) while the Chinese gauge has a collecting area of 314 cm^2 (diameter 20 cm or 7.9 in). Like the Hellmann gauge, the Chinese gauge is made of galvanised iron.

The remainder of the most commonly used world's gauges are as follows.

Russia:	13 620 gauges in seven countries, 200 cm^2, galvanised iron
The US:	11 342 gauges in six countries, 324 cm^2 (8 in), copper
India:	10 975 gauges just in India, 200 cm^2, fibreglass
Australia:	7639 gauges in three countries, 324 cm^2 (8 in), galvanised iron
Brazil:	6950 gauges just in Brazil, 400 cm^2, stainless steel
France:	4876 gauges in 23 countries, 400 cm^2, galvanised or fibreglass
Totalling:	55 402

The grand total of all these is 123 014, although Sevruk estimates there are about 150 000 gauges in all worldwide.

No doubt this has changed since the count was done and it probably missed gauges used by small organisations, but it gives a good idea of the situation. It does not include automatic gauges of any type, just manual ones.

Mechanical, chart-recording gauges

Recording raingauges are used mostly to supply information on the times when rain starts and stops and to give an approximate indication of the rate of rainfall. It is usual to operate a manual gauge nearby to act as a reference and to give precise totals.

Mechanical recording raingauges are of two main types – those that cause a pen to move across a paper chart through the movement of a float and those that use a balance to move the pen.

Float-operated recorders

In this design, the collected rainwater is funnelled into a container in which there is a float that moves upwards as the container fills, so that a pen also moves upwards across a paper chart turned by a spring-driven clock. The container is of smaller diameter than the collecting funnel, giving magnification to the pen movement. In early models, the container had to be emptied by hand when it was full, while in later designs mechanisms of various types were devised to make the pen fall back down to the bottom of the paper chart and restart its climb back up. It could do this several times before the container was full, but it still then needed emptying by hand. The gauges used today all have methods of emptying the container automatically each time it becomes full. Exactly how the container empties is the main difference between recorders, but all current models empty by siphoning.

The tilting siphon gauge

The most popular chart-recording raingauge in the UK, the tilting siphon gauge, was designed by Dines in 1920, the rain collected by a funnel being fed into a cylinder containing a float (Fig. 8.9 (*a*), (*b*)). The float rises as water enters the cylinder, moving a pen up a paper chart. When the float approaches the top of the cylinder it releases a catch that causes the cylinder, mounted on knife edges, to tip over to one side. The action of tilting causes the siphon tube to be suddenly flooded with water, kick-starting the siphoning process – which must start positively so as to prevent 'dribbling'. When nearly empty, and while still siphoning, the cylinder tips back to its vertical position (for it is counterbalanced) ready to repeat the process. Because it takes time to siphon the water out, the mechanism is designed to save the incoming water during the siphoning process.

Two models are made, one for temperate regions having a collector of diameter 28.73 cm (area 648.4 cm^2), which siphons for every 5 mm of rain, taking from 6 to 8 s to empty. The model for tropical, heavier rain is the same except that the funnel has only one-fifth the area of the temperate model and so siphons every 25 mm (taking the same 6–8 s, for the internal mechanism is the same). The tropical model is surrounded by a white outer case to shield it from the Sun. Charts can be for daily, weekly or monthly records.

The natural siphon gauge

The natural siphon recorder, designed by Negretti and Zambra, is very similar to the tilting siphon recorder and predates it by 20 years. The siphon consists of two

Figure 8.9. (a) A Dines tilting siphon recording raingauge. The paper chart mechanism, visible through the window of the gauge, records the movement of the pen. (b) Water enters the container on the right (through the small elongated funnel). As it accumulates it raises a float, connected directly to an arm that terminates in a pen in contact with the paper chart. The container is balanced on a knife edge and when full the float releases a catch that allows the container to tilt, starting siphoning abruptly. The counterbalance (bottom right) brings the cylinder back to its upright position when siphoning is almost complete.

coaxial tubes to the side of the float cylinder, constructed so that the annular gap between the inner and outer tubes is small, causing siphoning to start abruptly without any special mechanical trigger being necessary. When the water reaches the top of the outer tube, capillary action ensures that siphoning starts decisively, pushing all the air out and down the inner tube. Conversely, after siphoning, and once air gets to the top of the tube, siphoning abruptly stops. The simple construction of this instrument means that it is cheaper and there is less to go wrong than there is with the tilting siphon gauge; however, there is no means of saving the water that enters the funnel during siphoning.

As with the tilting siphon, there are two models: the temperate-climate design has a funnel of 8 inches diameter (20.3 cm, area 324 cm^2) and siphons for each 10 mm of rain, taking from 12 to 15 s to empty. The tropical model is the same but has a funnel of area 129.7 cm^2, that is, 2.5 times smaller, siphoning every 25 mm of rain.

The Hellmann siphon gauge

Hellmann designed a float gauge in Germany in 1897 using a siphon with a long tube that ensured accurate action. It has a collecting area of 200 cm^2. It became widely used in central Europe and, along with the Dines and Negretti recording gauges, probably accounts for most of this style of gauge in use today. As it is very similar to the natural siphon gauge discussed above, it needs no further description. Having a large container at the bottom to store the water after measurement (as back-up), it is bigger, standing 1.2 m high.

Weight-operated recording gauges

Weighing raingauges operate by recording the total weight of precipitation as it accumulates in a container, either by suspending the container on a spring or on the arm of a balance (Fig. 8.10). In both designs the weight of the water forces the container downwards, the vertical movement being magnified by lever linkages to move a pen. In some designs, a dual-traverse action is used so that the pen records half of the gauge's total capacity in the upward direction, the second half downwards. To prevent wind shaking the pen, and so introducing movement into the trace, the pen is damped by an oil dashpot.

Osler made one of the first weighing raingauges, in the UK in 1837, in which the container was counterbalanced by a weight on the opposite side of the fulcrum. A rod, connected to the weighted side of the balance, moved a pencil over a paper chart. The receiver had a siphon that emptied every 0.25 in of rain. Two dozen or so different types of weighing recorder have been made in the period since 1837, some being emptied by hand, some siphoning the water out, while others use containers that tip over when full, so emptying themselves.

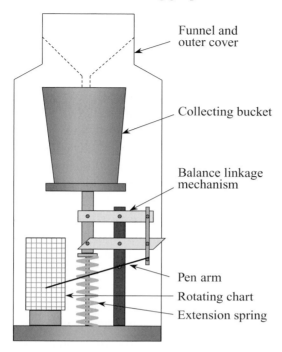

Funnel and
outer cover

Collecting bucket

Balance linkage
mechanism

Pen arm

Rotating chart

Extension spring

Figure 8.10. An alternative to float recorders is the weighing raingauge, illustrated here schematically to show the principle. This type can have the funnel removed to collect snow.

The advantage of periodic emptying is not just that operation can be continuous but that the balance mechanism can be made more accurate because it does not have to work over a wide range of weight. For example, in the case of an 8-in-diameter collector (area 324 cm^2), the weight of water representing 0.1 mm of rain is 3.24 g. If the gauge does not empty itself but instead has the capacity to collect up to, say, 12 inches (30.48 cm) of rain, the balance must be able to measure changes of 3.24 g in a total weight of 9884 g of water, or 0.03% of full scale. In reality only about 0.5% accuracy is achievable in a mechanical field balance, representing little better than a measurement of rain to 1 mm. In the case of float-operated recorders the same problem exists, but all modern float recorders siphon periodically, in such a way that the recorder needs to detect changes of 0.1 mm over a range of 5 mm, a resolution of only 1 in 50 (in the temperate models), or 0.1 mm over 25 mm, a resolution of 1 in 250 or 0.4% (in the tropical models). For anything other than approximate measurements, therefore, a weighing gauge, like a float gauge, must empty itself regularly.

The containers of weighing raingauges tend to be more open than in the case of float gauges, many using an ordinary galvanised bucket (US 'pail'), making them more vulnerable to evaporative loss, although this can be prevented by adding oil to the container so that it floats on the water.

Rate-of-rainfall recorders

There are two main types of raingauge that measure the intensity of rain – the fixed volume and the *calibrated orifice* (or *controlled orifice*) gauges. While they may be classed as rate recorders they can, with suitable integrators, also measure rainfall totals. The fixed volume type is more familiar as the tipping bucket (see below). The calibrated orifice type is much less common.

Calibrated orifice rate-of-rainfall sensors

There are three categories of this type of gauge. In the first, drops are formed that are then counted electrically, examples being gauges developed by W. Gallenkamp in Germany in 1887 and by W. J. Binnie in the UK in 1892. There are other more recent examples such as the gauge developed in the UK by Norbury and White (1971). Drop size is approximately proportional to the product of surface tension and the tube's bore (in the case sited, 2.8 mm diameter produces drops of 0.07 ml). Norbury and White state that, at a constant flow rate, the drop volume varies by up to 3%, although at high rainfall rates by up to 10%. Also, as drop size is affected by changes in surface tension and as this varies with the degree of pollution in the water and its temperature, additional errors occur. For the same gauge, over a temperature range of 5–35 °C, a change in drop size of up to 5% can occur. Clearly errors of this magnitude would be serious if precise totals of rain were required. However, for relative rather than absolute intensities, it is an attractively simple method with no moving parts.

In the second type of calibrated orifice gauge, the rainwater enters a container, from which it is also allowed to leave. The depth which the water reaches in the container is a combined function of the rates of input and output of the water. The input is controlled by the rainfall rate, the output is variable and controlled by a valve. In a gauge designed by Sprung in Germany in the early twentieth century, a weight-controlled valve over an orifice moves vertically as a function of rainfall rate; the higher the rate, the higher the valve and the more water is passed, the movement of the weight being recorded on a paper chart. The Spanish Jardi gauge of 1921 works similarly but the valve is controlled by a rising float.

In the third calibrated orifice type, the container has holes or a slit in its side, the amount of water retained being proportional to the rate of its input. The container is weighed, in the designs of F. J. Cornick in the US and C. Nell in the UK in 1950, but the depth could equally well be measured by a float.

Nevertheless, rate-of-rainfall sensors, apart from tipping buckets in electronic systems, are comparatively rare, rather specialised and little used.

Electronic raingauges

Although there are exceptions, which will be explained later, most electrical raingauges work on the principle of collecting the rainwater in a funnel and measuring it in some way.

Tipping bucket raingauges

Despite their problems, tipping bucket gauges are by far the most commonly used type of automatic raingauge in the world today.

Basic principles

The raingauge of Sir Christopher Wren's AWS (mentioned in the opening chapter) was single sided – it filled, tipped over and fell back again. Crossley, in 1829, was the first to use a double-sided tipping bucket. Other designs followed throughout the nineteenth century in Europe and America and they proliferated in the twentieth century. Less common has been the use of a water wheel, the principle being that, instead of a tipping bucket, a wheel rotates in discrete steps. But they are not used today.

Modern designs are all symmetrical (Fig. 8.11), although the shapes of the buckets vary from make to make. The size of bucket is important, for it is undesirable to attempt to make a bucket tip for very small amounts of water because the errors (described below) increase. Ten millilitres or more is an advisable minimum, and so to measure 0.2 mm of rainfall reliably requires a funnel of area at least 500 cm^2. Gauges are made with buckets that tip at intervals of 0.1, 0.2, 0.25, 0.5 and 1 mm. Which interval is chosen depends on the use of the data and on the conditions; the move today is toward 0.2 or 0.25 mm buckets.

The tipping of the bucket normally moves a permanent magnet, fixed to an arm on the bucket, past a magnetic reed switch, giving a brief contact closure. Reed switches can suffer from 'bounce' but, by ensuring that the counter circuits do not have too fast a response time, multiple counting can be prevented. Mercury-wetted switches also help to prevent this, as well as to reduce the other reed-switch problem, contact wear.

Invariably today a tipping bucket raingauge will be used in conjunction with a data logger, either built into the gauge as a single-channel recorder or as part of a multisensor system such as an AWS, which will include a multichannel logger. The contact closures of a gauge can be logged either as a total number over a

Figure 8.11. The most usual way to measure the water collected by a raingauge is by means of a tipping bucket. With a collector of 500 cm², 0.2 mm of rain produces 10 ml of water, which is adequate to tip the bucket reliably. As it tips, a magnet in each of the arms attached to the bucket closes a reed switch in each of the fixed arms of the pivot support, allowing two recording devices to be used with the gauge.

period or as the time and date of each tip, giving an indication of intensity and time distributions.

Strengths and weaknesses

The advantages of the tipping bucket are that it is simple, there is little to go wrong, it consumes no power and can be used with a variety of recorders. Its weaknesses are relatively few, but they can be a problem (Hanna 1995), as follows.

The record provided by a tipping bucket is not continuous, although if the tips are every 0.1 mm of rain this is close to being so. A UK Meteorological Office design overcomes this by weighing the bucket as well as counting its tips, giving an indication of the increasing amount of water on a continuous basis between tips (Hewston & Sweet 1989, Whittaker 1991). It operates by suspending the bucket on wires, the natural resonant frequency of which increases as the weight increases; the signal is picked off magnetically.

When the bucket has received the correct amount of water and starts to tip, water continues to enter the full side until it is in the level position, this rain being in excess and thus lost, although the error is only small. However, if the gauge is calibrated by passing water into it at a controlled rate, equivalent, say, to 1 to 20 mm of rain an hour, this error can be calculated for different input rates and automatic correction made for it during data processing (Calder & Kidd 1978). Dynamic calibration is preferable, in any case, to using a burette to pass water into the bucket until it tips once, since an average calibration for 50 tips will be more precise; drop-by-drop calibration is also difficult to do precisely. A carefully measured volume, of around one litre, is passed slowly into the gauge and the number of tips counted over a period of perhaps 60 min, the bucket being set to tip for the required amount by adjusting the pillars on which it rests. Since this can be a long process – of repeated adjustment and test – it can be preferable to set the bucket close to the desired sensitivity, of say 0.2 mm, and to measure exactly what it is by the passing of one litre of water. Thus a gauge may tip for each 0.19 mm, this being used as a bucket calibration factor.

Some water inevitably sticks to the bucket after it tips, this being lost if the bucket does not tip again until after it has evaporated. If it does tip again, slightly more water will be required to tip the second time because the wet side is now slightly heavier, but on third and subsequent tips, balance is restored. If the bucket is calibrated by passing a litre of water through the gauge, the wetting effect is automatically compensated for and the error only occurs in the event of one isolated tip. The bucket's surface will change with use, generally accumulating a thin film of oxide, if made of metal, or of adhering dust, and the amount of water sticking may thus change. There is, therefore, a case for calibration checks some months after installation. This is one reason why tipping buckets should be as large as possible, since in the case of small buckets the amount of water adhering is a larger percentage of their total capacity and so the error will be greater. To minimise the adherence of water, some tipping buckets have drip wires at the point where the water leaves the bucket, although this only reduces the problem at that one point.

Because rain rarely ceases just as a bucket tips, some water is usually left standing in the bucket until it next rains. If this is within a few hours, there is not much of a problem, but, if rain does not occur again for some time, all the water left standing in the bucket may evaporate. While this can amount to one full tip, on average over a long period it will average out at half a tip. Depending on the type of rain experienced this error can be large or small, light showers spaced by a day or so producing larger errors than long periods of heavy rain. This is another reason for having buckets that tip for small increments of rain (say 0.1 rather than 0.5 mm), for less will then be lost by evaporation.

Since the bucket must tip in response to a very small change in the amount of water in it, friction in the pivot or any variation of friction over time will affect performance. This is yet another reason for using as large a collector as possible. The type of bearing varies from make to make, including ball races, knife edges, rolling systems and wires under tension. Each performs differently.

Other types of electronic raingauge

Weighing gauges

By replacing the pen of the mechanical weighing raingauge by a mechanism that turns a potentiometer, an electrical output is obtained that can be logged. In electronic versions, the water is weighed by a straingauge load cell. Because electronic weighing is more precise than can be achieved in the field with mechanical springs and levers, greater accuracy is possible. One commercial raingauge senses increments of 0.01 mm of rain in a total of 250 mm – the capacity of its non-emptying container. Another model has a capacity equivalent to 1000 mm of rain.

An advantage of the weighing gauge is its ability to measure the weight on a frequent basis (just during rainfall), allowing the intensity of the rain to be measured in finer detail than by a tipping bucket. But for long-term unattended operation, some means of emptying the weighed container automatically is necessary, such as by siphoning.

Capacitance raingauges

An alternative to weighing is to allow the water to accumulate in a cylinder containing two electrodes that form a capacitor with the water as the dielectric (Fig. 8.12). As is shown in the measurement of soil moisture by the capacitance probe method, the high dielectric constant of water (about 80) makes it convenient for measuring water quantity by electrical means. In the case of the raingauge application, the situation is made much simpler because the water is in the form of bulk liquid, rather than as a very thin film of moisture on soil particles (see Chapter 9). As with weighing gauges, some means of periodically emptying the container is necessary for long-term operation.

Optical gauges

Optical raingauges are a completely different method of measuring rain, in the sense that they do not collect the water and measure it directly. They have evolved largely as a result of spin-off from visibility measurement (see Chapter 13) developments, raindrops and snowflakes being detected by their effect on a horizontal beam of

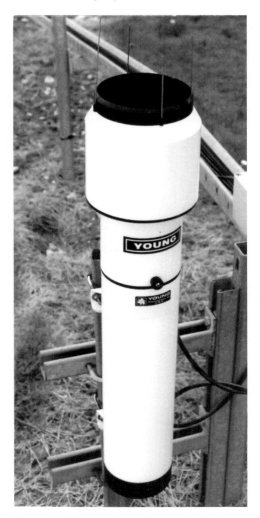

Figure 8.12. An alternative method of measuring the rain amount automatically is to accumulate the water in a container and to measure the change in capacitance between two electrodes submerged in the water (hence the depth of the outer casing). This is akin to the capacitance probe for soil moisture measurement, except that here liquid water is the only dielectric (no soil) and this makes the measurement simpler. As with the weighing raingauge, the container must be emptied periodically.

infrared light about one metre long (Fig. 8.13). An infrared receiver detects the passage of the particles through the beam, either by being in-line with the transmitter (as illustrated) or by detecting backscatter or sidescatter from the precipitation particles, the resultant scintillation waveform being analysed. It is essentially a rate-of-precipitation sensor although, by integration, hourly and daily means of

Figure 8.13. This optical raingauge at the Rutherford Appleton field site at Chilbolton is used to detect rain intensity and drops size in research into radio propagation. The large box on the right contains the light source (an IR LED) which produces a collimated beam detected by a diode in the small unit on the left. In clear conditions the beam is constant, but fluctuates (scintillates) when precipitation is passing through the beam. The amount and nature of the scintillation gives an indication of the type and intensity of the precipitation.

intensity can be obtained, thus giving the total amount of precipitation. The type of precipitation, whether rain, snow, a mix of the two or hail, can be detected by analysing the spectrum of the scintillation signal. This sensor also gives a yes/no indication of the precipitation state, making it a useful instrument at, say, an airport or an Antarctic base, where an observer can watch a screen and know whether it is raining or snowing and if so how intensely (another name for the instrument is a *present weather detector*). This can be particularly useful at night. It is also useful in cold regions where a large proportion of the precipitation falls as snow, which is difficult or impossible to measure with normal raingauges (see later), although the need to keep the optics ice-free means that power requirements might be high.

For indicating instantaneous precipitation rates, an optical raingauge has advantages over the tipping bucket. But for routine, long-term collection of precipitation totals, its high cost (about £7000 to £8000 currently) makes the tipping bucket raingauge, at one-twentieth the cost, inevitably preferable. Manufacturers of optical gauges (there are very few) quote an accuracy of from ±1% to ±10% (of actual intensity) depending on intensity and the type of precipitation.

An exception to the cost of optical raingauges being a deterrent might be their relative lack of sensitivity to non-stable deployment, which could be an advantage on ships or buoys.

Precipitation detectors

Precipitation detectors sense whether rain or snow is falling, but not its intensity. Another name for them is *ombroscopes*. A small sensing surface of somewhere between 10 and 50 cm^2 inclined at an angle (to allow water to run off) detects the presence of water on it by measuring either the change in resistance, or the change in capacitance, between interleaving electrodes, owing to the presence of water. The plate is heated if the temperature falls below +5 °C, to detect snow. It is also heated when there is water on it, so as to dry it off quickly after rainfall or snowfall stops. Lower-power heating prevents fog and dew from forming (and from being interpreted as rain). Ombroscopes have a shorter history than raingauges, having been made only since the start of the twentieth century. They find use particularly in operational situations where the occurrence of precipitation needs to be known in real time or where its duration needs to be known but its quantity is not required.

Impact-sizing sensors

Concern over drop size was mostly driven by an interest in cloud physics in the mid-twentieth century, but a few drop-size detectors were also developed for use on the ground to measure the energy contained in the impact of each raindrop as it hit a surface, as a means of measuring rainfall quantity. Nevertheless, the main application was directed more towards measurement of the size-spectrum of raindrops. A few instruments have, however, been developed to sense the size and number of the individual raindrops, as they hit a surface, and thus the rainfall quantity, but they have not found wide use.

One particular type was based on the measurement of small changes in the value of a capacitor made of interleaving electrodes (very similar to those of a precipitation detector), covered by a thin glass plate to protect them. This type of sensor detects the volume of each drop, the changes in capacitance that occur as each drop lands and spreads out on the glass modulating the frequency of an oscillator circuit (Genrich 1989). It is the sudden arrival of the drop that is detected, the water remaining on the plate afterwards presenting a DC component that can be ignored. But it is not a direct method and it is necessary to find a reliable relationship between the change in capacitance and the volume of the drop that produces it. It is unlikely that this can ever be as precise as measuring the volume of the accumulated drops collected by funnel. Genrich suggested that such sensors might find application in conjunction with a tipping bucket gauge, the impact-sizing sensor being able to detect small amounts of precipitation better than a conventional gauge, while the tipping bucket is better at all other rates.

Another type of impact sensor attempts to measure the energy in the impacting drop as it strikes a sensing membrane; the energy depends on the drop's volume (weight) and its speed of impact, larger drops falling faster than small ones. But again finding a convincing conversion factor from impact intensity to volume of water is not simple. The speed of impact is also affected by the speed of the wind, as is the angle of fall of the drop. Devices such as this may have a small place in applications requiring some knowledge of drop-size distribution, but they do not replace conventional raingauges.

Again, as with the optical raingauge, applications for impact sensors might be found at sea where conventional gauges are problematic. Indeed this is being investigated, but in relation to the sound-signature of the bubbles that raindrops produce in the water rather than the impact of the actual drops themselves (Chapter 17).

The measurement of snow

Snow is very difficult to measure, as will become clear.

Measuring snow as it falls

Wind effects

Snow is more easily blown by the wind than is rain, and this greatly worsens the aerodynamic problems. Above a wind speed of a few metres per second, the use of raingauges for measuring snow is not feasible without some form of screen. Pit gauges are obviously not practicable for anything but very small falls, although a possible alternative is the use of aerodynamically shaped collectors such as that shown in Fig. 8.7. After collection, wind eddies can lift out dry snow that has been successfully caught. This complication can be reduced by removal of the funnel, so presenting a deeper container, which lessens the reach of the eddies, or by introducing a cross-shaped vertical divider into the funnel, although this considerably increases evaporative loss when measuring rain.

As the wind speed increases so does the angle of fall of the snowflakes, becoming increasingly more horizontal; this causes the flakes to approach the gauge at such a small angle that the orifice presents too thin an ellipse for them to enter.

Figure 8.14. Dry snow and ice crystals, 'spindrift', blow like sand, as here on Deception Island on the edge of Antarctica. Caught by raingauges, spindrift can be mistakenly interpreted as snowfall.

After falling, dry snow can also be picked up by the wind and carried elsewhere as spindrift (Fig. 8.14), which if caught by gauges cannot be differentiated from true snowfall.

Measuring snow with manual raingauges

If the snowfall does not overcap or bury the gauge, and the wind is light, it is possible to use conventional manual raingauges to measure snow reasonably well. If the snow has not melted naturally at the time of reading, the procedure is to melt the snow in the funnel and to measure the resultant liquid water in the usual way.

If amounts of snow are greater, with a risk of over-capping the orifice, the funnel might be heated to melt the snow as it falls; alternatively, the funnel can be removed so that the snow falls directly into an open container, the rim being retained to define the catch area. Without some means of melting the snow, however, even a gauge of this design will become full quickly. Some open gauges use antifreeze to melt the snow but they suffer from greater evaporative losses than those with funnels, although this can be prevented by floating a thin layer of light oil on the melted snow.

Measuring snow with recording raingauges

If falls do not overcap the funnel or bury the gauge, and natural melting is relied on, siphoning, weighing and tipping bucket gauges will all measure snow satisfactorily, provided that their mechanisms do not freeze up. The timings of the record will indicate melting time, not falling time, even though the total may be correct. If their funnels, internal mechanisms and outflow drains are heated, all will operate as if they were measuring rain. But heating a funnel to supply the necessary energy to melt the snow takes from 250 to 1000 watts.

Measuring snow automatically with the funnel removed

By removing the funnel, just as with a manual gauge it is possible to collect the snow directly in a large, open container, the build-up of snow being weighed mechanically or electronically. Simpler designs do not melt the snow, they simply hold it, and these are limited in capacity, unless natural melting occurs. More complex types of this kind use antifreeze to melt the snow as it falls into the bucket, with a film of oil to prevent evaporation. A further refinement is that of emptying the container and automatically replenishing the antifreeze, retaining the oil film by stopping siphoning in good time. One such design (intended primarily for rain measurement) omits the oil, measuring the losses as well as the gains in weight, thereby allowing for evaporation. But unless the snow is melted, the collector will soon become filled with snow and without natural melting will be limited in its capacity.

In all open (funnel-less) gauges there must be a danger that snow will get between the fixed outer rim and the internal container, possibly bridging it and then freezing (blowing snow finds its way through the smallest of holes and cracks).

Because of their ability to measure snow more easily than can a siphoning rain-gauge, weighing raingauges are probably more commonly used in countries that receive a significant proportion of their precipitation as snow.

Measuring snow with an optical precipitation gauge

Optical gauges also measure snow. Since they do not catch or melt the snow, many of the problems are avoided and it could be seen as a possible answer to the snowfall measurement problem. And so it is, partly. But while the water content of raindrops is precisely known, from their size, thereby allowing a moderately plausible estimate of rain intensity to be inferred from the scintillation signal, snowflakes vary widely in shape and density, making the conversion from light signal to intensity of snowfall much less certain than for rain.

Snow shields

Because of the large wind errors in measuring snow, attempts have been made to shield raingauges when they are used for snow, the Alter shield being an early example. The best measurement of snowfall using raingauges was deemed by the WMO to be that obtained by having bushes around the gauge cut to gauge height, or to site the gauge in a forest clearing. But bush screens are only practicable if the bushes do not become full of, or covered by, snow. Also, because suitable

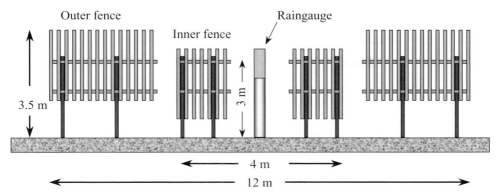

Figure 8.15. While wind effects can cause large errors in rainfall measurement, the problem is even greater with snow. To help mitigate these effects, the double-fence snow shield is the WMO-recommended international reference. A high outer octagonal slatted fence surrounds a similar but lower inner fence, the raingauge being at the centre, sometimes incorporating a further shield of the Alter or Nipher type (parts of the structure are omitted from the drawing for clarity).

forest clearings or bushes may well not be available at many cold sites, and as they will all be different, they were not selected as the reference. Instead, and after due consultation and review, the WMO designated the *octagonal, vertical, double-fence shield* as the international reference (DFIR) against which all snow gauges should be compared; it is a secondary standard (Fig. 8.15). Artificial screens can also be moved as the snow accumulates, whereas plants cannot. A comparison of the DFIR was made with bush shields at Valdai in Russia, the bush exposure catching most snow, the DFIR 92% of this. All other gauges caught significantly less, for example an unshielded 8-inch manual gauge caught 57% of the bush gauge while an identical gauge with an Alter shield caught 75%. Methods have been developed to correct the DFIR shield to the bush readings (Yang *et al.* 1994) and these can then be compared with the catch of national gauges in an attempt to correct the national readings. This was done for 20 different gauges and shield combinations (Metcalfe & Goodison 1993). An example of the findings of the comparison is that the catch of Hellman-type gauges can be as little as 25% of the corrected DFIR gauge reading in winds above 5 m s^{-1} (Gunther 1993). With correction procedures for raingauges (see earlier) it is necessary to know wind speed and drop size; corrections for snow readings are likely to be much less reliable than for rain.

Although correction procedures and field tests comparing national gauges with the DFIR are necessary, so as to be able to use the already existing data, the large size and expense of the double snow fence makes it impractical for general use – it is an equivalent, for example, to the pit gauge for rainfall. Since raingauges are, in so many ways, an unreliable means of measuring snow, alternatives have been sought.

Measurement of snow after it has fallen

Snow sections

After seven years of field tests 150 years ago, the only way that Huddleston could see of getting sensible readings was to take a 'cheese' of snow off the ground where it seemed to be of average depth and to melt it. This still represents one of the best ways of measuring snowfall – but it needs an observer. After small falls of snow it is possible to collect a sample by inverting a raingauge funnel and pressing it through the snow until it meets the ground. The snow in the funnel is then melted and the water measured. If there is doubt about whether the chosen site is typical, several samples can be taken some distance apart, the measurements being recorded separately and the mean quoted.

If the snow depth is greater than about 15 cm and an area can be cleared next to that which is to be measured, it is then possible to insert a thin metal or wooden sheet horizontally into the snow pack at a suitable height from the top and to press the inverted gauge down until it meets the plate. The process is then repeated as many times as necessary to reach the ground.

Alternatively, a longer tube of known diameter and with a sharp edge can be pushed down into the snow and withdrawn with the snow core in it. The whole is then weighed and the known weight of the tube subtracted, the weight of the snow being converted into the water equivalent. This method does not work too well with wet snow because some water may be pressed out in the process.

When new snow falls on old, a board, coloured white and with a slightly rough surface, is placed on the old snow just level with its surface and one of the above procedures carried out after the new snow has fallen. It is important not to trample the snow when collecting samples but to leave it undisturbed by following one path to the sample site every time. It is best to measure the collected snow soon after it has fallen to avoid evaporation.

Snow depth

It is relatively easy to measure the depth of snow manually, graduated poles fixed in the ground (or snowpack) simply being read at intervals. Automatic level sensors have also become available recently, sensing the top surface of the snow by downward-looking ultrasonic (echo) systems. They typically measure the depth to about ±2 cm, although different manufacturers quote different values. Falling snow and spindrift can interfere with the readings of these echo-sounders, spurious signals being received as noise from the flakes. The air temperature also has to be measured to compensate for the variation in the speed of sound with temperature.

However, the snow's density is unknown, varying considerably from place to place, with the type of snow and with time, and, as it is the water-equivalent of the snow that is usually required, without knowledge of snow density this information cannot be deduced from the depth. Various rules of thumb are used regarding the density, such as the one-tenth rule, but snow density can vary from one-tenth to one-fiftieth, an average being about one-thirteenth. Depth measurement alone is, therefore, only an approximate method. Taking test samples occasionally, to establish the density of the nearby snow, improves the estimate but is laborious and cannot be automated.

Snow weight

A way around not knowing the density of the snow is to measure the weight of the snowpack where it lies on the ground. Devices called snow pillows are used to do this. These are thin, flat containers made of flexible sheets of stainless steel, rubber or plastic, about a metre square and a centimetre or so thick, with a capacity of perhaps 50 litres of liquid. The liquid may be oil or a solution of antifreeze, the weight of snow being measured with a manually read manometer or automatically with an electronic pressure sensor or a float-operated level recorder in a standpipe.

However, the snowpack changes with time and in particular can become solid and form ice bridges so that gradually its weight no longer rests fully on the pillow. A larger pillow, sometimes in the form of an array of four operated in parallel, can reduce but not eliminate this problem. If the snow does not stay for long and is not deep, bridging and solidification may not have time to occur.

Gamma ray attenuation

If a collated beam of gamma rays passes through the snowpack, it becomes attenuated to an extent depending on the water-equivalent of the snow. This is an effective, if expensive, technique but even though the gamma source is low level (typically 10 mCi or 370 MBq of caesium-137), health concerns can cause problems. Consequently it is not in wide use. Natural background radiation is also sometimes used in a similar way.

Measurement of snow as it melts

Melting snow is of concern to hydrologists and to river managers, so information on melt rate is well worth having even if, due to evaporation and spindrift, it does not tell us how much fell in the first place, or when it fell. Snowmelt lysimeters collect and measure the meltwater flowing from the bottom of, or at some level within, a snowpack and take two forms – unenclosed and enclosed – the former being much the commoner (Fig. 8.16). In this design the collector has a shallow wall around it, while in the latter the wall completely isolates the snow column from its surroundings all the way up to the surface of the pack. In the latter design, the wall may

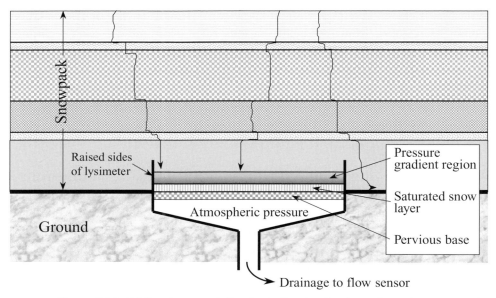

Figure 8.16. Meltwater percolating by indirect paths down through a snowpack, which comprises snow layers of varying texture and age. The volume of the meltwater can be measured by collecting it in a snowmelt lysimeter, the water being piped to a measuring device such as a tipping bucket. Provided that the sides of the lysimeter are higher than about 12–15 cm, the water is successfully collected, being unaffected by the pressure gradient introduced by the discontinuity that the presence of the collector generates. Because percolation can be spatially quite variable, the larger the area of the collector the better.

be raised slowly as the snowpack increases in depth, although this is not of course possible at an unattended site. Either type may operate at atmospheric pressure or under negative pressure (as can soil lysimeters for evaporation measurement).

Meltwater percolating down through the ice pack, which is normally layered through alternate melting and freezing with new snow being added periodically, follows anything but a direct downwards route, travelling laterally along the less permeable layers until it finds a way through (Wankiewicz 1979). The lysimeter intercepts this spatially variable flow but, as most lysimeters are small in area, what they collect may not be representative. The bigger the lysimeter, therefore, the better; the collecting area varies in practice, from less than 1 m^2 up to 100 m^2.

Meltwater flows mostly under the influence of gravity (Colbeck 1972, 1974), capillary effects being negligible although these may increase at discontinuities (Wankiewicz 1979). But the snowmelt lysimeter itself is just such a discontinuity, the layer at atmospheric pressure within the snowpack causing an end effect. This results in a saturated layer of up to 2–3 cm thick (Kattelmann 1984), where water is held in the pore spaces, between the snow grains, by capillary force; once formed, this saturated layer remains fairly constant. Above it is a pressure gradient zone, which decreases from atmospheric pressure at the lysimeter's supporting base to a negative pressure not far above. Percolating water may avoid this region and go round the lysimeter, but because the pressure gradient zone is only a few centimetres deep it can be contained within a lysimeter wall of from 12 to 15 cm, and so the problem can be avoided. However, the presence of this zone affects the timing of meltwater outflow from the lysimeter, since the saturated zone has first to form – a process that can take up to several days initially. Thereafter, the response to percolating water can be within hours or even minutes. The end-effect problems can be largely overcome by using tension lysimeters, in which a porous plate is held at a negative pressure approximating to that of the capillary pressure of the overlying snow. This much reduces the start-up and response times.

Enclosed lysimeters are useful for more precise work since they eliminate the lateral flow of meltwater at natural discontinuities within the snowpack. However, they are much less commonly used and their own, specialised problems will not be discussed here.

The meltwater collected by a lysimeter, piped away to a suitable site, can simply be read manually as it collects in a container, but more usually it will be measured automatically. The level of water in a container, perhaps sited slightly downhill if the installation is in mountains, can be measured with a float or pressure sensor, the water being siphoned off periodically. A weir tank with a water level recorder is an alternative if amounts of water are large, while weighing methods are also used, but a tipping bucket raingauge is the most common method. In all cases it is necessary to prevent freezing of the outflow pipes from the lysimeter, and of the measuring system, and this is helped by using large-diameter pipes and electrical heating if necessary.

Precipitation measurement using radar

Its place in environmental monitoring

Remote sensing is dealt with in Chapter 19, but while radar measurements rightly belong under that heading, it is better to deal with this particular RS technique in this chapter. Radar measurements of rainfall from satellite are included in Chapter 19.

Radar was developed for the detection of aircraft, and the effects of precipitation on the received signal were originally seen as an inconvenient source of interference. However, even 50 years ago it was recognised that radar could be used to detect precipitation. Indeed Kurtyka (1953) remarked that 'In the last five years, the necessity of adequate precipitation instruments to calibrate radar for precipitation measurement has pointed to the primitiveness of the present-day rain gauge.' He went on to say that 'In all likelihood, the rain gauge of the future may be radar, for even in its present developmental stage, radar measures rainfall more accurately than a network of one raingauge per 200 square miles'. Well, radar has not replaced raingauges, and while it has advanced greatly over the years, it still benefits from, and relies on, in situ data from telemetering raingauges to calibrate the system, although this need may eventually be overcome.

Advantages of radar

One of the great advantages of radar is that it gives an areal estimate of precipitation rather than a single-point measurement, and the area covered is quite large, typically about 15 000 km^2 or more for each station. It also has the advantages of giving data in real time and of not needing anything to be installed in the area, or even access to the area.

Principle of operation

A classic weather radar system, which had been very thoroughly evaluated by a consortium of interested groups (North West Water, the Meteorological Office, the Water Research Centre, the Department of the Environment and the Ministry of Agriculture, Fisheries and Food) and which was the first unmanned weather radar project in the UK, was installed on Hameldon Hill about 30 km north of Manchester during 1978. This brought within range the hills of North Wales, the southern Lake District and the Pennines and so gave the opportunity to evaluate a system in different terrains (Collier 1985). Much of the information below refers to this particular installation, but it is representative of the genre. Currently in the UK there are 12 weather radars spread fairly evenly from the south Cornish coast to the Outer Hebrides.

Weather radar uses a conical radar beam of 1 or 2°, usually in the C-band (3–6 GHz), and measures the energy reflected and scattered back from the precipitation, converting this into an estimate of surface precipitation intensity. Estimates of the rainfall rate R are derived from the returned energy, or radar reflectivity Z, using an empirical equation of the form $Z = aR^b$. Both a and b have many possible values, the most usual being $a = 200$, $b = 1.6$. b is usually left at about this figure while a may lie anywhere from 140 for drizzle, through 180 for widespread rain to 240 for heavier showers. These are figures for the UK. In other situations they can be different, for example 500 for thunderstorms in an Alpine setting. These values are quoted to illustrate the magnitude of the variation of the reflected signal from different rain types and at different places; yet other values of a apply to snow.

Causes of uncertainty of measurements

The uncertainty in converting from reflectivity to precipitation intensity, and its variation from one place to another, is due to many factors, including the variation in drop-size spectrum and the presence of hail and snow. In addition, melting snow (snow turning into rain) increases the reflectivity, giving what is called the *bright band*, leading to overestimation of the rain reaching the surface. Conversely, drizzle tends to be underestimated, owing to the absence of large drops.

The second main cause of uncertainty is due to changes in precipitation intensity between the radar beam and the ground. A low beam elevation is necessary to detect rain as near to the ground as possible and to cover as long a range as possible, but hills can interfere with too low a beam. In the case of the Hameldon Hill installation, four elevations of beam were used: 0.5, 1.5, 2.5 and 4°. The two lower angles sense surface precipitation, the 0.5° beam having a range of 24 km and the 1.5° beam being used for distances beyond. The two higher elevations are for the detection of the bright band (melting snow) and for other studies.

The antenna dish rotates at a speed of about one revolution per minute, scanning each elevation in turn and so giving about one scan of the surface rainfall every 3 min. However, the need to direct the beam slightly upwards can result in the beam missing events below it, such as shallow precipitation or the orographic enhancement of rainfall due to the presence of hills. Furthermore, even with an appropriate upward elevation of the beam, reflections from the ground or sea can still occur, particularly in the presence of a strong hydrolapse (dry air over moist air), which causes the beam to bend downwards (Fig. 8.17). Since the beam scans round, it does not take continuous readings at any one point but samples each direction and distance briefly as it passes. The measurements are thus spot instantaneous readings of intensity taken about once every 3 min over any one particular location. This snapshot view inevitably adds (perhaps 10%) to the uncertainty of conversion from spot observations to mean values, which arises from the temporal and spatial variability of rainfall.

Corrections can be made for permanent ground obstructions and spurious reflections (ground clutter) since they do not move and are present in dry weather, and for the reduced beam depth caused by intervening hills (although what is happening beyond and below the hills cannot be seen). Allowance is also made for a reduction in outgoing and returned signals due to their absorption as they pass through heavy rain. And, of course, allowance has to be made for the reduction in signal due to the inverse square law. At longer ranges, allowance also has to be made for Earth curvature and the fact that a significant part of the beam may, therefore, be above the rain.

An improvement in the conversion from reflectivity to intensity can be obtained by adjusting the value of a, in the empirical formula, through the use of in situ

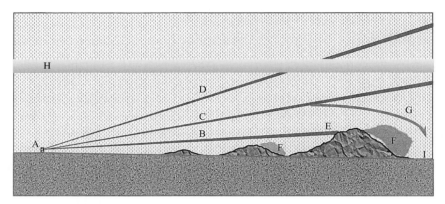

Figure 8.17. In this schematic of a weather radar system, *A* is the radar transmitter, which radiates a beam of width 1–2°. Beams of a low angle of elevation (0.5°, *B*, and 1.5°, *C*) sense precipitation close to the ground (*B* up to, and *C* beyond, 24 km), although they can miss orographic rainfall caused by hills below the beam (*F*). As distance increases, the fall-off of the ground due to the curvature of the Earth (*I*) increases the height below which events cannot be measured. The beams can also be obstructed by hills, which cause shadows beyond them (*E*). In the presence of a hydrolapse, the radar beam can be bent towards the ground, giving false reflections (*G*). A higher beam (*D*) is used to detect the bright band (*H*) caused by melting snow, which gives anomalously high reflected signals.

raingauges telemetering an input of ground-truth. Over the last ten years, however, there has been extensive development of adjustment techniques and, while there are still problems, the reliability of estimates has much increased.

Accuracy and corrections

The accuracy achieved by the radar system operated on Hameldon Hill for rainfall measurements was tested by comparing the hourly values of the radar estimate, for the areas containing a raingauge, with the hourly raingauge readings (Collier 1986). The radar estimate was the average of 12 instantaneous readings of intensity integrated to give hourly totals. Individual differences between the radar and rain-gauge measurements varied very considerably from the mean. When observations influenced by the bright band were excluded, the average difference, under conditions of frontal rain, was 60%, and 37% for convective rain. If bright band episodes were not excluded, the difference could be 100%.

To help reduce the error, ground-truth from telemetering raingauges can be used. However, if only a few gauges are used – to give an average correction for the whole area – it is unlikely to help, because of spatial and temporal variations. An alternative is to use a large number of gauges to sense more widely the spatial variations in the difference between radar and raingauge readings, but this requires too many gauges.

A less direct method is based on the fact that automatic detection of the type of rainfall allows the most appropriate calibration procedures to be used. Detection of rainfall type can be achieved if the temporal variability of the difference between readings is measured at a small number of raingauge sites, for it has been found that this variability can identify the type of rainfall well enough, and so can be used to choose the best correction factor. This approach lessens the difference between radar estimates and raingauge measurements from 60% to 45% for frontal rain and from 37% to 21% for convective rainfall. Over the last decade much work has been done to improve the conversion from radar signal to estimated rainfall. For those needing to know more, there is a detailed review by Collier (1996).

More sophisticated techniques could improve the performance of radar estimates, such as the use of beams polarised alternately in the vertical and horizontal directions, whereby it is possible to detect the raindrop-size distribution more precisely. The principle is that raindrops are nearly oblate spheroids, with their axes of symmetry aligned close to the vertical. The degree of oblateness can be detected by the polarised beams and this relates to drop size. However, there are problems with this method, not the least being an increase in the complexity of data processing and of hardware, and thus of cost. So far this type of radar has not been introduced operationally. There are a number of other new techniques using polarisation-diversity radar, the most promising seeming to be the use of differential attenuation and differential phase (Collier 1997).

Nimrod and Gandolf

Another approach to the enhancement of radar rainfall measurements was begun some time ago under the name of FRONTIERS (forecasting rain optimised using new techniques of interactively enhanced radar and satellite data). Now known as *Nimrod*, a more recent version of this nowcasting technique, combines IR and visible Meteosat cloud images (Chapter 19) with surface weather reports, including telemetering raingauges, along with numerical models. The system collects the measurements from 15 radars around the UK with resolutions of 2 and 5 km every 5 and 15 min. This combination enables judgement to be made (automatically) as to whether a radar echo is from rainfall or from spurious echoes due to ground clutter or 'anaprop' (anomalous propagation, for example due to a hydrolapse), allowing the radar data to be more accurately converted to rainfall intensity than when used in isolation. Nimrod also goes some way towards compensating for the gradual increase of the height of the beam with increasing range, caused by the fall away of the horizon due to Earth curvature, and because the beam has to be angled slightly upwards. Estimates of surface rainfall from Nimrod are regularly compared with measurements from hourly raingauge records and adjustments made to the radar data, but to avoid making detrimental adjustments, adjustments are made on a weekly or monthly basis to exclude short-term variability introduced by the different sampling characteristics of the radar and the raingauges. Remaining differences in monthly rainfall maps are due to residual errors in the correction process, actual radar faults and to the inherent shortcomings of the radar technique (as described earlier).

An extension of the Nimrod system is *Gandolf* (Generating Advanced Nowcasts for Deployment in Operational Land-surface flood Forecasts) in which satellite images are used to differentiate between convective and frontal rain clouds, a separate forecast being produced for the convective rain. This is then combined with a forecast of frontal rain using Nimrod techniques, resulting in a forecast for rain rates and totals for each 15 min for up to three hours forward.

The combination of several different methods of measurement, as represented here by Nimrod and Gandolf, is seen again later in the correction of remotely sensed measurements from satellites (Chapter 19). Such combinations will undoubtedly be an important development in all remote sensing (see Chapter 21, Forward look).

Snowfall measurement by radar

Since snowfall is very difficult to measure by other means it would be useful if radar could do it better. The same empirical equation is used for snow as for rain, and as in the case of rainfall there is considerable variability in the values of a and b, a lying somewhere between 2100 for wet snow and 540 for dry. However, the same problem exists for radar as for optical raingauges – the size and density, and thus the water equivalent, of snowflakes vary so much that there is a considerably larger margin of uncertainty than for rain. Just as spindrift is a problem when trying to measure snowfall with collectors, it presents a problem when using radar since the subsequent movement of fallen snow by wind takes place below the beam and so goes undetected.

There is also the additional problem that because it is extremely difficult to measure snowfall by any conventional means there are no reliable measurements against which to compare the radar's snow readings, either for checking performance during development or as real-time ground-truth in the operational state. It is possible, however, to use measurements of snow made after it has fallen to assess the radar's performance in retrospect. Because snowflakes are not spheroid-shaped, polarisation techniques are inappropriate. Snowfall measurement is just a very intractable problem, however it is approached.

Displaying the rainfall measurements

Although radar positions are given in polar co-ordinates, they can be converted to Cartesian co-ordinates for display on a monitor screen in the form of a square grid. A typical size of grid (resolution) will vary from 1 km for ranges up to 75 km to 5 km for ranges from 75 to 200 km, and proportionally between. Modern radars allow the user to retain polar grid displays if preferred. Levels of precipitation intensity are displayed using different colours. In addition, rainfall totals over areas of interest, such as river catchments, can also be calculated over a period of, say, 15 min.

Cost

The capital cost of a weather radar system is high – hundreds of thousands of dollars – and being complex it needs high-level technical expertise to maintain it. Weather radar is thus best suited to situations where measurements are needed from a large area on a permanent basis in real time, such as for a national weather service or a large river authority. Radar is not suited to measuring precipitation over smaller areas or for short durations, although this kind of information may be purchasable by smaller organisations from those operating large radar networks.

References

Alter, J. C. (1937) Shielded storage precipitation gauges. *Mon. Weather Rev.*, **65**, 262–5.
Calder, I. R. & Kidd, C. H. R. (1978) A note on the dynamic calibration of tipping bucket raingauges. *J. Hydrol.*, **39**, 383–6.
Colbeck, S. C. (1972) A theory of water percolation in snow. *J. Glaciol.*, **11**, 369–85.
Colbeck, S. C. (1974) The capillary effects on water percolation in homogeneous snow. *J. Glaciol.*, **13**, 85–97.
Collier, C. G. (1985) Accuracy of real-time estimates made using radar. In: *Proceedings of the Symposium on Weather Radar and Flood Warning*. University of Lancaster 1985.
Collier, C. G. (1986) Accuracy of rainfall estimates by radar, part 1: calibration by telemetering raingauges. *J. Hydrol.*, **83**, 207–23.
Collier, C. G. (1996) *Applications of Weather Radar Systems. A Guide to Uses of Radar Data in Meteorology and Hydrology*, second edition. John Wiley/Praxis, Chichester, 390p.
Collier, C. G. (1997) Private communication.
Danby, H. (1933) *Translation of 'The Mishnah'*. Oxford University Press, Oxford, UK.
Dobson, D. (1777) Observations on the annual evaporation at Liverpool in Lancashire; and on evaporation considered as a test of the dryness of the atmosphere. *Philos. Trans. R. Soc. London*, **67**, 244–59.
Essery, C. I. & Wilcock, D. N. (1991) The variation in rainfall catch from standard UK Meteorological Office raingauges: a twelve year case study. *J. Hydrol. Sci.*, **36**, 23–34.
Folland, C. K. (1988) Numerical models of the raingauge exposure problem, field experiments and an improved collector design. *Q. J. R. Meterol. Soc.*, **114**, 1485–516.

Genrich, V. (1989) Introducing 'electronic impact sizing' as a new, cost-effective technique for the on-line evaluation of precipitation. In: *Proceedings of the WMO/IAHS/ETH International Workshop on Precipitation Measurement.*, St Moritz, pp. 211–16.

Grew, N. (1681) *Musaeum Regalis Societatis*, London pp. 357–8.

Gunther, Th. (1993) German participation in the WMO solid precipitation intercomparison: final results. In: *Proceedings of the Symposium on Precipitation and Evaporation*, Bratislava 1993, pp. 93–102.

Hanna, E. (1995) How effective are tipping-bucket raingauges? A review. *Weather*, **50**, 336–42.

Heberden, W. (1769) Of the quantities of rain which appear to fall at different heights over the same spot of ground. *Philos Trans. R. Soc.*, **59**, 359.

Hewston, G. M. & Sweet, S. H. (1989) Trials use of weighing tipping-bucket raingauge. *Meteorol Mag.*, **118**, 132–4.

Huddleston, F. (1933) A summary of seven years' experiments with raingauge shields in exposed positions, 1926–1932 at Hutton John, Penrith. *Br Rainfall*, **73**, 274–93.

Hughes, C., Strangeways, I. C. & Roberts, A. M. (1993) Field evaluation of two aerodynamic raingauges. *Weather*, **48**, 66–71.

Jevons, W. S. (1861) On the deficiency of rain in an elevated raingauge as caused by wind. *London, Edinburgh and Dublin Philos Mag.*, **22**, 421–33.

Jones, S. B. (1983) The estimation of catchments average point rainfalls. *Inst. Hydrol. Rep.*, No. 87.

Kattelmann, S. C. (1984) Snowmelt lysimeters: design and use. In: *Proceedings of the Western Snow Conference*, pp. 68–76.

Koschmieder, H. (1934) Methods and results of definite rain measurements III. Danzig Report (1). *Mon. Weath. Rev.*, **62**, 5–7.

Kurtyka, J. C. (1953) *Precipitation Measurement Study*. [Report of Investigation No. 20]. State Water Survey, Illinois, p. 178.

Metcalfe, J. R. & Goodison, B. E. (1993) Correction of Canadian winter precipitation data. In: *Proceedings of the 8th Symposium on Meteorology, Observation and Instrument 1993*. Anaheim, California, pp. 338–43.

Nipher, F. E. (1878) On the determination of the true rainfall in elevated gauges. *Am. Assoc. Adv. Sci.*, **27**, 103–8.

Norbury, J. R. & White, W. J. (1971) A rapid-response raingauge. *J. Phys. E. Sci. Instr.*, **4**, 601–2.

Reynolds, G. (1965) A history of raingauges. *Weather*, **20**, 106–14.

Robinson, A. C. & Rodda, J. C. (1969) Rain, wind and the aerodynamic characteristics of raingauges. *Meteorol Mag.*, **98**, 113–20.

Rodda, J. C. (1962) An objective method for the assessment of areal rainfall amounts. *Weather*, **17**, 54–9.

Rodda, J. C. (1967a) The rainfall measurement problem. In: *Proceedings of the Bern Assembly IASH.*, pp. 215–31.

Rodda, J. C. (1967b) The systematic error in rainfall measurement. *J. Inst. Water Eng.*, **21**, 173–9.

Sevruk, B. (1996) Adjustment of tipping-bucket precipitation gauge measurements. *Atmos. Res.*, **42**, 237–46.

Sevruk, B. & Hamon, W. R. (1984) International comparisons of national precipitation gauges with a reference pit gauge. WMO Instruments and Observing Methods, Report No 17. WMO/TD, No. 38.

Sevruk, B. & Klemm, S. (1989) Types of standard precipitation gauges. In *WMO/IAHS/ ETH International Workshop on Precipitation Measurement.* St Moritz.

Sevruk, B., Hertig, J. A. & Spiess, R. (1991) The effect of a precipitation gauge orifice rim on the wind field deformation as investigated in a wind tunnel. *Atmos. Environ.,* **25**, A(7), 1173–9.

Shamasastry, R. (1915) Translation of *'Arthasastra'* by Kautilya. Government Oriental Library Series, Bibliotheca Sanskrita, No. 37, part 2, Bangalore.

Stevenson, T. (1842) On the defects of raingauges with descriptions of an improved form. *Edinburgh New Philos J.,* **33**, 12–21.

Strangeways, I. C. (1984) Low cost hydrological data collection. In: *Proceedings of the IAHS Symposium on Challenges in African Hydrology and Water Resources.* Harare, Publ. No. 144, 229–33.

Strangeways, I. C. (1996) Back to basics: The 'met. enclosure': Part 2(b) – Raingauges, their errors. *Weather,* **51**, 298–303.

Symons, G. J. (1864) Rain gauges and hints on observing them. *Br Rainfall,* 8–13.

Symons, G. J. (1866) On the rainfall of the British Isles. In: *Report on 35th Meeting of the British Association for the Advancement of Science.* Birmingham, 1865, pp. 192–242.

Thiessen, A. H. (1911) Precipitation for large areas. *Mon. Weather Rev.,* **39**, 1082–4.

Townley, R. (1694) Observations on the quantity of rain falling monthly for several years successively. A Letter from Richard Townley. *Philos Trans. R. Soc. London,* **18**, 52.

Wankiewicz, A. (1979) A review of water movement in snow. In: *Proceedings of the Modelling of Snow Cover Runoff.* US Army Corps of Engineers. Cold Regions Research and Engineering Laboratory, Hanover, US, pp. 222–52.

Whittaker, A. E. (1991) Precipitation rate measurement. *Weather,* **46**, 321–4.

Wiesinger, T. (1996) Aerodynamically superior precipitation gauges. [Report to CEN/TC 318 Working Group 5]. British Standards Institute, London.

Yang, D., Sevruk, B., Eloma, E., Golubev, B., Goodison, B. & Gunther, Th. (1994) Wind-induced error in snow measurement: WMO intercomparison results. In: *Proceedings of the 23rd International Conference on Alpine Meteorology.* German Weather Service, **30**, pp. 61–4.

9

Soil moisture and groundwater

The cart-way of the village divides, in a remarkable manner, two very incongruous soils. To the south-west is a rank clay, that requires the labour of years to render it mellow; while the gardens to the north-east, and small enclosures behind, consist of a warm, forward, crumbling mould, called black malm, which seems highly saturated with vegetable and animal manure; and these may perhaps have been the original site of the town; while the woods and coverts might extend down to the opposite bank.

Gilbert White *The Natural History of Selborne* (published 1798).

Subsurface water processes

Soil moisture

Soil moisture (or soil water) refers to the water that occupies the spaces between soil particles. It is at its maximum when the soil is saturated, that is when all the air between the particles is replaced by water but, if the soil can drain, the spaces will normally also contain air, the water then forming a thin film on and between the soil particles, held by capillary attraction. As the soil dries out this film becomes thinner and progressively less easy for plant roots to extract. The water is free to move through the soil, up or down, by gravity and by capillary attraction; it is taken up by plant roots, evaporates at the surface or recharges the groundwater.

However, there is also water present in soil which is not free to move or to be taken up by plants but which may nevertheless be detected during measurement – but not differentiated from the free water, depending on the measurement technique used. This is the *water of crystallisation* or *water of hydration* – water that is chemically bound to minerals within the soil such as gypsum ($CaSO_4\ 2H_2O$); water may also be bonded to organic material to varying degrees of strength. In making measurements of soil water content, soils of great variety are encountered and any instrument or

Measuring the Natural Environment, second edition, Ian Strangeways. Published by Cambridge University Press. © Ian Strangeways 2003.

measurement method must be able to handle them all, from pure peat or sand to silt and clay or a mix of them, all varying in pore size or having a variety of pore sizes combined, and all varying in the extent of chemically bonded water. Anomalies, such as stones dispersed at random in the soil profile, can also affect measurements since stones do not usually hold any water and if not detected can give rise to large errors, again partly depending on the method of measurement. The terms 'soil moisture' and 'soil water' are generally interchangeable although the latter could be interpreted as including the chemically bound water, which, as explained above, is not in the form of moisture.

Soil tension

The spaces between the soil particles contain air, water and roots. When the soil pores contain only water then the water can move freely, the largest force acting on it being that of gravity which causes it to move downwards provided there is drainage. It then drains away until an equilibrium is reached known as the *field capacity*, at which point any further water added cannot be retained and will drain away. As the soil dries out (for whatever reason), the water in the pores decreases in volume and retreats to increasingly small spaces between the particles, becoming held more and more tightly by capillary attraction. Under anything less than field capacity there is said to be *soil moisture deficit*. As drying continues, the suction that plant roots must apply to extract water increases. This suction or *soil tension* (or *soil water potential* or *soil matrix potential*) is measured in the same units as barometric pressure. At saturation (no air spaces), the tension is zero while at field capacity it is around -0.3 bar (-300 hPa). As drying continues, the remaining water is eventually held so tightly that it is not possible for plants to apply enough tension to extract it, this state being known as the *permanent wilting point*, occurring at about -15 bar. The exact value differs between plant types, but at this point they die. However, most of the water available to plants is held at pressures of less than -1 bar, although negative pressures can rise as high as -60 bar. In soil physics, a non-SI unit of pressure, *centimetres of water*, is often used as it is more directly meaningful (1 bar $= 1017$ cm of water; see Chapter 6).

Knowing the strength of soil tension is of value to agriculture, since it acts as a warning of when it is necessary to start irrigation as well as when it is wasteful to continue. But the significance of soil tension is also due to the control it exerts on the movement of water within a soil profile. If the combined upward tension exceeds the pull of gravity, it is possible for water to move upwards and be evaporated or transpired. If gravity and downward tension combined are the stronger force, water moves downwards, eventually joining the saturated zone of groundwater at the water table. But temperature gradients within the soil also cause water to move

(Gilman 1977), liquid water moving from colder to warmer zones, vapour from warm to cold (by evaporation and condensation). This was understood many years ago (Patten 1909).

The zero-flux plane

The point in a soil profile at which the water above flows up and the water below flows down is known as the *zero-flux plane* (ZFP) – the plane through which no soil moisture is passing at the time of the measurement (Fig. 9.1). The position of the

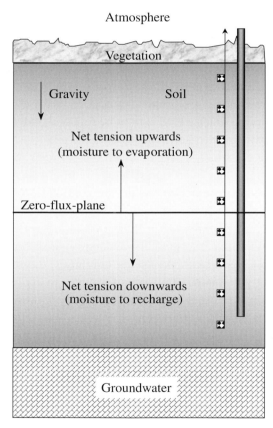

Figure 9.1. The zero-flux plane. Depending on conditions, there can be a depth in the soil through which the flux of soil moisture is zero, moisture above moving upwards to evaporation and moisture below moving downwards to groundwater recharge. The presence and position of the plane depends on the net force on the soil water of gravity and soil tension. The depth of the plane can be established by installing tensiometers at several depths (the vertical array of squares), and the amount of soil moisture at these depths can be measured by neutron probe (the access tube is shown on the right of the array of tensiometers).

plane moves up and down depending on conditions, or it may not exist at all (for example after heavy rainfall in summer, or throughout the winter – all movement then being downwards). Knowing the plane's position is useful since it divides the soil moisture into that which is being evaporated or transpired and that going to recharge the groundwater and eventually to contribute to river flow. The position of the ZFP can be measured by sensing the relative values of tension in the soil at different depths. Further, if the ZFP is measured in addition to the moisture content in the same profile, the amounts of water going up and down can be quantified.

Groundwater

The part of the soil moisture which does not rise upwards and evaporate instead percolates downwards and joins the groundwater – a region in which the soil or permeable rock is saturated with water without airspaces. From here, in due course, the water finds its way into a drainage ditch, stream, river or lake and eventually to the sea. Measuring the depth of the water table, changes in its level and the quality of the groundwater is of importance not just to researchers investigating hydrological processes but to those concerned with the practical matters of water resources, civil engineering, agriculture and pollution.

Units and terms

The simplest method of measuring soil moisture is to take a sample of soil, weigh it, dry it and weigh it again. This basic method is described more fully later, but is introduced here because it allows the units of measurement to be understood more graphically.

There are two ways of expressing the amount of moisture present in a sample of soil, one giving the value in terms of the weight of the moisture, the other in terms of its volume. Suppose a sample is taken roughly with a spade – with no knowledge of the volume it occupied in situ – and that this sample is weighed before it has lost any moisture. It is then dried in an oven and weighed again. The loss of weight (loss of water) is divided by the dry weight of the soil and the result is known as the *moisture dry-weight fraction* (MDWF). By multiplying the fraction by 100, the result can be given as a percentage.

But it is generally more meaningful to express the amount of moisture in terms of the volume, not the weight. The result is then expressed as *the moisture volume fraction* (MVF) or *volumetric water content* (VWC), the volume of the water divided by the volume of the wet soil. Again if multiplied by 100 the result is expressed as a percentage – the *moisture volume percentage* (MVP). In these terms, 100% indicates pure water with no soil, while 0% is pure soil containing no water. In the same terms, when all air spaces are occupied by water the MVP is about 35% to 40% for sand, through 65% for clay and as high as 98% for peat. The reason the MVP is often preferable to the MDWF is that it is then in units compatible with related variables such as rainfall and evaporation – that is, they are all expressed as a thickness (millimetres) of water gained or lost over a period. This helps in calculating a water balance.

As an example, if measurements have shown that most of the change in soil moisture occurs within the top metre of a particular soil, and by measurement it is found that the average MVP has changed by +5% over a month, then the change can be expressed as 5% of 1000 mm (1 m) in a month, or +50 mm of water; that is, the soil now holds 50 mm more water than it did. Perhaps it has retained 50 mm of the rainfall, or some moisture has moved upwards from below under the influence of tension. This can be compared with the rainfall, of say 200 mm, and with the measured evaporation of, say, 30 mm. From this, the amount left that will appear as runoff can be calculated:

$$\text{Runoff} = 200 - 30 - 50 = 120\,\text{mm}$$

If the area of the catchment is 1 km^2 then the volume available as runoff is $10^6 \times 0.12\,\text{m}^3 = 120 \times 10^3\,\text{m}^3$. This amount will not run off into the river at a constant rate over the period of, say, one month, but will follow a curve known as a hydrograph (a plot of runoff against time), the shape of which will depend on how the 200 mm of rain fell (the time and intensity) and on the physical characteristics of the ground. The point to note is the advantage of expressing soil moisture as a fraction of the volume of soil, not as its weight.

Measuring soil moisture

The gravimetric method

The gravimetric method was the only way of measuring soil moisture until the development of the electronic methods to be described later; it is now mainly used to calibrate the electronic methods, although it is still in routine use where costs must be kept to a minimum, or where only a few measurements for a short time are needed and the high cost of electronic instruments would be unjustified.

The basic procedure

A soil sample is taken, stored so that water cannot evaporate from it, returned to a laboratory, weighed and dried in a *ventilated* oven at 105 °C until its weight is constant. It is then cooled and reweighed (hence the more correct name is the *thermogravimetric* method); it can take from 0.5 to 2 days to complete the drying process. This gives a measure of the weight of (free) water which the sample contained in the field. Because the density of water in c.g.s. units is as good as unity for all practical purposes, knowing the weight of water driven off by heating tells us the volume; for example, if the weight loss is 20 grams, the volume of water loss is 20 ml.

However, measuring the volume of the soil is more difficult. The usual method is to collect a sample with a *soil corer* – a sharp-edged cutting cylinder, pushed into the soil to a precisely known depth and then withdrawn with the soil inside it. But if the sample does not slide easily into the tube, soil can be pushed sideways and lost as the sample becomes compacted. Stones can prevent a sample being taken or they can deflect the corer. If the corer is moved from side to side to get it to go down against resistance, spaces can be introduced into the sample and small stones in the sample can introduce a large random error. It is also difficult to insert the tool to exactly the correct depth. When the corer is removed, a clean break (level with the end of the corer) rarely occurs. While some of these difficulties can be prevented by careful work and repeated attempts, even then the error in the volume, and so in the MVP, can be $\pm 10\%$.

A more precise method

If an error of this magnitude is unacceptable, more complex procedures can be adopted. One such involves the careful digging of a small hole of known dimensions (say 10 cm in diameter by 15 cm deep), by hand using a trowel, the soil being stored in a plastic bag for later weighing, drying and reweighing. The volume of the hole is then measured by lining it with a thin plastic bag into which water is added until level with the top of the hole, the volume of water poured into the bag being equal to the volume of the sample of soil (Bell 1996). This is very time consuming, but it has the added advantage that in addition to measuring the MVP it also allows the *dry bulk density* σ_d of the soil to be established at each depth at the site:

$$\sigma_d = \text{mass of the dry soil in grams} \div \text{volume of the dry soil in ml}$$

This needs to be done only once because σ_d is fairly constant for most soils, clays excepted; it ranges from just below 1.0 for topsoil to 1.75 for silts, sands and gravels. The advantage of knowing σ_d is that it can then be used in the third and most practical way of using the gravimetric method, which we now discuss.

The screw-auger method

Soil samples are collected, using a screw auger, at the various depths required (at which σ_d has already been established). The MDWF (moisture dry-weight fraction) is then determined. In this method, the volume of the sample is not needed, for if σ_d is known then the MDWF can be converted to the MVF, since MVF $=$ MDWF $\times \sigma_d$.

For example, if the MDWF is 0.30 and σ_d is 1.5 then the MVF $= 0.30 \times 1.5 = 0.45$ (or 45% MVP).

Auger holes are easy to make down to over two metres in depth, and need only be 2.5 cm in diameter. Using this technique errors are most likely to occur in the course of establishing σ_d, so care is needed in that crucial initial measurement. Errors may also occur if σ_d varies much around the point at which measurements will be made, so several initial measurements of σ_d are a wise precaution even if time consuming and tedious. Errors in weighing will be very small in comparison.

The advantages of the gravimetric method are the cheapness of the equipment and the simplicity of the field work (after the initial careful measurement of σ_d), so that unskilled workers can be used. It is also an easy-to-understand method and there is nothing to go wrong with the equipment.

Limitations and sources of error

The method assumes that the properties of the soil are fairly consistent within the area around the sampling point, but as with other variables there is always some spatial variation. Because the method is destructive, a different sample of soil is measured each time and so there will be some error due to this variation – it could be that the reason for a difference between measurements from one observation and the next is that the soil properties are different, not that there are moisture changes. Repeated measurements over time are normally what are required, because it is the changes in moisture content that are usually significant, for example in an estimate of catchment water balance or a measurement of the use of water by a particular crop. Some of the new electronic, non-destructive methods have the advantage of measuring the same volume of soil each time, and so they overcome the spatial variation problem – at least as far as comparing changes at the one point is concerned. (They do not overcome the problem that the sampling point may not be representative of the area. That is another problem, akin to that encountered with all variables when selecting representative sites, but just plain common sense goes a long way in this.)

To quantify the problem that spatial variation causes uncertainty about the interpretation of differences between consecutive readings at the site, it is possible to do a preliminary evaluation of the extent of this variability. About 10 sampling points are selected and samples taken at each depth at which measurements will routinely be made. Using the dry bulk density method, the MVP is calculated for each point and for each depth and the standard deviation determined for the readings at each depth. These can then be used either to estimate the accuracy that one single sample will give or the number of samples that needs to be taken to achieve the accuracy required (Bell 1996).

For better results, however, there are more complex (and more expensive) methods of measuring soil moisture, as follows.

The neutron probe

The neutron probe is arguably the best method available for the measurement of soil moisture, although two others, described later, the *capacitance probe* and the *time domain reflectivity* method (TDR), are also now available with their own particular advantages and problems.

Principle of the neutron probe

A source of radioactive material, generally americium 241-beryllium, with an intensity of between 30 and 100 millicuries (1 millicurie (mCi) = 0.037 GBq or gigabecquerels) and contained within a tube, emits fast neutrons into the soil (Fig. 9.2); americium–beryllium is used because it emits little gamma radiation and so is safer. As the fast neutrons move through the soil they progressively lose energy and become slowed and scattered as they collide with hydrogen nuclei. Most of the hydrogen atoms in the soil are those contained in water molecules. After many collisions and direction changes, some of the fast neutrons find their way back to the tube containing the radioactive source, as slow or 'thermal' neutrons (Fig. 9.2). The soil itself and some elements within it, such as cadmium, have a similar, though slight, effect. Also in the sensor tube is a slow-neutron detector, usually a boron trifluoride sensor, owing to its cheapness, stability and reliability, which produces a pulse output for each slow-neutron that enters it, the number of such encounters being proportional to the number of water molecules in the soil.

The rate at which thermal neutrons are returning is measured and displayed by a *rate-scaler* over a period, typically of 16 or 60 s, the rate being displayed as counts per second. For more accurate readings the counting time can be extended to minutes. (Because the radioactive source emits neutrons randomly, the count rate will vary from one measurement to the next, even if the water content did not change, but by counting over a longer period this effect is minimised.) The half-life of the source is 450 years, so its gradual decline in intensity is not a factor. The count rate is displayed on a LCD screen, but can also be stored in the memory within a *recording rate-scaler* for later downloading, the site location of the reading also being stored in the memory.

For health safety, the tube containing the radioactive source is housed in a protective spherical shield made of a moderating material such as polypropylene. This reduces the radiation level outside to a value that is well within safety limits (less than 0.5 mrem h^{-1}), allowing the instrument to be carried by an operator all day without danger. Since the instrument often has to be carried across rough country,

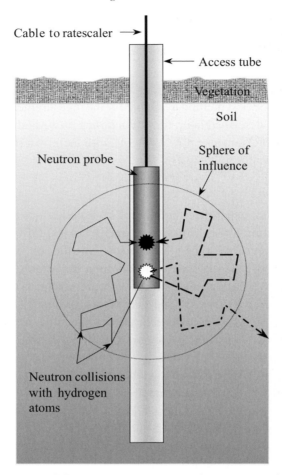

Figure 9.2. A neutron probe contains a source of fast neutrons (white sunburst), which are emitted into the surrounding soil and collide with the hydrogen atoms of soil moisture; some return, much slowed, to a slow-neutron detector in the probe (black sunburst). The resultant pulses are counted by a rate-scaler (above ground). Depending on the amount of soil moisture present, the sphere of influence varies from a radius of 15 to 30 cm. The solid, broken and broken-and-dotted lines show the paths of three neutrons.

its weight needs to be minimised and a typical complete probe will weigh around 10 kg (Fig. 9.3).

Operation in the field

The probe is placed on an *access tube*, permanently installed in the ground to a depth appropriate to needs, which might be from near-surface to several metres (Bell 1976). The tube is usually of aluminium alloy, which is fairly transparent to

Figure 9.3. In the foreground, a neutron probe is seated on an access tube, the probe having been lowered into it; a rate-scaler (folded open, to the right) indicates the rate of return of slow neutrons from the moist ground. Beyond, against the wall, is a tensiometer manometer board; see Fig. 9.11(*b*).

slow neutrons (stainless steel absorbs them more and so reduces the count rate, as does brass although less so). A cap at the base prevents water or soil entering from below and aids installation, while a removable rubber bung at the top protects the probe from water, debris ingress and insects. The tubes are installed by first auguring out the soil to a diameter that allows the tube to fit as tightly as possible, in order to prevent water running down the outside of it, which would change the soil moisture below. As with all soil observations, it is important not to tread the ground down, which would compact it around the access tubes; wooden walkways are used to minimise the problem.

The probe is lowered from its shield into the access tube on a cable to each depth at which a reading is to be made, a counter indicating the depth that the detector has reached. At each depth, the rate-scaler is set counting and at the end of the period it displays the count rate, this being written down by the operator or automatically stored in the memory.

Calibration

The MVF, MVP or VWC are defined, for the purpose of calibrating a neutron probe, as the volume of water that would be removed from the soil by heating it to 105 °C – as in the gravimetric method. But, as noted earlier, not all the water is driven off by heating, some remaining as water of crystallisation and bound to organic material. This water affects the count rate as if it was free water but, as the amount is fairly constant for a given sample of soil, changes in count rate may be attributed to changes in water content. It is largely the changes, rather than an absolute measure of water content, that matter in practice.

Calibration can be done theoretically, or in a soil sample set up in the laboratory or in the field. Theoretical calibrations are based on a precise chemical analysis of the soil for such materials as cadmium, boron and chlorine that also scatter neutrons, but it is not much used since the analysis is expensive. Laboratory calibrations require a soil sample, weighing several tons, to be carefully dug from the field and repacked into a drum at least 1.5 m diameter and 1.5 m deep with an access tube down the centre. But it is only practicable if the soil is homogeneous and can be repacked to something resembling field conditions. In practice this is only useful for gravel, sand or silts.

The simplest method is field calibration, although it is slow and laborious. Near to the permanent access tube, a temporary tube is installed and, at half a dozen points around it, gravimetric measurements are made at each depth at which the probe will be used. This number of points is necessary to obtain a linear regression, because of the variability of the soil. Probe readings are taken at the same time at each depth. Because the readings must cover the full moisture range of the soil, readings need to be taken periodically throughout a full season and so it can take up to a year to obtain a full calibration.

The gravimetric soil moisture measurements are then plotted against the ratio of the probe count rate in the soil, R, to the count rate in a standard moderator, usually water, R_w. The latter is used instead of simply R in order to compensate for variations, from probe to probe, in the sensitivity of the detectors, the intensity of the neutron source and other characteristics. Figure 9.4 shows the type of calibration obtained. The lines are almost linear and this allows changes in MVP to be estimated reasonably accurately by the probe. The lines are also nearly parallel, and so one composite representative graph can be used for all soils.

The value of R_w is obtained by constructing a cylindrical container at least 60 cm deep by 50 cm diameter, with an access tube into which the probes can be lowered,

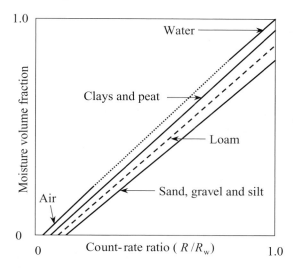

Figure 9.4. It is better to calibrate a neutron probe not in terms of the count rate in soil, R, but as the ratio of R to the count rate in water, R_w. Calibrations will vary according to the soil's chemistry and density, roughly as illustrated here. The lines do not go through zero because there will always be a small residual count.

to the centre. The container is filled with water. To minimise the random error a long count is made (perhaps 8 minutes). As a further check, a quick test is made at each field site before taking the actual soil readings, by taking a 16 second count with the probe in its shield (which acts as a moderator in a similar way to water – but less consistently).

Volume of soil measured by a neutron probe

The radius of the sphere over which the neutrons have influence depends not upon the strength of the radioactive source but upon its energy, so any probe using americium–beryllium will cover the same range. Tests have shown that, using this type of source, the effective radius of the probe is about 15 cm in wet soil and perhaps 30 cm in dry soil (Fig. 9.2). So the measured value of MVP is the average over a sphere with these diameters. This also means that greater resolution cannot be obtained by taking readings at levels closer together than about 15 cm, since the sampling volume then overlaps and some of the same soil is included in two adjacent readings. It also means that it is difficult to make reliable measurements in the top 20 cm of soil, as part of the neutron cloud is then in the air above ground. Ways around this include the use of 15 cm-deep *surface extension trays* – circular, open-bottomed trays, which are filled with the same soil and vegetation as the surroundings and which can be lifted from an open-bottomed shallow storage pit and placed around the access tube during readings.

Access-tube networks

Apart from situations where soil moisture is to be measured at one localised point for a specific reason, the aim is usually to obtain an areal estimate of soil moisture – just as in the case of rainfall – and the methods used in deciding the siting of access tubes can be the same as those used for selecting raingauge sites (Chapter 8). One of these techniques is to install as many access tubes as can be conveniently visited in, say, a day; readings are taken quickly (in 16 seconds), high precision at a few points being less useful than lower precision at many, because of the great spatial variability of soil moisture. After a year or more, *index sites* may be identified that correlate closely with the network mean, and if this is the case many of the access tubes can then be discontinued and just the key sites used (as with raingauges). At these key sites, measurements can then be made more often and with longer counting periods (1 min instead of 16 s, for example); see Bell (1987).

Problems

The neutron probe is a manual instrument carried to a site. The method could not be automated without developing a technique to move the probe up and down the access tube, and although this was considered in the early days of probe development, it would not be economic to commit a probe to every access tube in a network, made all the more expensive by the inclusion of automation. It might have been an option at a few selected sites, but safety considerations – with regard to leaving radioactive sources unattended at field sites – were a further deterrent. Safety considerations also make the probe difficult to transport, with warnings required on cars and problems when the probe has to be transported by air. There is also always the danger that in a technologically backward country a probe could be abandoned and later dismantled, the source being left unprotected. Over the decades, since its first development, health considerations have made it increasingly difficult to use the neutron probe. Because of these limitations and also because of the difficulty of measuring the top 20 cm of soil with a neutron probe, alternatives were sought and developed. Nevertheless, the neutron probe continues to be used, where possible, because it is such an excellent instrument.

The capacitance probe

Techniques to determine the water content of soil by measuring its dielectric constant had been investigated before the neutron method was developed (Debye 1929), and these became the basis for new instruments in several guises, one being the capacitance probe.

Principle

The *relative permittivity*, ε_r, of a material (also known as its *dielectric constant*) can be defined as the factor by which capacitance is increased over the free-space value (or free-air value, for all practical considerations). Since the value of ε_r for water is about 80 (it varies from 81 to 78 over the temperature range 15 °C to 25 °C), while for air it is 1 and for soil between 4 and 6, the capacitance value of a capacitor with soil as its dielectric is decided primarily by the amount of moisture it contains.

However, care is needed with regard to how this capacitance is measured, because ε_r for many substances, including water, is not a constant but frequency dependent, having different values as the frequency rises from sub-audio up through the audio and radio bands to microwave. The reasons for this are complex and do not need to be fully explored here. Superficially, they involve the dipole nature of the water molecule and its tendency to rotate so as to line up with an electric field applied across it, such as that between the plates of a capacitor, and to reverse its orientation when the field reverses, for example when an AC signal is applied across the plates; the speed at which it does this (its *relaxation frequency*) is also a factor (Smith-Rose 1933, Hasted 1973, Hoekstra & Delaney 1974).

As the amount of water in soil reduces, it concentrates largely where the particles touch, owing to capillarity, being perhaps only one molecule thick over surfaces not in contact. Under such conditions, bonding of the water molecules to the soil can be tight, restricting the movement of the water dipoles under the influence of an electric field. This affects the capacitance in an uncertain way, particularly when measured with a low frequency (tens of kHz). To minimise the problem, the frequency at which the capacitance of soil containing moisture is measured must be greater than about 30 MHz. Early attempts probably failed because audio frequencies were used (Anderson 1943, De Plater 1955). Later designs (McPhun 1979) used frequencies in excess of 30 MHz. By the 1980s, modern electronics and a better understanding of the physics involved had made the technique more practical, and portable capacitance probes became a reality.

Practice

Bell *et al.* (1987) and Dean *et al.* (1987) describe a probe based on a modified Clapp transistor oscillator running at about 150 MHz (in air), housed in a plastic tube of 4.4 cm diameter, by 29 cm long, with two concentric cylindrical capacitor plates (Fig. 9.5(*a*), (*b*)), the whole fitting into an access tube with an internal diameter of 5.0 cm. The access tube is similar to those used for neutron probes, except that it is made of PVC to allow the passage of the electrical field into the soil. The probe is kept centred in the access tube by a soft nylon fabric ring at each end. The sensor tube can be moved up and down the access tube, in this case by attachment

(a)

(b)

Figure 9.5. (*a*) A capacitance probe being placed in an access tube, with the hand-held frequency meter (*b*) connected by fibre optics to the electronics and the sensing electrodes in the lower (white) section of the probe. The grey upper tubing is used to locate the probe precisely at the required depth. (*b*) With the capacitance probe in place at the required depth, the frequency indicated on the meter is used to determine the moisture content of the soil immediately adjacent to the capacitor electrodes.

tubes rather than by cable. At each required depth, a reading of the frequency of the oscillator is taken by a hand-held frequency meter (Fig. 9.5(*b*)) at the surface; it is connected to the probe by a fibre-optic cable rather than by copper wire, so as to prevent stray capacitance from influencing the frequency of the oscillator.

More recently, a probe has been developed with two short rod electrodes that can be inserted directly into the soil for the measurement of the top 10 or so centimetres, without the need for an access tube (Robinson & Dean 1993; see Fig. 9.6). In this, it is similar to the *time domain reflectometry* instruments described below.

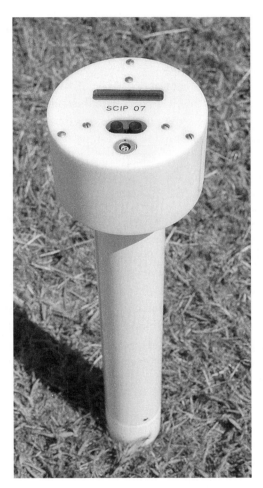

Figure 9.6. An important advantage of the capacitance probe is its ability to measure the upper few centimetres of the soil – for which the neutron probe is less well suited. To take advantage of this, a hand-held probe with short, external electrodes (Fig. 9.8) is an ideal tool for such measurements, being easily inserted and quickly read and removed.

Because of the dipolar nature of water molecules, it may be that the capacitance probe senses well the water that is free enough to respond easily to the reversals of the electric field, because it is not too tightly bonded, and less well to that more strongly bound by surface tension in the pore spaces and as thin films on particles, especially in the case of clay. Because of this uncertainty, there may be no simple relationship between the capacitance probe frequency and the MVF as measured by the gravimetric method or by the neutron probe method, each of which responds differently to the several types of water bond in the soil. There is probably no way of assessing with certainty which water molecules are sensed in the capacitance probe method. Nevertheless some reference is necessary for calibration, and the gravimetric method is the one used, because it is the most fundamental.

The bulk of the capacitance probe's response is derived from a volume of soil between 4 and 8 cm in thickness with a diameter of about 13 cm, the shape being approximately a vertical ellipsoid, which is much smaller than the sphere of influence of the neutron probe. While this may have the advantage of giving greater depth resolution, it also means that small anomalies in the soil can produce greater errors (Whalley *et al.* 1992).

Installation

Extreme care is necessary when auguring the holes and installing the plastic access tube so as not to allow an air gap to form between the tube and the soil, since this would dominate the frequency response. A perfectly installed access tube is essential and a gap as small as 0.5 mm affects the response; one or two millimetres is unacceptable. To achieve this standard of installation, special tools are needed and this puts up costs as well as demanding skilled and careful work. Air gaps can also be a problem when trying to work with shrinking clays, as the gap may form later, some time after a perfect installation. In the case of the hand-held probe, in which the electrodes are inserted directly into the ground, the problem of an air gap is less, but not absent.

Calibration

A capacitance probe typically operates in the frequency range of 80 MHz in water to 150 MHz in air. (In Fig. 9.5(*b*), the meter indicates 13437. This is the frequency in Hz divided by 8000, meaning that the probe output is actually 107 MHz.) This reading is converted into the relative permittivity, ε_r, using an empirically derived equation of the form

$$\varepsilon_r = a/(R^2 - b) - c, \tag{1}$$

where R is the reading and a, b and c are constants specific to each probe design and to each individual probe of any one design (be it the direct-insertion type or one that is lowered down an access tube).

Conversion from ε_r to MVF is, however, dependent on soil type and, for best results, the probe should be calibrated in the soil of interest. The much smaller volume of soil sampled by a capacitance probe means that the technique of calibrating a neutron probe, by taking soil samples from somewhere in the vicinity of a temporary access tube, is not applicable – since the scatter of points is too great. Instead, known-volume samples have to be collected as the actual access tube that is to be used is installed. This requires care so as to ensure that the samples, and the readings taken by the probe, refer to the same depth down the tube. Even this can give a spread of points because the probe is still not measuring the actual sample. The samples' water contents are then measured by the gravimetric method to determine the MVF and plotted against the probe's frequency readings; the MDWF can also be calculated from the soil samples.

The procedure gives only one point on the calibration curve – the MVF at the time of installation. To get around this, several access tubes, close to the original tube, have to be installed at different degrees of soil water content and the calibration process repeated each time; this method assumes lateral homogeneity of the soil, which may not be the case. Graphs produced in this way are approximately linear over the restricted range of soil moisture changes normally associated with each particular soil type, but graphs from different soils can have very different slopes and intercepts. If the various graphs from different soils are combined over the full range of water content, the overall relationship is non-linear, and some soils may not fit neatly onto this combination graph (see Fig. 9.7). In the case of a hand-held probe a similar procedure is necessary, the soil to be measured by the probe being removed for measurement by the gravimetric method. If this lengthy procedure is not possible, or pending its being done, a rough estimate of the MVF can be obtained by using another empirical equation; in the case of the Institute of Hydrology probe, and for typical UK loams and brown-earth types of soil, the equation takes the form

$$\text{MVF} = 0.0977\sqrt{\varepsilon_r} - 0.167 \tag{2}$$

(Robinson *et al.* 1998).

The overall procedure, therefore, is to take a probe reading at the field site, convert it to ε_r by eq. (1) and then to convert this to the MVF by curves obtained over a period from field calibration or by using eq. (2).

Effect of soil conductivity

An *equivalent circuit* of soil consists of a capacitor in parallel with a resistor, the resistor representing solutes in the moisture. It is necessary for the

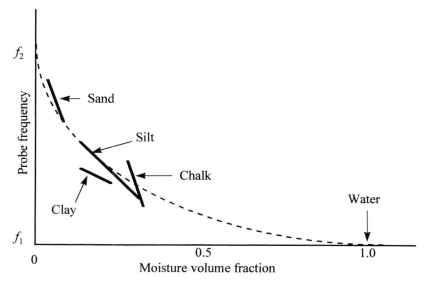

Figure 9.7. Calibration curves of a capacitance probe for different soils may be nearly linear, but each may have a very different slope and intercept, as this combination graph shows. The frequencies f_1 and f_2 will depend on the make of probe, falling roughly into the region 100 to 150 MHz. Thus, unlike the neutron probe with its linear calibration, the capacitance probe is far from linear. Further, as soils may often be layered, with one type overlying another, or mixed, the situation is further complicated.

capacitance-measuring circuits to be able to cope with this shunt resistance, which can be quite low in some soils. For capacitance methods to be successful, the effect of the conductivity of the soil must be understood. When the shunt resistance is high, the capacitance measured by the electrodes is correct, but as the resistance falls the oscillator response is damped, the oscillator frequency reduced, increasing the apparent ε_r and so giving an overestimate of the water content. If the conductivity is high (2000 µS cm^{-1}, equivalent to a salt solution of 1000 ppm – see Chapter 10), ε_r can be overestimated by as much as 30 (Robinson *et al.* 1994). For such soils, a relationship existing between the true capacitance, the apparent capacitance, the resistance and the angular frequency (the frequency in hertz multiplied by 2π) can be used to derive the correct capacitance (Dean 1994), although work remains to be done on the effects of temperature on the conductivity. However, with a conductivity below about 500 µS cm^{-1} there is little effect, and most soils are not conducting enough to cause problems, although some clays and saline or fertilised soils can put the measurement badly in error.

Comparison of the capacitance probe with the neutron probe

The care required in the installation of access tubes in order to avoid air gaps, compared with the relative ease of neutron-probe access-tube installation, is a considerable disadvantage. This, taken along with the need to calibrate the capacitance probe's response in different soils for precise work, the uncertainty over exactly which water molecules the instrument responds to and the much smaller volume of soil measured, considerably offsets the method's originally perceived advantages over the neutron probe, of low cost and simplicity. Neither advantage has yet materialised, the cost of a capacitance probe being not very different from that of a neutron probe; both are priced at between £5000 and £7000, in part owing to the equipment required to install the access tubes accurately. Indeed the neutron probe, despite its main disadvantage of using a radioactive source, must still be seen as the reference, the more accurate and preferred method (Evett & Steiner 1995). But with research and development continuing, the capacitance method is likely to become more sophisticated, or perhaps it will be superseded by the time domain reflectrometry method of measuring a soil's dielectric constant.

Time domain reflectrometry

Time domain reflectrometry (TDR) is another way of measuring the dielectric constant of soil – and thereby its moisture content. It is not a new concept; references to its application to soil moisture measurement go back over the last three decades (Hoekstra & Delaney 1974, Davis & Chudobiac 1975, Topp *et al.* 1980, Dasberg & Dalton 1985, Nadler *et al.* 1991, Whalley 1993).

Principle

A fast rise-time (200 picosecond) electromagnetic step pulse, launched down a transmission line (such as a coaxial cable or twin feeder), travels not at the speed of light but at something less, the actual speed depending on the value of the relative permittivity ε_r of the material filling the space between the conductors. The larger the value of ε_r, the slower the speed of travel. Because the value of ε_r for water is about 80 while for soil it is less than 6, the speed of travel of a pulse down a transmission line with soil as its dielectric is largely dictated by the soil moisture content (the MVF). To be precise, the combination of soil, air, water and roots has what is known as an *apparent dielectric constant* because it is a mixture of dielectric materials. The TDR method operates by measuring the time of propagation of the pulse along the line to the end, where it meets an open circuit; this causes the energy to be reflected back up the line, returning to its origin. (If the line were terminated in a load, such as a matching impedance of the same value as the source and the transmission line, the energy would be absorbed and there would be no reflection.) It is the round-trip time of the pulse that is measured, an average of perhaps 1000 readings being taken by the instrument. Times as short as 2 to 3 ps (10^{-12} s) have to be measured with short transmission lines in dry soil, but typically the round-trip time is several nanoseconds. In addition to a reflection from the end of the line, if there are changes in ε_r along the length of the transmission line then the consequent change in impedance causes a fraction of the signal energy to be reflected at each discontinuity back up the line. By an analysis of the resultant waveform, further information can be obtained about the changes in the moisture content profile, in addition to that obtained by simply measuring the round-trip time for the pulse.

For laboratory experiments investigating the TDR method, it has been more convenient to fill a coaxial transmission line with soil, but for field use a parallel transmission line, consisting of two or more parallel metal rods similar to the capacitance probe (Fig. 9.8), is more practical and the most widely used. A coaxial cable with a characteristic impedance, Z_0, of 50 Ω is the most convenient type of transmission line to interconnect between the pulse generator and the parallel transmission line in the ground. But the two cannot be connected together directly, since a parallel transmission line has a value of impedance different from that of a coaxial feeder. A coaxial cable is unbalanced; its inner cable carries the signal and its outer screen is 'earthed', the field being contained within the screen. A parallel transmission line is balanced, both lines carrying the signal, the field being between and around the conductors. At the point where the balanced meets the unbalanced line, they must be interconnected through an impedance-matching transformer (known as a *balun* – an abbreviation for balanced–unbalanced). This applies to those sensors that are made up of two parallel conductors (Fig. 9.8).

There is an alternative design of sensor based on three (or more) conductors and, as this acts as an unbalanced transmission line like the coaxial cable, no balun is then needed; the inner conductor is connected to the inner coaxial conductor and the outer conductors to the coaxial screen (Fig. 9.9). The sphere of influence is roughly the volume of an ellipsoidal cylinder with a greater diameter of about 8 cm and a length of 5 to 15 cm, which is akin to the capacitance probe's (and considerably less than that of the neutron probe), with the same problems that this causes for the capacitance probe.

If the soil moisture contains dissolved salts then although the amplitude of the signal is reduced the travel time of the pulse is not affected, but there is a limit when the conductivity rises towards 2000 μS cm^{-1}, because the waveform becomes so attenuated and smoothed that it is difficult to detect the pulses reliably. But, if not too large, the attenuation is not entirely a disadvantage since it is possible to use it to measure the conductivity of the sample.

Figure 9.8. In common with both the capacitance probe and TDR for near-surface measurements, the sensing electrodes take the form of two parallel rod electrodes from 5 to 15 cm in length. The ones illustrated are those of the hand-held capacitance probe of Fig. 9.6.

Practice

TDR is not the exclusive preserve of soil moisture measurement, indeed it is a specialised application of it, a more general use being the detection of faults in long cables. For convenience, some workers have simply used cable testers for the measurement of soil moisture and these have built-in oscilloscopes which aid interpretation of the trace, but today most TDR systems, with and without oscilloscope displays, are custom built for measuring soil moisture and there are now several makes of instrument available. Some simply measure the transit time of the pulse and convert this into ε_r and thus into MVF; others also include an oscilloscope that allows the reflected waveform to be displayed and thus more information to be gleaned from the reflections due to changes in ε_r at different depths, as mentioned earlier.

At its simplest, a measurement is made manually by pushing the sensor into the ground (as shown in Fig. 9.6 for the capacitance probe), taking a reading on a built-in meter, withdrawing the sensor and moving on to the next site. If measurements were taken of the same volume of soil each time, then reinsertion of the probes into the same holes could give rise to air gaps around the probe rods. This procedure might also allow more rainwater to enter that area (via the holes left after taking a reading) than the untouched ground nearby. To avoid these problems, and yet also guard against spatial variation, several readings are usually taken at different nearby points and an average produced.

Manual insertion of a probe into the ground only provides a reading for the top 5 to 15 cm of the soil. If deeper measurements are required, one option is to bury a sensor permanently, a coaxial cable connecting it to the surface where either a manual reading can be taken with a hand-held meter or the measurement can be logged automatically. If several probes are installed permanently, then they are connected by cables to a multiplexing switch that allows them to be scanned and their readings logged.

If a profile of soil moisture is required, one option is to bury a sensor at each required depth but this can be expensive and, where probes contain the electronics built-in, the price can be four times as much. The presence of a buried probe can also affect the behaviour of the soil being measured. There may also be a period of stabilisation after installation. Nor is it possible to obtain a profile at precisely one point, since if several probes are to be installed at different depths, they must be displaced laterally from each other. This can introduce uncertainty unless the soil is homogeneous. If a pit is dug adjacent to the point where the profile is required, probes can be inserted sideways into the ground from the pit, immediately over each other. But this has its disadvantages, since it disturbs the ground even more than augering access holes from the surface.

To avoid these problems, an alternative to burying several probes at one site is available in the form of a probe that can be moved up and down an access tube, as with the capacitance probe (Fig. 9.5(*a*), (*b*)). This probe is constructed with its transmission line in the form of two vertical metal strips on opposite sides of the probe, much as the capacitance probe has two horizontal rings as capacitor plates. However, because the plan view of the sphere of influence of the vertical plates is elliptical (Fig. 9.9), it is necessary to take several readings at each depth, rotating the sensor so as to give all-round coverage. This is also true for the portable probes

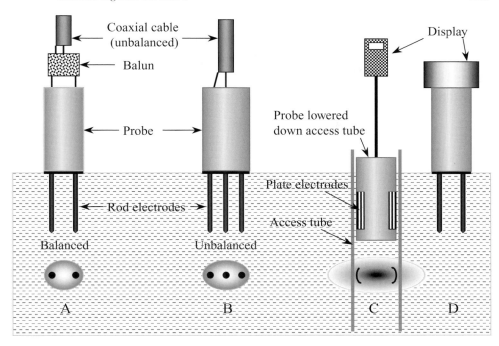

Figure 9.9. TDR probes can take several forms. In A, two short rods are inserted into the soil (or the whole can be buried in the ground at any required depth). Because the probe's two electrodes form an electrically 'balanced' system, a balun must be interposed between the probe and the 'unbalanced' coaxial cable transmission line to match their impedances. The coaxial cable can run some distance to a data logger. B is similar except that the probe's three electrodes form an unbalanced system and so can be connected directly to a coaxial cable. In C, a supporting cable allows movement of the probe up and down an access tube, as with the capacitance probe (Fig. 9.5). However, in this case, because its field of influence is an ellipsoid, the probe must be turned several times at each depth to get a fully representative reading (whereas the capacitance probe uses two ring electrodes one above the other, resulting in an equal response all round). D is the same as A, except that it contains the display built into the top, making it a portable system similar to the capacitance probe of Fig. 9.6. In cases A, B and C the measured volume, an ellipsoid, is shown below the electrodes.

with external rods that are simply pushed into the soil by hand, a plan view of their sphere of influence also being an ellipse.

Furthermore, just as it is essential to ensure that there are no air gaps between the access tubes and the soil when installing a capacitance probe, it is equally important with TDR sensors; the sensitivity decreases exponentially into the medium so that the volume nearest to the tube has greatest influence. Similar special equipment is, therefore, advisable to enable the (fibreglass) access tubes to be installed with minimal air gaps.

Calibration

Whereas each capacitance probe requires a calibration of its frequency against ε_r to be obtained, with TDR this is not necessary since the travel time of the pulse is related directly to ε_r. However, the conversion from ε_r to MVF requires exactly the same procedure as that described for the capacitance probe (Robinson *et al.* 1998).

Modified TDR

A modification to the TDR method (which we might dub CW TDR) has been developed in which a continuous wave of about 100 MHz is emitted, instead of a single fast-rise pulse (Gaskin & Miller 1996). The continuous wave is reflected from the end of the line just as is the pulse. The outgoing and returning signals interfere, producing a composite standing wave, the amplitude of which depends on the phase and amplitude of the reflected wave (which either adds to or subtracts from the source wave). This in turn depends on the length of the probe and the soil's relative permittivity, and thus its moisture content. The ratio of the outgoing wave to the composite standing wave is calibrated against MVF, using gravimetric techniques similar to those developed for the capacitance and TDR instruments. The advantage of the method is that it does not rely on the detection of pulse thresholds, which can be somewhat uncertain, nor on measuring the very short travel times of pulses. In consequence it is cheaper. Otherwise it is similar to TDR, with the same advantages and disadvantages.

There seems to be some uncertainty whether to call this method TDR or a capacitance probe. Indeed workers in the field are hesitant to pigeonhole it, saying that in reality both capacitance probes and TDR overlap in principle. This is certainly true in that both measure the dielectric constant of the soil, albeit in different ways.

TDR compared with the capacitance and neutron probes

Because the TDR method measures the soil's ε_r, it is subject to many of the same limitations as the capacitance probe. Air gaps around the access tubes introduce errors. There is the same uncertainty as to which water molecules are measured (the question of their levels of binding and water of crystallisation), although this is less of a problem if it is changes in water content that matter. The same very careful and lengthy methods are required in calibrating a TDR probe against readings obtained using the gravimetric method (for precise scientific work, that is, although perhaps not for less demanding applications). The relatively small volume of soil sampled by the probes (their sphere of influence) makes the method more vulnerable than the neutron probe to local anomalies within the soil such as stones and other (unknowable) spatial variations.

In consequence, for applications such as estimating irrigation needs, which do not require precise quantitative measurements, for measurements near to the surface and when automatic logging is necessary, dielectric methods have much in their favour. But for accurate, precisely calibrated data, collected manually, the neutron probe is still difficult to better.

By radio propagation

Yet another way of measuring the dielectric constant of soil, and so its moisture content, is to measure the effect of soil moisture on the passage of a radio wave (Hasted 1973).

When a transmitter launches a radio wave towards a receiver some distance away, the received signal strength is dependent on the relative permittivity of the ground between the two antennae; the presence of soil moisture increases the field strength in direct proportion to the MVF, linearity being maintained over a wide range of moisture content. The effect can be visualised as being like that of a large capacitor, the antennae being the plates, the ground the dielectric. Vertical polarisation is best, since a horizontally polarised wave is very dependent on antenna elevation. The main problem with vertical polarisation is that the signal is attenuated by thick green vegetation, although low-growing plants such as grass, or sparse vegetation, have little effect. Because radio waves are attenuated exponentially with depth into the ground, it is the top metre or so of ground that has most effect on signal strength.

The method has been tested successfully at frequencies ranging from 30 to 150 MHz. There must be a minimum of eight wavelengths between the transmitter and receiver antennae, allowing distances of from 15 to 300 m or more to be measured (Chadwick 1973); this gives an integrated value of soil moisture over a large area, unlike the other methods, which give localised, spot measurements on small volumes of soil at specific depths. Perhaps because of its relatively broad-brush picture, the method has not been exploited in the way that the methods that produce spot readings have, although its large areal coverage could be seen as an advantage.

By gamma ray attenuation

Just as snow can be measured by its attenuating effect on gamma rays, so too can soil moisture. The principle is identical: a source of gamma rays is placed in a tube below the surface, a Geiger–Müller detector being sited some distance away to measure the level of gamma rays transmitted through the soil. Scattering and absorption of the rays occur in their journey from source to receiver, and although the soil is the principal determinant of these processes, variations in gamma ray

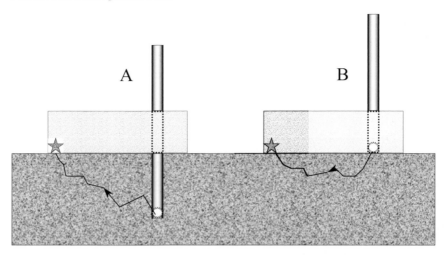

Figure 9.10. In A, the gamma-ray path is directly from source (white sunburst) to detector (shaded star); in B the path is by back-scatter. The former requires an access hole to be made, although this gives control over the depth to which the soil moisture is measured. In the latter case, the instrument can be simply stood on the ground, although only the top surface is then sensed.

intensity over time are due largely to changes in water content, provided that the soil is not subject to shrink–swell behaviour.

The method is little used for soil moisture measurement, the other options generally being preferred; however, it is more widely used for measuring soil bulk density (see the gravimetric method above) in civil engineering applications. For this type of measurement, the source (typically caesium-137 with a strength of 10 mCi) is housed in a portable unit from which it can be lowered incrementally to a depth of about 30 cm, the detector being housed in the base of the unit. An alternative mode of operation is to sense the back-scatter of the gamma rays, as Fig. 9.10 shows, removing the need to install an access tube.

By heat conduction

The rate of heat dissipation in soil is a function of the soil's thermal conductivity, which increases with water content. It is thus possible to measure moisture content by measuring the rate at which an electrical heat source, buried in the soil, warms up – the greater the water content the slower the rate. The idea was first proposed in the 1930s (Shaw & Baver 1939), but because of the problem of maintaining good thermal contact between the sensor and the soil, it has not found the wide use that the other methods have. If it could be made reliable it would be a low-cost and simple method, requiring only a heat source and a means of measuring temperature (Bloodworth & Page 1957).

Measuring soil tension

As the level of soil moisture changes, so does its tension – the dryer the soil the higher the tension. The significance of a combined measurement of soil moisture and soil tension is that not only is the amount of water known but its direction of flow can also be inferred – up, in the case of evapotranspiration, and down in the case of groundwater recharge. Soil tension can be measured in two main ways, by tensiometer or by gypsum resistance block.

Porous-pot tensiometers

Manually read tensiometers

A porous pot, filled with water, is connected by a plastic tube, also filled with water, to a mercury manometer (Fig. 9.11(*a*), (*b*)). A stopper at the top of the tube to which the porous pot is fixed has a smaller bung, which is used to allow the system to be filled with water and purged of air. If the pot is held in air and at the same level as the mercury, the mercury level in the tube remains at reservoir level. If the pot is lowered, mercury is drawn up the manometer tube by an amount equal to the pressure-head (suction) caused by gravity. This suction due to gravity increases the further the pot is lowered. This also applies when the pot is in the soil, and so needs to be subtracted from any suction caused by the soil. Since it is entirely a matter of height difference, it can be calculated.

If the pot is now installed in the soil, soil tension adds its negative pressure to that of gravity and so draws the mercury further up the manometer tube, by an amount additional to that due to gravity – unless the soil is saturated, when there is no soil tension. The installation procedure, to ensure good hydraulic contact with the soil, is the same as the procedure, explained later, for gypsum blocks. Several pots can be installed at one site at different depths, usually corresponding with those at which the MVP is measured, each with its individual manometer tube. Manometer readings are taken at the same time as the MVP is measured. This might be daily, weekly or monthly, depending on the conditions and requirements.

Manual tensiometers are also manufactured in which the mercury manometer is replaced by a mechanical vacuum dial gauge, with the means to fill the system with water and to ensure that no air is present in it. These are useful for obtaining rapid readings in the top soil layers during a field visit; they are not left in situ.

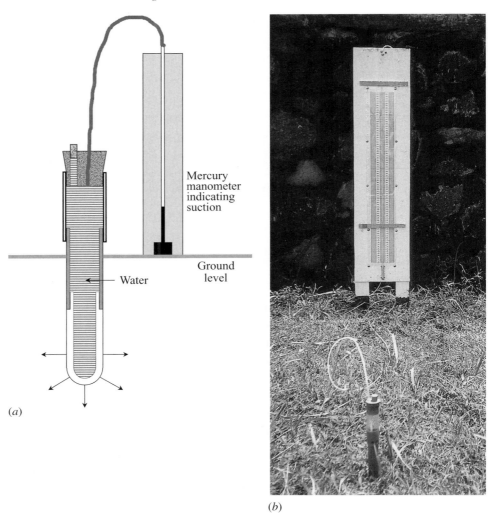

Mercury
manometer
indicating
suction

Water

Ground
level

(a)

(b)

Figure 9.11. (*a*) The unglazed ceramic U-shaped porous pot of a tensiometer is
fixed to a plastic tube (dark shading), the top few centimetres being made of
transparent material to allow the water inside it to be seen. The top of the tube is
sealed with a rubber bung, through which passes a flexible small-bore plastic tube,
connected at its far end to a mercury manometer. The small bung is removable to
allow the system to be purged of air, using a syringe. The arrows below the porous
pot show the effect of tension plus gravity on the water. (*b*) With the porous pot
installed in the ground, the negative pressure produced by soil tension and gravity
draws mercury up the manometer tube, which is read manually at the required
interval, often at the same time as the moisture content is read with, for example, a
neutron probe (Fig. 9.3). The top of the tensiometer is seen in the lower foreground,
the flexible tube connecting it to the manometer scale board in the background.

Unfortunately, this type of tensiometer can only function over a small range of tensions, from 0 to about −0.8 bar, because, as the vacuum pressure increases, air bubbles start to form in the water, drawn out from crevices or from the pot itself. (The water used should be boiled to remove the air dissolved in it in order to lessen the effect, but this does not fully prevent it.)

Electronically measured tensiometers

It is possible to connect an electronic pressure sensor (see also Chapter 6, the section on electronic barometers, and Chapter 10, the subsection on sensing water level by pressure) to the same type of porous pot as is used in the manual tensiometer, and so to measure the pressure automatically (Fig. 9.12(*a*), (*b*)). The engineering of this type of instrument makes it much more complex than a manual tensiometer and therefore much more expensive. And again these tensiometers can only operate up to about 1 bar suction.

Gypsum blocks

While most of the water useful to plants is in the region covered by porous-pot tensiometers (up to −0.8 bar), it is important to be able to measure well beyond this range because the processes of evaporation, transpiration and the downward percolation of water continue at much higher tensions.

Principle

The electrical conductivity of soil varies with its water content because the water contains dissolved salts, making it much more conductive than the soil particles. But because of the highly variable nature of soil, it is better to measure the conductivity not of the soil itself but of a porous block kept in close contact with the soil. The water content of such a block tends to be in equilibrium with the surrounding soil and so measurement of its resistance provides a measure of soil tension. Such sensors are known as *resistance block tensiometers*. Variations in water content of the block result also in a capacitance change but it is usually the resistance that is measured because it changes more than the capacitance in response to soil water potential, from around 500 Ω at −0.4 bar to 10 000 Ω at −10 bar.

Because the conductivity of soil also depends on the quantity of dissolved material it contains (as well as on how much water it contains), it is desirable to minimise the effect of the dissolved material so that just the quantity of water is sensed when the resistance is measured. While soil tension resistance blocks have been made of various porous materials, those made of gypsum (or Plaster of Paris – the commonly occurring mineral hydrated calcium sulphate) are preferred because the water they contain is always saturated with calcium sulphate. In non-saline

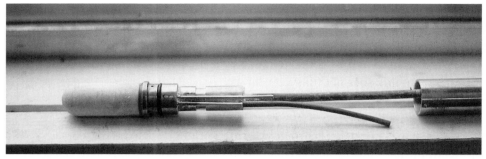

(a)

(b)

Figure 9.12. (*a*) The negative pressure sensed by porous pots can also be measured electronically using a similar type of pressure sensor to that used to measure water level (Chapter 10); it is seen in the figure affixed immediately to the right of the porous pot. (*b*) At the surface, automatic tensiometers are filled with water and purged of air in a similar way to the manual type, but with valves rather than a simple bung. The cable from the sensor carries the signal to a data logger, and usually several such tensiometers are operated together to give a depth profile. Within the cluster of tensiometers shown (on the right) is a neutron probe access tube, allowing the amount of moisture to be apportioned, upward to evaporation and downward to percolation.

soils, the concentration of natural salts is likely to be much less than this and so using gypsum buffers the conductivity of the water they contain against variations in the soil's conductivity. (In acidic soils the blocks quickly dissolve.)

Since the 1940s, gypsum blocks have been used to measure both soil moisture and soil tension but because of the confusion that arose between the two uses, they fell into disfavour in the 1960s and 70s and were looked at with suspicion until the situation was clarified (Wellings *et al.* 1985). Resistance blocks are unreliable as soil moisture content sensors, owing to the hysteresis of the relationship between soil moisture content and the potential of the soil and that of the block itself. But if blocks are individually calibrated on a drying cycle and are used for tension measurements only on a drying cycle, good results can be obtained. On a wetting cycle results are only approximate. Calibration is done in a ceramic pressure-plate apparatus, the blocks being covered in a thick slurry of, for example, chalk. Instead of applying suction to the outside of the ceramic plate, pressure is applied to the inside, this having the same effect (Wellings *et al.* 1985). It is necessary to put blocks through several wetting and drying cycles before calibration, however, as their resistance characteristics change considerably at first. Figure 9.13 illustrates a typical resistance versus pressure curve of a gypsum block calibrated in this

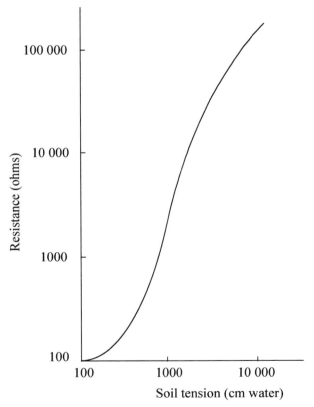

Figure 9.13. A gypsum block of the cylindrical type shown in Fig. 9.14 will have a response roughly of the shape shown here. Other designs, such as those with parallel plates, have slightly different responses.

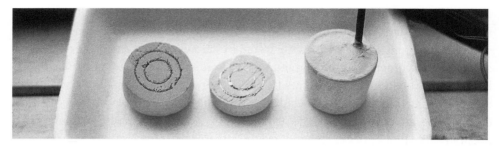

Figure 9.14. Much simpler and cheaper than porous pots, and with certain technical advantages (a much greater tension range), gypsum blocks are constructed of (typically, but not exclusively) two concentric stainless steel mesh cylinders embedded in gypsum, here sectioned to show the interior construction.

way. Because calibration by this method takes several days, just three points are measured as an adequate compromise.

Construction and installation

Gypsum blocks are probably the simplest of all of the sensors described in this book, although manufacturers may challenge this and point to the many pitfalls of do-it-yourself construction. They are quite right, since they are cheap to buy and consistency from one to another is much more certain in the bought-in item. All designs consist of Plaster of Paris moulded around two electrodes. While some are flat, a common design is cylindrical, with concentric stainless steel mesh electrodes (Fig. 9.14).

Installation is similar to the installation of a neutron probe access tube, by auger. When the required depth is reached the block (previously well soaked in water) is lowered to the bottom of the tube and pressed into good contact with the ground. Thereafter the hole is back-filled with soil slurry for about 50 cm, which is consolidated by gentle pressure with a rod or tube. The remainder of the hole, if it was deeper than 50 cm, is filled with dry soil sieved to pass a 2 mm mesh. It is difficult to retrieve blocks at a later date if they are much below 50 cm, the only way being to dig them out, but as they are low cost this is not essential.

Accuracy and corrections

These are not precision sensors. If semiquantitative data are all that is required then resistance blocks can be used without calibration, but if more is needed the calibration briefly described above must be used (Wellings *et al.* 1985). In addition to calibration, temperature correction is needed and for this thermistors can be installed along with the blocks to monitor the temperature at the time the blocks are read; errors are in the order of 1.5% K^{-1}, although temperature changes are slow and small below ground. Because the sensors can change through chemical reaction with the soil, their calibration will drift with time. This can be detected by comparing data a year apart, at similar soil water contents, and this allows some correction to be applied. Depending on the chemistry of the soil, blocks will last for three to five years. While the absolute values of matrix potential may be in error by ±20% after a year or more, if only the depth of the zero-flux plane (ZFP) is required this may be acceptable. Response times vary from minutes at low tensions to hours at 5 bar and up; this is adequate to follow the changes that occur in natural soil conditions.

Measuring block resistance

A DC bridge or digital multimeter is not suitable for measuring the resistance of gypsum blocks because electrolysis occurs and the blocks polarise, causing a rapid rise in resistance due to gas bubbles forming on the electrodes. Bridges for manual use operate at about 1.5 kHz, the operator balancing the bridge to a null point on a meter by hand. For automatic logging, various methods can be used to prevent polarisation. In one design (Strangeways 1983) a single, short (120 ms) DC pulse is applied to the block via a resistor, the voltage across the resistor being logged. The block is then short circuited until the next reading to discharge any slight polarisation that has occurred. In all methods, the block's resistance must be measured in isolation from the other blocks of a profile since there is a resistance path through the soil between blocks that will otherwise affect the readings.

Accessing groundwater

Since instruments for ground- and surface-water measurement, with respect to both depth and quality, are similar in principle, they are most logically dealt with together in the next chapter, on surface water measurement. Here we look at how access is gained by drilling to the groundwater, in order to measure it.

Drilling methods can be divided into rotary and percussive, either of which may or may not use circulating fluids (Nielsen 1991).

Rotary drilling

Hand augering

Hand augers offer a simple method of drilling holes in the top few metres of ground provided it is of unconsolidated material. Several types are available, from the familiar helical auger to those for soft fine soils and for gravel; the cutting tool is mounted on a rod, the operator pressing down while turning a handle until the auger is full, when it is withdrawn. Holes of 5 to 10 m depth or more can be augered in this way, with diameters up to 20 cm.

Continuous-flight and hollow-stem augers

A central shaft about 1.5 m long, known as a *flight*, has a steel helix wound along its full length. Flights can be joined together, forming a continuous helix that acts as a conveyor belt to the surface (Fig. 9.15), torque being transmitted from the surface through the flights. The auger's cutting head makes a hole 10% wider than the diameter of the helix, to ease withdrawal and to reduce wear and friction. Such augers are generally used to drill unconsolidated formations, although rock can sometimes be penetrated if the appropriate bit is used. Diameters range from 10 to 60 cm with depths in excess of 25 m.

In *hollow-stem* augers, the rod up the centre of the helix is replaced by a tube, the drill bit being in two parts, one on the helix and one on a rod that passes down the centre of the tube, both rotating together. (Alternatively the central bit can be run down the stem on a wire cable and locked onto the auger.) The advantage of hollow-stem augers is that the central rod, or wire cable, can be withdrawn, leaving the tube and helix in place; this allows a water sampler, measuring instrument or casing (see later) to be inserted. It also allows a change in drilling technique to be made at an interface between unconsolidated formations and rock. Cutting speeds vary from 30 rpm for heavy clay to 75 rpm for sand (Clayton *et al.* 1982).

Percussive drilling

Here the method of advance is by hammering, striking or beating the formation. There may also be some rotation, but this will be mostly to maintain the roundness and straightness of the borehole.

Driven wells

The simplest way to install a shallow, small-diameter borehole in unconsolidated formations free of large stones is to drive a *well point* (metal cone) into the ground using a pneumatic or hydraulic jackhammer or a sledge hammer, a tube behind the point containing holes (a *screen* – see later) to allow water to pass into it from the surrounding formation. To prevent damage to the screen during installation, a rugged outer tube (or casing – see later) can be used to carry the driving force to the tip, the casing being partly or completely withdrawn on completion of installation, in order to expose the screen to the water. Alternatively, the driving force can be applied directly to the point using a down-hole weight, raised and lowered on a rope.

Figure 9.15. One of two principal ways of making a borehole for monitoring groundwater is to drill the hole using a *continuous-flight* auger, which acts as a conveyor belt to the surface. In this case the auger is of the *hollow-stem* type, allowing access to the cutting surface down its centre, perhaps to take water samples or so that casing can be inserted. Torque is generated by the hydraulic equipment at the top. Several flights can be joined to reach greater depths.

Cable tool drilling

The earliest drilling rigs were probably made about 3000 years ago in China using bamboo. In the present-day equivalent, the rig is of steel and the *tool string* is raised and lowered at a short distance from the bottom of the borehole, breaking off fragments of rock or loosening unconsolidated material. The string comprises a drill bit, a heavy length of steel (the *drill stem*) to give the necessary mass to break the material (and also to keep the hole vertical), a *jar* to help if the tool string becomes jammed (by back-hammering) and a *swivel socket* that allows the string and thus the bit to rotate at random by virtue of the lie of the *steel wire cable* (or *rope*) on which the tool string is suspended. The *cable* is moved up and down by a *spudding arm* fixed to the rig.

Various bits are used depending on the formations being drilled, although they are mostly chisels of various types. To remove the cuttings, water is added (if the hole is above the water table) to form a slurry which is then periodically bailed out using a *shell* or *bailer* (see next section). For gravel, a bit with 'orange peel' jaws may be dropped through the gravel, the jaws closing as the bit is withdrawn. Temporary casing is inserted behind the bit to prevent collapse.

Although this is one of the oldest drilling methods, it is still widely used because it is simple and cheap and does not need large volumes of water. It is, however, slow (perhaps 5 to 10 m a day) and holes smaller than about 15 cm diameter are not practicable because of the need to use relatively large and heavy bits and drill stem to make adequate progress. The method is, therefore, less used for drilling monitoring-boreholes than for production-boreholes, but there is some overlap between drilling techniques and borehole use (Nielsen 1991).

Light percussive drilling

This type of drilling is also known as *shell and auger* (originally non-cohesive formations were shelled and cohesive formations were augered using this method). It is a scaled-down form of the cable tool method and is used mainly for unconsolidated formations (Fig. 9.16). In place of the spudding arm, the reciprocating motion is here imparted manually by using a clutch on a simple diesel engine, which raises the string and then lets it drop. The same cutting tools are used (with a *sinker bar* to give the necessary weight – compare with the *drill stem* of the cable tool method). When a clay cutter is used, the clay or soil usually sticks inside the cutter and can be drawn up and removed manually. With gravel and sand, a *shell* (or *bailer* or *sand pump*) is used. This is a heavy tube that is surged up and down about 30 cm each second, the slurry and suspended matter being forced by this action past a non-return *clack valve* in the base of the tube, the tube then being drawn

Figure 9.16. Light percussive drilling is a simple, if slow, way of producing a well. In the figure, the *tool string* is made up of, from top to bottom, a *swivel, sinker bar* and *clay cutter*.

to the surface and emptied. The casing must be advanced while drilling to prevent collapse (Dixon 1998).

Drilling with circulated fluids

The drilling techniques described so far, both rotary and percussive, have not involved the use of fluids pumped down the boreholes. The following methods use circulated fluids – water, air or 'mud'.

The fluid is usually pumped down a central drill pipe, emerging out through the cutting bit and returning back up the annular space between the borehole wall and drill pipe. The purpose of the fluid is to cool and lubricate the bit and to carry cuttings back to the surface. To achieve the latter action, a certain minimum velocity is needed, depending on the fluid and on the particle size. For water this is 1 m s^{-1}, and for air 15 m s^{-1}. Because air is everywhere, and water may have to be transported, air is often the more convenient, but air compressors are expensive both to buy and to operate. However, air has the advantage that it does not contaminate the borehole to the same extent that water can, although even the compressed air will contain hydrocarbon lubricants.

A thin, mud-like, mixture of water and bentonite (a naturally occurring clay of hydrous aluminium silicate mineral) is sometimes used in the place of plain water, helping to raise particles more easily. A mixture of air and water, as foam, is another commonly used fluid. Alternatively polymers may be used as the 'mud', although these can affect the chemical make-up of the water in the borehole even more than bentonite can. Indeed the use of fluids to aid drilling is always liable to contaminate the borehole and its contents, whatever the fluid used, and this is not ideal when the measurement of the quality of the groundwater is the purpose of drilling the borehole. But it is often unavoidable, particularly for the deeper boreholes.

Direct-fluid circulation with rotary drilling

A drill pipe with an attached bit is rotated continuously against the face of the borehole while fluid is pumped down the pipe. The down-force required on the bit is typically between 200 and 400 kilograms force for each centimetre of the bit diameter, with rotary speeds of 30 to 250 rpm depending on the formation being drilled. The cutting tool may be a three-toothed cutter of hardened steel or tungsten carbide, which rotates on its own bearings (a *rock roller*), or for soft formations a three- or four-blade *drag bit*. To keep the bit straight, a heavy *drill collar* can be fitted above it. The drilling fluid is pumped into the top of the drilling tube and emerges through the drill bit, returning to the surface up the annular space between the hole wall and the central tube. Problems can occur, however, if the borehole diameter is large, since the pump may not be capable of producing enough updraft in the larger annular space between borehole and drilling tube, and so may fail to raise the cuttings. To overcome this problem, reverse circulation is used.

Reverse-fluid circulation with rotary drilling

In this method, the drilling fluid is simply allowed to fall by gravity down the annular space until the hole is full of water up to the surface. It is then pumped back up the drilling tube (in the reverse direction to the previous method). Because the diameter of the drilling tube (and thus the pump power required to achieve the minimum velocity and so lift the cuttings) is a constant, larger-diameter boreholes can be drilled. Although this method is not much used for drilling boreholes for monitoring applications, since large-diameter holes are not required, a modified form of it, known as *centre-stem recovery* or *dual-pipe drilling*, is useful for making monitoring boreholes. Here two concentric drill pipes are used, the fluid being pumped down the annular space between the two tubes to the bit, which cuts the hole in such a way as to deflect the fluid back up the central tube. The spoil is either collected as a sample or allowed to pile up on the ground.

Top hammer drilling

Also known as *drifter drilling*, this method combines percussion, rotation and circulating fluids and although it has been used to install access tubes for neutron probes (see Chapter 9) it is used mostly for drilling in mines and quarries for blasting. The hammer is a scaled-up pneumatic or hydraulic road-breaker, with the addition of rotation, all mounted on a rig. Air is the usual drilling fluid but water can be used. Penetration rates are high but diameters are limited.

Down-the-hole hammers

Having a pneumatic hammer immediately behind the bit avoids the necessity of a top hammer, when the impact energy has to be passed via drill rods or tubes, in which it will be partly dissipated. In a down-the-hole (DTH) hammer, blows are delivered to a tungsten carbide *button bit* at 500 to 1000 blows a minute, while the whole drill is rotated at 10 to 30 rpm. The exhausted air from the hammer cools the bit, and also brings the cuttings to the surface up the annulus between the drill and the borehole wall, the updraft being about 15 metres a second. The method works well in hard formations but is unsuitable for unconsolidated material.

Casing and screening

Having drilled the borehole, it must next be *developed*; this comprises a cleaning-up process to rid the borehole of sediments or contaminants introduced during drilling.

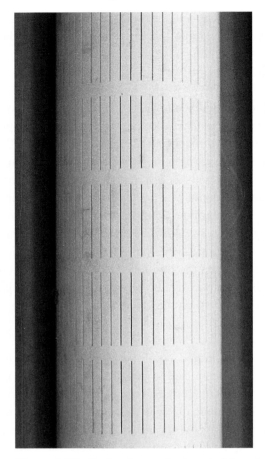

Figure 9.17. Casing has to be inserted in a borehole after or during drilling to prevent collapse; part of the casing has to be in the form of *screening*, in order to allow water to enter. Here the screening takes the form of a plastic tube with fine slits, but it can also be of metal mesh.

It may be done in several ways but a typical procedure is to stir up the water at the base of the hole by raising and lowering a *surge block* on the end of the drill rod and then pumping the water out, the process being repeated until clear water is pumped up. Alternatively compressed air can be used to stir the water.

Unless the borehole is in rock, its walls will have to be supported by *casing-tube* inserted either during or after drilling the borehole, depending on the formation and the drilling method. In order to allow water to enter the well, part of the casing has to be in the form of *screening*: this is a tube containing fine slits (Fig. 9.17) or in the form of a metal mesh. The materials used for both casing and screening are either plastics such as PVC or ABS, or some form of steel or fibreglass-reinforced

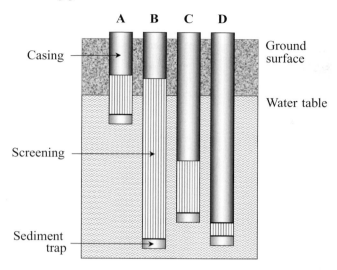

Figure 9.18. Four examples of how casing and screening may be installed are shown here. In *A* the screen extends across the full range of water-table changes. *B* covers the full saturated layer, *C* monitors a specific zone, while *D* is for a specific point.

epoxy or plastic. It is not necessary here to go into the intricacies of how the casing and screening tubes are placed in the borehole, but four typical end-products are illustrated in Fig. 9.18.

To ensure a tight seal at the required depth, the casing and screen may be installed centrally in the borehole, leaving an annular space around them (Fig. 9.19). Below the level of the screen the annular space is filled with a sealing compound, usually the clay bentonite (see the discussion of drilling fluids), inserted as pellets down the annular space to the point required, where, in contact with water, it swells and forms a watertight seal. (However, chemists may object to the use of bentonite as a seal on the grounds that it can contaminate water samples.)

Above the seal, a *filter pack*, usually of medium-to-coarse quartz sand (which is chemically inert) is introduced via a pipe from the surface. The pack prevents fine particles of the natural wall of the borehole from entering the well via the screen slots, while also maintaining good hydraulic conductivity between the borehole wall and the screen. Above the pack is introduced a second seal of bentonite, thereby isolating the screen, hydraulically, at that precise depth.

When just a measurement of water level is required from a borehole, dedicated *piezometers* can be used. These are tubes of small diameter (2 to 5 cm) with a conical metal tip, behind which is a short tube of porous material, with a pore diameter of around 60 microns, acting as a screen. This is attached to a tube that can be extended to any length required, giving access from the surface. These are much

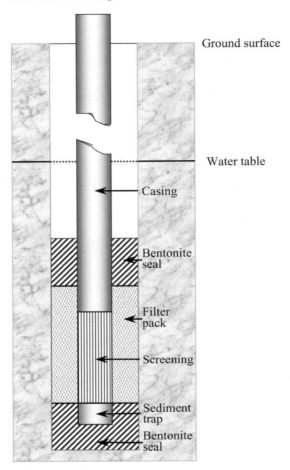

Figure 9.19. To ensure that only groundwater at the required depth is obtained for monitoring, the screen can be centred in the borehole and sealed above and below the required level with bentonite, with a chemically inert filter pack in between, to prevent particles from the borehole wall entering the screen.

smaller, and so cheaper, than well casing and screening; they also require smaller-diameter, and so cheaper, boreholes. In complex situations, nests of piezometers are installed at different depths to detect vertical hydraulic gradients. Screens, or piezometers, may be installed at several depths in the same borehole to obtain a profile of water quality and piezometric head. Some piezometers are designed to be hammered into the ground (see 'driven wells', in the subsection on percussion drilling), which saves the cost of drilling a borehole, although this is only practicable under some circumstances such as near the surface or at the bottom of an existing borehole.

Costs

A drill rig with the necessary minimum of two operators will cost upwards of £300 a day depending on the type of rig. It can take from one hour to several months to drill a borehole, depending on circumstances. Boreholes are not, therefore, usually cheap and may well cost as much to drill as the cost of the instruments that are subsequently used in them.

References

Anderson, A. B. C. (1943) Method of determining soil moisture content based on the variations of the electrical capacitance of soil at low frequency with moisture content. *Soil Sci.*, **56**, 29–41.

Bell, J. P. (1976) Neutron probe practice. Institute of Hydrology Report No. 19, p. 63.

Bell, J. P. (1987) Neutron probe practice. Institute of Hydrology Report 19 (third edition).

Bell, J. P. (1996) Determination of soil water content by the thermogravimetric method. Private communication .

Bell, J. P., Dean, T. J. & Hodnett, M. G. (1987) Soil moisture measurement by an improved capacitance technique, Part II. Field techniques, evaluation and calibration. *J. Hydrol.*, **93**, 79–90.

Bloodworth, M. & Page, J. (1957) Use of thermistors for the measurement of soil moisture and temperature. *Proc. Soil Sci. Soc. Am.*, **21**, 1056–8.

Chadwick, D. G. (1973) Integrated measurement of soil moisture by use of radio waves. Utah State University, Logan, College of Engineering Report PRWG 103–1.

Clayton, C. R., Simons, N. E. & Matthews, M. C. (1982) *Site Investigation*. Granada Publishing, ISBN 0 246 11641 2.

Dasberg, S. & Dalton, F. D. (1985) Time domain reflectometry field measurements of soil water content and electrical conductivity. *J. Soil Sci. Soc. Am.*, **49**, 293–7.

Davis, J. L. & Chudobiak, W. J. (1975) *In situ* meter for measuring relative permittivity of soils. Geological Survey of Canada. Paper 75–1, Part A, Project 630049.

Dean, T. J. (1994) The IH capacitance probe for measurement of soil water content. Institute of Hydrology Report 125.

Dean, T. J., Bell, J. P. & Baty, A. J. B. (1987) Soil moisture measurement by an improved capacitance technique, Part I. Sensor design and performance. *J. Hydrol.*, **93**, 67–78.

Debye, P. (1929) *Polar Molecules*. pp. 77–108. Dover, New York.

De Plater, C. V. (1955) Portable capacitance-type soil moisture meter. *Soil Sci.*, **80**, 391–5.

Dixon, A. (1998) Light percussive drilling. Private communication.

Evett, S. R. & Steiner, J. L. (1995) Precision of neutron scattering and capacitance type soil water content gauges from field calibration. *J. Soil Sci. Soc. Am.*, **59**, 961–8.

Gaskin, G. J. & Miller, J. D. (1996) Measurement of soil water content using a simplified impedance measuring technique. *J. Agr. Eng. Res.*, **63**, 153–60.

Gilman, K. (1977) Movement of heat in soils. Institute of Hydrology Report 44.

Hasted, J. B. (1973) *Aqueous Dielectrics*, p. 302. Chapman & Hall, London.

Hoekstra, P. & Delaney, A. (1974) Dielectric properties of soils at UHF and microwave frequencies. *J. Geophys. Res.*, **79**, 1699–708.

McPhun, M. (1979) Soil moisture meter. Report to the Department of Engineering Science, University of Warwick, UK.

Nadler, A., Dasberg, S. & Lapid, I. (1991) Time domain reflectrometry measurements of water content and electrical conductivity of layered soil columns. *J. Soil Sci. Am.*, **55**, 938–43.

Nielsen, D. M., ed. (1991) *Practical Handbook of Ground-water Monitoring*. Lewis Publishers, ISBN 0 87371 124 6.

Patten, H. E. (1909) *Heat Transference in Soils*. US Department of Agriculture Bur. Soils Bulletin. 59.

Robinson, D. A., Bell, J. P. & Batchelor, C. H. (1994) Influence of iron minerals on the determination of soil water content using dielectric techniques. *J. Hydrol.*, **161**, 169–80.

Robinson, D. A. *et al.* (1998) The dielectric calibration of capacitance probes for soil hydrology using an oscillation frequency response model. *Hydrol. Earth System Sci.*, **2**, 111–20.

Robinson, M. & Dean, T. J. (1993) Measurement of near surface soil water content using a capacitance probe. *Hydrol. Processes*, **7**, 77–86.

Shaw, B. & Baver, L. (1939) An electrothermal method for following moisture changes in the soil *in situ*. *Proc. Soil Sci. Soc. Am.*, **4**, 78–83.

Smith-Rose, R. L. (1933) The electrical properties of soils for alternating currents at radio frequencies. *Proc. R. Soc. London*, **140**, 359.

Strangeways, I. C. (1983) Interfacing soil moisture gypsum blocks with a modern data logging system using a simple, low-cost, DC method. *Soil Sci.*, **136**, 322–4.

Topp, G. C., Davis, J. L. & Annan, A. P. (1980) Electromagnetic determination of soil water content: measurement in coaxial transmission lines. *Water Res.*, **16** (3), 574–82.

Wellings, S. R., Bell, J. P. & Raynot, R. J. (1985) The use of gypsum resistance blocks for measuring soil water potential in the field. Institute of Hydrology, Report 92, p. 26.

Whalley, J. R. (1993) Considerations on the use of time-domain reflectrometry (TDR) for measuring soil water content. *J. Soil Sci.*, **44**, 1–9.

Whalley, J. R., Dean, T. J. & Izard, P. J. (1992) Evaluation of the capacitance technique as a method for dynamically measuring soil water content. *J. Agric. Eng. Res.*, **52**, 147–55.

10

Rivers and lakes

Going up that river was like travelling back to the earliest beginnings of the
world, when vegetation rioted on the earth and the big trees were kings. An
empty stream, a great silence, an impenetrable forest. The air was warm, thick,
heavy, sluggish. There was no joy in the brilliance of sunshine. The long
stretches of the waterway ran on, deserted, into the gloom of overshadowed
distances. On silvery sandbanks hippos and alligators sunned themselves side by
side. The broadening waters flowed through a mob of wooded islands, and butted
all day long against shoals, trying to find a channel till you thought yourself
bewitched and cut off for ever from everything you had known once –
somewhere – far away – in another existence perhaps.

> Joseph Conrad *Heart of Darkness* (steaming up the Congo River).

Sensors for measuring the quality and quantity of surface water, including the
oceans, and groundwater are similar in principle, and so it makes for greater clarity
if this chapter is organised by sensor type rather than by application.

Measuring water level

Staff gauges

Graduated staff gauges are widely used for the manual measurement of rivers, lakes
and sea level. They are usually installed vertically in the river bed or fixed to a weir
(Fig. 10.1(*a*)), bridge or harbour wall. Boards are made in 1-m and 2-m lengths
and are about 15 cm wide, fixed one above the other to cover greater depths, and
marked to span up to 12 m, or more. Alternatively, several may be installed, each
progressively higher up the bank of a river if there is no structure to which to fix
them and the river is deep and wide (Fig. 10.1(*b*)). They are graduated in a variety
of ways, some every centimetre, others every 10 or 20 cm – as in the case of some

Measuring the Natural Environment, second edition, Ian Strangeways. Published by
Cambridge University Press. © Ian Strangeways 2003.

(a)

(b)

Figure 10.1. (a) Staff gauges, as here at a trapezoidal flume on the Red River in Southern Papua New Guinea, are the commonest means of measuring water level, a manual reading being taken periodically. Some care is necessary not to misread the scale. (b) In the larger Iguacu River near Curitiba in Brazil, one staff gauge cannot cover the full range of level change and several must be stepped up the bank. This is a common procedure.

sea level gauges. Boards are also available for fixing at an angle of 45 or 30°, laid flat on river banks, their markings being stretched to compensate. Most are graduated from bottom to top, but others are made with an inverted scale for situations where levels below a reference point are needed. Gauging boards have the advantage of cheapness and simplicity, although care is needed in reading them. Observers may simply write down the readings or send them by telephone or radio to a distant base, providing a simple form of telemetry.

Peak gauges

While water level recorders, of the types to be described, give a detailed plot of the rise and fall of river level on each side of the peak flow, a device that records just the peak level of flow can be much simpler. Knowledge of the peak alone is valuable, but a fairly good estimate can also then be made of the shape of the hydrograph (from prior knowledge and experience) and thus of flow. This can be particularly useful when measuring events in ephemeral streams in the wadis of arid areas. Here flow events are rare, but when they do occur the amount of water can be extreme and the speed of the event dramatic, often missed by observers who have insufficient time to get there. The amount of water flowing is important because it sinks into the ground, recharging the groundwater, which is often the only supply of water available in arid countries.

The sticking-float method

A float, through which passes a vertical rod, rises as the river level rises, but as the level falls, the float is prevented from falling back down by a ratchet fixed alongside it. The float has a pointer attached to it which moves up a gauging board, the peak level being read after the flow has subsided; the float is then reset to the bottom.

The powder-in-a-tube method

An even simpler arrangement can be made in which a tube has powder, such as talcum or chalk dust, placed in its base (Fig. 10.2). As the water level rises, entering through a small hole at the base, the powder rises in suspension up the inside of the tube. As the level falls, a deposit of the powder is left sticking to a rod fixed within the tube, giving an indication of the highest level reached; the rod is then removed for reading, cleaning and replacement ready for the next event.

Stilling wells

Staff gauges stand unprotected in the water. But if a float is to be used to measure water level it must be protected from wind and waves. There are two main types of stilling well. In one a hut stands on the bank some distance from the river,

Figure 10.2. Set in the bed of a wadi in northern Oman, the tube records the maximum level reached during the previous flow event (the flow can be short lived, and sudden in onset). Powder placed in the base of the tube rises to the flood level and sticks to a rod extending up the centre of the tube as the water subsides.

protecting a well dug down to the lowest level the river will reach (Fig. 10.3). At its lowest level, a pipe connects the well to the river, the level of the water in the well slowly following the level changes in the river, with surface irregularities filtered out. This tube can become blocked, and care is needed to detect and prevent this. In an alternative design, often constructed of corrugated iron, the well stands directly in the river with a housing on top and walkway to it from the bank (Fig. 10.4).

Figure 10.3. In the Red River of Papua New Guinea (see Fig. 10.1(*a*)), this stilling well is set back from the river, to which it is connected by an underground tube; this arrangement allows the hut to be accessed even at times of maximum water level.

Electrical contact gauges

Known as *contact meters* or *well dippers* (Fig. 10.5), an electronic probe is lowered by hand down the well (or borehole), detecting the water surface by completing an electrical contact through the water between the case and a point electrode. The probe is lowered on a reel of cable, marked in metres and centimetres, a unit at the surface emitting a light and/or a sound indication when the electrodes contact the water, the depth being read off the graduated cable against a datum point on the top of the borehole. It is possible to read depths to within ±1 cm at 100 m depth, and dippers are made for use at up to 500 m depth in boreholes. To support the weight of a long length of cable in the deeper boreholes, the wire is of steel.

Some dippers also contain a resistance thermometer, allowing a temperature profile to be measured, while others may also include a conductivity, pH or

Figure 10.4. In the Aquidauana River in the Mato Grosso area of Brazil, this stilling well stands in the river. It is accessed by foot at normal flows, over the walkway, and by boat if necessary when the river is in flood. Since it houses automatic instruments, access is less essential.

dissolved oxygen sensor (see later). It is also possible to include a *ground detector*, which indicates when the probe touches bottom, the difference between the two readings giving a measure of water depth as well as the distance of the water surface from the top. Their use is not limited to boreholes, since they are just as functional in a river, lake or the sea. In some situations, a stilling well may not be necessary.

Sensing water level by float

A range of methods, mechanical and electrical, of sensing and recording the position and movement of a float has been developed over the last 150 years.

Figure 10.5. A well dipper is a simple means of measuring water depth manually, in stilling wells and boreholes. The probe (right) is lowered from a reference point down to the water. Upon contact, it sends a signal up the graduated cable, the depth being read from the cable. The reel shown here holds 200 m of cable.

Paper chart level recorders

Together with staff gauges, instruments which measure water level by float, moving a pen across a paper chart, account for most of the world's past river data and they are still in wide use today. The various designs are distinguished by three main features, as follows.

The paper chart can either have the water level axis horizontal (Fig. 10.6) or vertical (Fig. 10.7). The chart can wrap once around a drum (Fig. 10.7) or take the form of a long chart wound from one reel to another (Fig. 10.6).

The third main difference concerns how the pen moves across the chart. In the simplest design, the full range of level change is represented by the width (or height) of the chart. To record different level ranges, different gear ratios or pulley ratios are used. A typical chart is 25 cm high and is usually marked in 2 mm divisions. If the gear-reduction ratio is 1 : 1, then every 2 mm movement of the float will move the pen 2 mm across the chart, covering a total level change of 0.25 m. With a reduction ratio of 20 : 1, the range is extended to 5 m, but each 2 mm movement of the pen now represents 4 cm float movement and this may not be good enough for some purposes; this is why the reversing-pen mechanism was developed.

By introducing a double-spiral groove in the rod that moves the pen (Fig. 10.6), the pen reverses its direction when it reaches the limits of the chart, allowing an unlimited range of level change to be accommodated without sacrificing level discrimination. The penalty for this is a record that is not immediately understandable

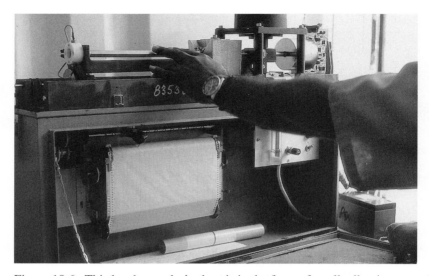

Figure 10.6. This level recorder's chart is in the form of a roll, allowing extended recording times. The level is recorded horizontally (the time vertically), the pen being moved by a shaft (visible at the top of the chart) with a double-spiral groove, allowing the pen to reverse direction at the edges of the paper. This allows a wide range of level changes to be recorded to high resolution. Although this method is frequently and most commonly used with float recorders, the one illustrated here is of the bubble type. The pressure of the gas (representing the depth of water) is converted into shaft rotation by a servo-motor controlled by a mechanical balance that senses the bubble pressure via a bellows.

and so, to aid in its interpretation, another pen marks the chart whenever a reversal of pen direction occurs, indicating if the level is rising or falling (Fig. 10.8). A reversal mechanism requires a larger float to overcome its extra friction. With a reversal mechanism, the range can still be varied through a selection of gears.

Accuracy starts at about ±2 mm for the small-range instrument, falling as the gear ratio (level range) increases. A well-maintained mechanism will perform better than a neglected instrument since friction will increase if it is not kept clean. (As with all pen recorders, the friction of the pen on the chart and the friction of the mechanical linkages have to be overcome.) The zero reference needs careful specification, but it is often arbitrary, such as a mark on a pipe, making it difficult to relate to levels at other points along a river. Height above mean sea level is preferred.

Time resolution depends on the speed of the chart. In the case of a drum, one revolution can be adjusted (again by changing gear ratios) to represent a time ranging from one day to one month, or longer, while for a reel-to-reel chart, from 2 to 60 mm per hour is selectable. To ensure appropriate timing accuracy, a third pen may be included to give a time-mark on the chart, triggered by the clock mechanism for, even though the chart is time-marked, the paper can change dimensions slightly

Figure 10.7. In this much simpler instrument, water level is recorded on the vertical axis (and time on the horizontal axis), the chart being a single sheet wrapped once around the rotating drum. The float moves the pen up and down the chart, the height of the paper (by choice of gear ratio) representing the maximum level change expected. When this is large, the resolution is low.

with humidity (this can also cause jamming if sprocket holes on the chart do not align precisely with sprocket wheels, for example through dampness). A refinement to the time-mark is to incorporate an independent electronic timer to trigger the pen.

A float turning a potentiometer

The simplest, and cheapest, way of converting float-movement into an electrical signal is to make it turn a potentiometer, which produces a resistance output proportional to water depth, this being converted into a voltage for logging. River level

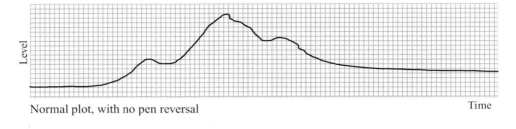

Normal plot, with no pen reversal   Time

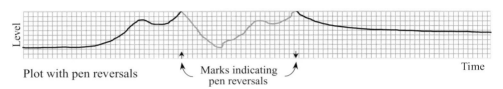

Plot with pen reversals   ↖ Marks indicating ↗   Time
                                     pen reversals

Figure 10.8. By reversing the direction of movement of the pen when it reaches the edges of the chart, a wide range of level changes can be recorded to high discrimination on a chart of practical dimensions. The mechanism that achieves this is a double spiral cut in the rod that moves the pen (visible in Fig. 10.6).

is one of the few variables that may justify a high-resolution sensor and logger because it is often necessary to measure water level changes of 1 or 2 mm over a range of 5 or 10 m. Even 1 mm in 1 m represents a resolution of 1 in 1000 (0.1%); 1 mm in 5 m is 0.02%, representing 12-bit resolution (Chapter 11).

A potentiometer can take several forms. It can rotate just once, through about 358°, or it can have a wiper that crosses the small gap (between the ends of the resistance track) and rotates continuously, or it can be of the multi-turn (10 to 20 turns) helically wound type. But a potentiometer in which the full range of level change is covered by just one rotation will only be able to achieve acceptable resolution when the change in level is small, say 1 m or less.

Where higher resolution is required, the same principle can be adopted as in the mechanical reversal of the pen of a chart recorder. In its electrical equivalent, the potentiometer rotates continuously, the wiper contact moving across the gap in the winding (Fig. 10.9). The float, in the example shown, turns a pulley with a circumference of 20 cm, which rotates a potentiometer once for each 200 mm change in level. As the potentiometer rotates its output rises until, as it approaches the gap in the winding, its reading is full scale. Thus, even at only 8-bit resolution, the 200 mm change can be measured to a resolution of better than 1 mm. After the gap is crossed, the reading returns to zero and starts rising again (if the water level is still rising), this going on indefinitely; thereby an infinite range is covered. To keep track of the overall (absolute) level, a second potentiometer, driven through a

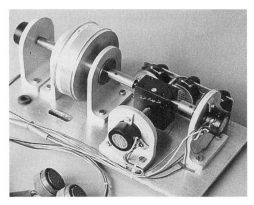

Figure 10.9. The simplest way of converting float movement into electrical form for logging is to make it turn a potentiometer. In the instrument shown (one of my earlier developments), the pulley (left) is 20 cm in circumference, causing the first potentiometer (right) to turn once for a level change of 20 cm. At the far left, a second identical potentiometer is also directly turned, but the gap in its winding is 180 degrees out of phase with the first, to allow the sensor to operate over the full 360 degrees. A 20 : 1 gearbox (the black cube in line with the pulley) drives a third potentiometer (foreground) once for every 20 × 20 cm or 4 m level change, so as to prevent errors if the level changes by more than 20 cm between recordings.

reduction gear box, rotates once over the full range of level change – or sufficiently slowly to avoid more than one rotation between loggings. A third potentiometer can be included with its gap 180° out of phase with the first potentiometer, to avoid any error near the gaps in the windings. By using three logger channels, a high level-resolution is obtainable over a large level change. Such a technique works well even with loggers using only 8-bit analogue-to-digital converters.

However, if shaft rotation can be converted directly into digital form within the sensor, all the analogue errors disappear. This is the rôle of *shaft encoders.*

Floats turning a shaft encoder

Some shaft encoders are 'incremental', meaning that as the shaft rotates a switch opens and closes, the angle between operations of the switch being proportional to a change in float level (of say 1 mm, or 1 cm). To take account of whether the float is moving up or down, two switches are needed, 180 degrees out of phase. Which switch operates before the other indicates the direction of rotation, the sensor electronics making the necessary computation and giving an output of absolute level in straight binary, binary-coded decimal (BCD) or the Gray code. (These codes, and their strengths and weaknesses, are described in the section on analogue-to-digital conversion in Chapter 11.) Because such sensors measure just *changes* in level, not absolute level, they need to be programmed at installation with information as to the current water level, thereafter keeping track of the changes automatically, unless the power is removed, when they must be reprogrammed. The switches can be mechanical, but most modern encoders sense rotation magnetically or optically, these methods having less mechanical resistance to float movement and a longer life. An incremental encoder, with electronics to keep count of the absolute level, can cover as wide a range of level change as required, to high discrimination, the shaft simply rotating many times.

Some shaft encoders, however, use a disc encoded directly in the Gray code. In such designs the level changes which can be resolved for a 360° rotation of the encoder shaft depend on the number of (concentric) tracks on the optical disc. If the number is large, say 16, a wide level range can be covered in just one revolution of the shaft, to high resolution. In such a case, a reduction gearbox may be necessary between the float-pulley and the shaft-encoder, so that only one revolution of the encoder results from the full range of level change. This is of no matter, except possibly for backlash in the gears. It is also possible to use a Gray-coded sensor, in a way similar to the multi-turn analogue potentiometer level sensor, by allowing it to rotate many times over the full range of level change, but this then loses track of the absolute level.

Tone transmitters

Little used today, but useful in the past, and still needed in some situations, is the float-operated sensor that generates a series of tones for transmission by radio or telephone, decoded by ear by an operator in a distant office. In 1960s and 70s models, the float turned cylinders that were mechanically encoded with 0 to 9, one cylinder driving the next via a 10 to 1 gear reduction. Typically there would be four cylinders, representing centimetres, decimetres, metres and tens of metres. A microswitch, moved by a motor, sensed each coded cylinder in turn, generating a string of four groups of pulses, converted into voice-frequency tones for transmission. For example, a depth of 13.65 m would be represented by bursts of 1 tone, 3 tones, 6 tones and 5 tones, with spaces between each group.

More recently it has become possible to convert the BCD, or similar, codes generated by absolute shaft encoders into tones, thereby avoiding the complex mechanics of the early designs. Simulated voice announcements can also now be generated for transmission. In situations where just a few levels are to be measured and where there is no need for computer-based systems, this kind of instrument is an effective and simple way of using modern technology.

Combined methods

It is quite common to see two of the above techniques combined in one instrument, typically a shaft encoder or potentiometer being added to a paper chart recorder, thereby producing an on-site recording as well as being able to log or telemeter the measurements. Sometimes a shaft encoder and a pressure sensor are operated at the same site, one backing up the other in the event that one sensor fails, or the river freezes.

Sensing water depth by pressure

The bubble method

Nitrogen from a cylinder (or air compressed from a pump) is allowed to bubble slowly out of the end of a tube, the end being fixed in the river at the lowest flow level or suspended in a borehole below the lowest water level. The pressure of gas that causes gentle bubbling is equal to the pressure of the water above the outlet of the tube. By measuring the pressure, the depth of water can be calculated. The nitrogen gas must first be reduced in pressure from the high pressure in the cylinder. An adjustable valve then controls its speed of flow, the rate of flow being checked visually by bubbling the gas through a small water container on the instrument before it is piped to the vent point under the water. There are a number of methods of sensing the pressure and they fall into two groups – electromechanical balances and electronic diaphragm sensors.

Figure 10.6 shows an electromechanical balance in which the gas pressure, representing the river depth, operates a bellows that applies pressure to one side of a balance arm, while on the other side a weight is moved along the arm (by servo-motor) until balance is achieved. Balance is detected by an electrical contact on the arm, which stops the servo-motor. Movement of the servo-mechanism is transferred, via gears, to a pen which records a trace on a paper chart (with a reversal mechanism to enlarge the scale). A damping device, with a time constant of about 1 minute, is included in the tube to the bellows, to prevent waves on the water from causing the gas pressure to fluctuate rapidly. The accuracy claimed by manufacturers ranges from ±0.07% to ±0.1% of full range. These pressure sensors are not, therefore, as good as the better float-operated instruments, although when measuring level changes of a few metres the resolution is generally adequate, and even over large ranges this sort of accuracy is often sufficient. In a similar design, a mercury manometer is used in place of the bellows. A self-cleaning facility may be included that periodically applies a high gas pressure to the tube to blow out sediments.

By electronic pressure sensor

An alternative, and simpler, way of measuring water pressure is offered by electronic sensors, fixed in the river (or borehole or sea) at the lowest water level expected.

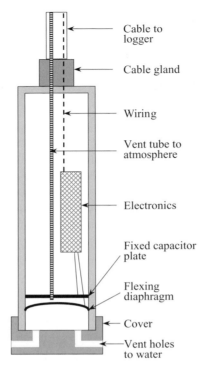

Cable to
logger

Cable gland

Wiring

Vent tube to
atmosphere

Electronics

Fixed capacitor
plate

Flexing
diaphragm

Cover

Vent holes
to water

Figure 10.10. The pressure of the water overhead, and thus its depth, can be measured by sensing the deflection of a diaphragm, one side being exposed to the water pressure, the other to atmospheric pressure. In this schematic figure, the movement of the diaphragm is sensed by measuring the capacitance between it and a fixed plate, the associated electronics being housed close by in the body of the sensor. Some sensors use different methods of measuring the deflection.

Figure 10.10 illustrates the principle of this method, in which a diaphragm, in a tube, has the water pressure applied to one side, the other side being vented to atmospheric pressure (through its cable to the water surface). If the pressures on both sides of the diaphragm are the same (when the water level is zero), the diaphragm is flat and unstressed. As the depth increases, the diaphragm flexes under the increasing pressure, the deformation being measured by straingauges fixed to the surface of the diaphragm – generally in the form of a bridge circuit. Today, the diaphragm is often of silicon, the straingauges being etched directly onto the surface using microelectronic techniques. Some sensors now use ceramic as the diaphragm material, which has better elastic properties than metal and a minimum of hysteresis, the deflection being measured capacitively (as Fig. 10.10 illustrates). Figure 10.11 shows an actual sensor. To prevent damage due to shock waves in the water and to remove very rapid fluctuations in level, the nose cap may contain a

Figure 10.11. With its protective cap (left) removed, the diaphragm of this pressure sensor is visible (right); this flexes under the changing weight of the water above it, giving an electrical output proportional to depth. The cap contains vent holes to allow the water pressure to be sensed while protecting the diaphragm from damage.

damper to suppress pressure transients, and filters to prevent sediment getting into the diaphragm cavity.

An advantage of using either electronic pressure sensors or bubble methods is that a stilling well is not essential and they can also be used in boreholes of small diameter and of any depth. Electronic pressure sensors, however, have the advantage over bubble-type systems of much greater simplicity. Some instruments, however, combine the two techniques, the gas pressure of the bubbler system being measured by an electronic sensor instead of mechanically. The advantages of this hybrid design are that a pressure sensor is simpler than a mechanical balance and that the electronic sensor is housed safely above water in the recorder housing, in the dry rather than in the river. Being close to the logger or chart recorder, long (expensive) cables are also avoided. The analogue voltage output from diaphragm sensors can, after suitable processing, drive an electrical chart recorder, display the level visually, be recorded on a logger or be telemetered to a distant base – or a combination of these possibilities.

There are, however, disadvantages to the diaphragm sensor. At best an accuracy of ±0.05%, and more typically ±0.1%, of full range is achievable. (This makes them very similar in performance to mechanical bubbler instruments.) If additional errors are introduced in the sensor electronics, and in the analogue-to-digital conversion process, a more realistic accuracy might be closer to ±0.3%. Thus, while the resolution of a flexing diaphragm is infinite, if the level change to be measured is, say, 10 m, the best accuracy achievable is about ±5 mm, and a more realistic performance is ±20 mm. Pressure sensors are thus not as good as the best float-operated instruments, and, although it is less true of today's pressure sensors, there might be a long-term calibration drift. Nevertheless, because of their convenience and relative cheapness they are becoming widely used.

Sensing water level by ultrasonics

Not widely used for sensing water level in environmental applications are ultrasonic sounding systems (compare the method for measuring snow depth in Chapter 8), which sense the time of flight of an ultrasonic pulse from a transmitter to the surface of the water and back to a receiver, the sensor either being above the water looking down or in the water looking up. More details of this are given later in this chapter under ultrasonic flow gauging.

Measuring water current velocity

A brief history

The first river current meter was probably made by the Italian physician, Dr Santorio, around 1610, and consisted of a plate held vertically in the stream on a balance arm, the pressure of the water on it being measured by a movable weight (Frazier 1969) although the mathematical tools to convert the pressure into velocity had not yet been devised. Robert Hooke also designed a current meter, about 70 years later, for measuring the speed of ships through the sea, as well as the current of rivers. This was based on a propeller, reconstructed by Frazier (1969), which looked remarkably like a present-day current-meter propeller. Paolo Frisi, professor of mathematics at the University of Milan in the eighteenth century, described a paddle wheel for measuring surface velocity (Leliavsky 1951, 1965). It should be remembered that at this time even the origin of rivers was uncertain; Frisi did in fact believe that they derived their origin from the rain, although he was wrong in the belief, shared by others, that river currents changed in speed parabolically with depth (Herschy & Fairbridge 1998).

The Pitot tube (Fig. 5.8) was developed specifically to measure river velocity (De Pitot 1732). However, Pitot was unable to calculate correctly the velocity of the current from the height to which the water rose in his instrument, because the value for g had not yet been determined and so the correct equation $v = (2gh)^{1/2}$ could not be derived for the Pitot tube. Nevertheless the instrument was very useful in determining the relative velocities of currents at different depths, and proved the fallacy of the formerly held view that it changed parabolically with depth; exposing the error of this view was itself an important contribution made by Pitot.

Finally, Reinhard Woltman, a German engineer, designed a current meter in 1790 in which two angled plates on a cross-arm turned a shaft connected to a revolution counter (Frazier 1969). According to Biswas (1970), for many years after Woltman's death each improved version was still called a Woltman current meter out of courtesy. It was, in effect, a crude propeller.

Mechanical current meters

A modern mechanical current meter (Fig. 10.12), senses flow through the water impinging on a propeller, or impeller – a better word might be 'helicoid' since it is helix-shaped. The helicoid rotates a shaft on which there is a rotation detector – usually a magnet operating a reed switch, although optical methods are also now used. The pulses produced by the meter are counted, either by an electromechanical counter (which is limited in the maximum speed to which it can respond – this is around 10 counts per second) or by an electronic counter (which has no speed limit). In a design that is an alternative to pulsed outputs, but is again comparable with a wind speed sensor, the propeller rotation is converted into a varying DC voltage by a small generator. This allows a continuous strip-chart record to be made of the flow. Although current meters can be permanently installed in a river, they are normally used as portable instruments operated manually – hand-held on graduated *wading rods* while the operator walks across a shallow river, or for deeper waters suspended from a jib on a bridge or boat or from a cableway.

Depending on whether light or strong currents are to be measured, the size of the instrument can be varied, the smaller measuring currents up to about 5 m s^{-1}, the larger up to 10 m s^{-1}, typically with accuracies of about $\pm2\%$ of full scale. The helicoid propellers range in diameter from 5 to 12 cm and are made of metal or plastic. In addition they vary in pitch, expressed in metres. Pitch is not a characteristic that can be measured physically on the helicoid itself; a pitch of, say, 0.1 m

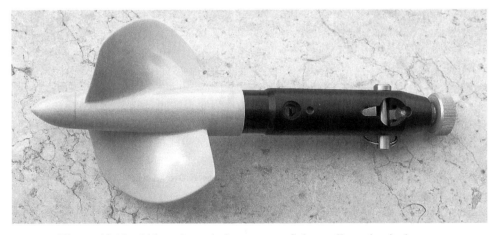

Figure 10.12. Although made in a range of sizes, all mechanical current meters sense the movement of water by means of a propeller in the shape of a helicoid, generally with two or three blades, the pitch varying to suit different velocities. A magnetic reed or an optical switch produces pulse signals as the blades rotate.

expresses the fact that one revolution will be produced by the passage of 10 cm of water. One revolution per second would, in this example, mean that the current velocity is 0.1 m s^{-1}. Pitches vary from 0.1 to 1.0 m and have to be calibrated in currents of known velocity. Pitch is a characteristic of the shape of the helicoid and once a particular shape has been tested it is assumed that all helicoids of that design will behave in a similar way. It is usually possible to interchange impellers on the same instrument to measure currents of different speeds. There is a minimum speed below which the instrument will not respond, typically around 0.03 to 0.05 m s^{-1}, limited by friction at the bearings. Also, as with helicoid wind sensors, the response of a current meter is dependent on the angle of attack of the current and can be assumed to have a cosine response over a small off-centre alignment. When suspended from a jib or cableway, meters are provided with a tail that points them into the current. In stronger currents it is also necessary to add weight to the meter so as to keep its suspension cable reasonably vertical. The *sinker weights*, which vary from 5 to 10 kg, fit just below the meter and are streamlined.

Ultrasonic current meters

It is possible to measure the flow of water in a river by the Doppler shift in frequency of a continuous sound wave emitted upstream, reflections from bubbles and small particles in the water providing a returned signal of shifted frequency. Hand-held instruments using this principle are now available, in the style of a propeller meter, although they are less common than the electromagnetic meters described next. Because the method operates by emitting a cone of sound into the stream of water, it senses the velocity of the cone of water, not, like the propeller-type, just the water with which the blades come into contact. Typically an ultrasonic meter will sense water velocity up to about one metre upstream within a cone of about 15 to 30°. It is not, therefore, so position specific as a propeller. Because flow in rivers contains eddies, an instrument of this type is designed to take many readings, up to 1000, and to determine the mean velocity from a potentially quite wide velocity spectrum. In contrast, propeller sensors have a natural averaging response, being driven by the sum total of the eddies impinging on the blades, rapid changes in rotation being prevented by the inertia of the mechanism.

The accuracy quoted by manufacturers is between ±2% and ±5% of the measured velocity (not of full scale) and the range covered is typically from 0.1 to 5 m s^{-1}. The resolution is around 1 mm s^{-1}.

Impeller and Doppler current meters are not intended for permanent installation in a river. Where continuous current metering is useful, however, a larger design of Doppler instrument is now available for permanent installation on the river bed. As there are no moving parts in ultrasonic systems and thus no mechanical wear-and-tear, they are better suited to continuous operation than their mechanical counterparts.

Electromagnetic current meters

Michael Faraday showed that a conductor moving in a magnetic field has a voltage induced across it proportional to the length and speed of the conductor and to the strength of the field, such that $E = BVl$, where E is the induced voltage (emf), B is the strength of the magnetic field in tesla (symbol T; 1 tesla = 1 weber m^{-2}), V is the average velocity of the conductor in metres per second and l is the length of the conductor in metres. By making the conductor the river water, the voltage induced in it through its motion enables its velocity to be measured. Several hand-held current meters are now based on this concept.

Figure 10.13 shows how water 'filaments' cut through the magnetic lines of force as the water moves, and so generate a voltage across its length. In electromagnetic flowmeters, the magnetic field is generated by a coil powered by a square-wave

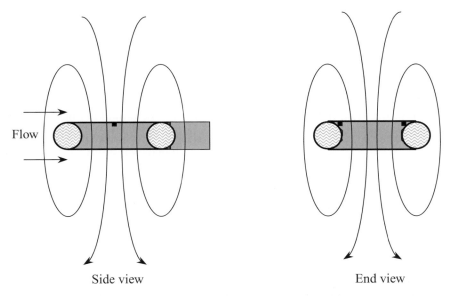

Side view   End view

Figure 10.13. In this electromagnetic current meter, an epoxy resin body, about 2 cm thick, 5 cm wide and 20 cm long, is held so as to face into the water flow. The rounded end contains a coil (light shading), in the plane of the flow, that induces a vertical AC magnetic field through which the water flows, inducing a voltage across the conducting 'filaments' of water; this is picked up by the two titanium sensing electrodes (small black squares) and amplified, giving a measure of current velocity.

current with a frequency of about 130 Hz. By using an AC signal, and thus a varying magnetic field (rather than a DC current and a fixed field), it is easier to amplify the resulting AC signal generated by the movement of the water through the field. Small DC signals are more difficult to handle than are AC, suffering more from zero drift and random noise.

Readings are taken several times a second and averaged over periods from 5 to 100 seconds. At the upper end of the averaging periods, the accuracy is in the region of $\pm 0.5\%$ of the reading (not of full scale). The range usually covered is up to ± 5 m s^{-1}, although instruments measuring up to ± 10 m s^{-1} are available (forward and backward directions are measurable because if the direction of flow reverses so too does the polarity of the induced voltage). Although the more usual instrument has just one pair of sensing electrodes, by introducing a second pair at right angles to the first it is possible to separate the velocity into two vectors at 90° to each other and thus to measure both the velocity and the direction of the current. Sensors to measure current in three dimensions are also manufactured.

As with ultrasonic meters, the volume of water being measured is less clearly defined than in the case of a propeller sensor, since the conductor (the water), unlike a piece of copper wire, is not clearly bounded but extends out beyond the electrodes into the surrounding water; this is in part dependent on the conductivity of the water. So it is necessary to calibrate sensors of this type in the laboratory in order to arrive at a relationship between current velocity and induced voltage; once this is done for a particular design, however, it is the same for all instruments of that design, although conductivity-dependent.

Because the voltage generated by the movement of the water is measured by an amplifier with high input impedance, no electrical load is placed on the source of voltage so very little current is drawn from it and, as a result, it is the open-circuit voltage that is measured. In consequence, the conductivity of the water (which can be visualised as a resistor in series with the source) has no effect on the readings, although below a conductivity of about 5 μS cm^{-1} its functioning may be impaired. As the current velocity approaches zero, the generated voltage becomes exceedingly small and at velocities below a few millimetres a second the exact zero point becomes uncertain.

Measuring river discharge

Apart from knowing the depth of water in a river, which is an important variable in itself (for example, in flood warning), the quantity of water flowing, the *volumetric flow* or *discharge*, is equally important. Discharge is expressed in cubic metres per second (m^3 s^{-1}), abbreviated to 'cumecs', although at very low flows in small streams litres per second might be more convenient.

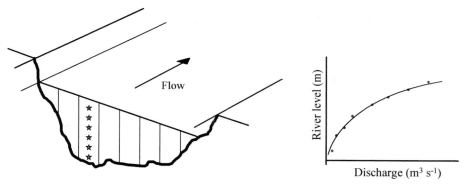

Figure 10.14. To produce a stage-discharge rating curve (right-hand figure), the river is divided into about 20 to 30 verticals (only eight are shown here, left figure), and in each segment the velocity is measured at several depths (star positions) with a current meter, producing an average discharge estimate for each segment. These are summed to give a total discharge for the river. This procedure must be repeated for a range of river levels. From the readings, a plot of discharge against depth – a rating curve – is drawn.

Stage-discharge rating curves

If the cross-sectional profile of a river is known, a measurement of the depth of water (the *stage*) at any moment will give the area of water flowing and, if the velocity is also known, the discharge can be calculated by multiplying the cross-sectional area of water by its mean velocity. In practice this is done by using a current meter to measure the velocity at several depths at each of a set of *verticals* across the width of the river (Fig. 10.14, on the left). In all but the smallest rivers, from 20 to 30 verticals are selected, the spacing between them generally being chosen so that they each represent a segment of about equal flow. No segment should contain more than about 10% of the overall flow. In rivers wider than 300 m, the verticals are usually spaced equidistantly.

The total discharge is equal to the mean velocity in the segment times the width of the segment times the depth of the water. The total discharge of the river is the sum of each segment's flow. From the measurement of discharge and depth, a *rating curve* is drawn smoothly through the points obtained by current metering (Fig. 10.14, on the right), allowing discharge, thereafter, to be measured simply by measuring the depth of the river (Herschy 1995).

As explained in the previous section, depending on the size and depth of the river the current meter may be hand-held, the operator wading across the river, or it may be deployed from a bridge or a boat, the meter being lowered by hand or from a winch with a jib. Measurements can also be made by suspending the current meter from a permanent cableway installed across the river from bank to bank (Fig. 10.15).

Figure 10.15. If wading in a river is not possible and there is no boat available or bridge conveniently in place, current metering must be done by suspending a current meter (Fig. 10.12) from a cableway. The meter is suspended from the fixed upper cable on a crane trolley, the two tow cables drawing the meter across the river, where it is stopped at several points and lowered to several depths. When suspended rather than being hand-held on a rod, the meter is fixed to a sinker with a stabilising tailpiece to keep it at the required depth and facing into the current.

If the profile of a river changes with time, the rating curve also changes. This is particularly the case after flooding, when the banks and bed may be suddenly and perhaps considerably altered. When this occurs, the field work has to be repeated and a new curve produced, which can be an expensive process involving a team of field workers at remote sites for days. Normally the relationship should be checked several times a year.

While it is fairly easy to obtain a rating curve for low to medium flows, it is often impossible to get to a river to make current meter measurements when flooding is occurring, and so it is usually necessary to extend the stage–velocity curve up to flood stage, beyond the level to which current metering has been possible. This procedure is based on a study of flow characteristics (Corbett *et al.* 1945), which includes allowances for the slope of the river, the presence of a contraction or a widening downstream, bends in the river, downstream tributaries, dams and plant growth. Even so, at very high flows, when the river overflows the banks, it may be difficult to establish what the flow is.

An alternative to actual current metering is to make measurements after the flood event. The most commonly used technique, the *slope-area* method (Dalrymple & Benson 1967, Riggs 1976), involves surveying the slope of the debris left by the flood (the *silt line*) and obtaining an estimate of the *bed-roughness coefficient*, through a knowledge of the bed material and the size distribution of the bedload. To compute river discharge, the *Manning formula* is often used (Barnes & Davidian 1978, Limerinos 1970). Here the discharge $Q = AR_h^{2/3}S^{1/2}n^{-1}$ where A is the cross-sectional area, R_h is the hydraulic radius (the cross-sectional area of the river divided by its wetted perimeter), n is the bed-roughness coefficient and S is the friction slope (the change of water-surface level over a distance). As this method does not involve the use of instruments, it will not be expanded on further.

Weirs and flumes

To overcome the problem of changing profiles, it is sometimes possible to route the river through an artificial structure, the shape of which does not change with time. Knowledge of the precise shape also makes it possible to calculate a theoretical rating curve (Horton 1907).

Weirs

A weir is a dam constructed across a stream or river, controlling the depth rather than the width of the channel; the water flows either through a hole cut in a plate across the channel or it may flow across the full width of the channel, as Fig. 10.16 illustrates. The shape of the outflow hole is usually either rectangular or V-shaped, the angle of the triangular V-notch varying from 90° downwards. The most usual alternative angles are those known as 'half-ninety' (53° 8′) and 'quarter-ninety' (28° 4′), the half and quarter referring not to the angle but to the area of the notch relative to one of 90° – and thus to the relative flows they can deal with. V-notch weirs allow small flows to be more accurately measured than rectangular-notch weirs. Although provision is usually made to prevent the build-up of debris, any blockage of the outflow will affect the readings and the narrower-angled notches are more prone to this. The measurement of flow is achieved by measuring the depth of water behind the weir plate, the velocity of the outflow being theoretically equal to $(2gh)^{1/2}$, where g is the acceleration due to gravity and h the measured depth of water above the crest of the weir. As the area of the notch through which the water flows is known precisely, the flow can be calculated from a measurement of the head only (Ackers *et al.* 1978, Herschy 1995).

The most commonly used weirs are the broad-crested (square-edged) weir, the rectangular-profile weir and the triangular-profile weir, these being designed individually to suit the channel and flow conditions. Although weirs are more suitable for small streams than large rivers, several designs are available for larger channels and natural rivers; they are usually constructed in concrete.

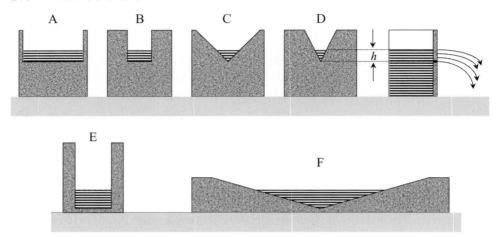

Figure 10.16. The height of water *h* behind a weir (A to D) gives an accurate measure of the outflow velocity and, since the area of the notch is known, the discharge can be calculated. A is a weir extending across the full width of the rectangular approach channel while B, C and D are contracted weirs with different cross-sectional shapes to suit different flow rates (see text). For correct operation there must be atmospheric pressure under the sheet of water (the nappe) flowing over the weir. E and F are flumes, rectangular and trapezoidal respectively, which control the shape of the channel, stabilising the rating curve. The advantage of the trapezoidal design is that the range of measurable flows is wide.

Flumes

A flume is a channel through which a river passes. A river bed is a natural flume, but the term is usually applied to an artificial channel. The cross-section of a flume is usually either rectangular or trapezoidal (Fig. 10.16) and as Figs. 10.17 and 10.18 show, they vary in size and circumstance considerably.

Even though laboratory rating curves have been developed for flumes and weirs, it is usual to confirm their actual performance by field tests. These tests might use current meters or other methods such as dilution gauging, which we now discuss.

Dilution gauging

Where river flow is very turbulent, such as in steep streams in mountains, the use of a current meter may be difficult or impossible and may give unreliable estimates. In such situations, *dilution gauging* may be more suitable (Water Research Association 1970, Gilman 1977).

A chemical tracer, in the form of a solution of a salt, for example an iodide or a lithium salt, of known concentration C_1, is injected into the river at a known and constant rate q. After a time, sufficiently far downstream to allow thorough

Figure 10.17. This small flume at the Institute of Hydrology experimental catchment in Plynlimon, central Wales, is designed to cope with the fast-flowing steep streams encountered in upland areas, with provision to slow the flow temporarily and to deal with bedload movement.

mixing, the tracer concentration will reach a plateau C_2, which can be measured. Knowing these values allows the quantity of water flowing in the river, Q, to be calculated from $Q = qC_1/C_2$. If the tracer occurs naturally in the river, its natural concentration C_o must be subtracted from the measured concentration C_2 of the samples. This may only need to be done once, although it is wiser to check it at different flows by sampling upstream of the injection site. To ensure a constant injection rate, a 'Mariotte' bottle is used; this gives a constant head of liquid as the level in the bottle falls (Fig. 10.19). The flow rate is checked by weighing a timed release. To ensure that the plateau of concentration has been reached, samples need to be taken at intervals over a sufficiently long period (a few hours), this either being done manually or by automatically timed samplers. The full process of injection and sampling can be completely automated so that when the river reaches a preset level the injection and sampling can be started. At the same time, the level of the

Figure 10.18. In contrast to the previous figure, this trapezoidal flume in the path of a wadi in Oman has to measure over a range of flows, from the low level seen in the figure up to large, sudden, flood events. The instrument stilling well is visible on the end of the wall of the flume (to the left of the two men).

river must also be measured so that a rating curve can be derived. When the rating of the river is completed, the dilution gauging can be stopped. An automatic dilution meter, which uses common salt as the tracer and which can measure up to 4 m^3 s^{-1}, is now available.

Ultrasonic flow gauging

As well as being the basis of hand-held current meters, ultrasonics can also be used in permanent installations in large rivers to give a continuous measurement of volumetric discharge. But instead of sensing the velocity at a single point, it is measured across the full width of the river, in this case not by Doppler shift but by the flight time of an ultrasonic pulse. In its simplest configuration, two piezo-electric ultrasonic transmitter–receivers are fixed in the river on opposite banks, one upstream of the other (Fig. 10.20(a)). Both simultaneously emit an ultrasonic pulse and then immediately switch functions to receive the pulse arriving from the transducer on the opposite bank, the pulses propagating diagonally across the river in opposite directions. The pulse travelling downstream, with the current, arrives more quickly than that going upstream. The flight times (t_d and t_u) are measured.

If the length of the sound path is L metres, then $t_u = L/c - v$ and $t_d = L/c + v$, where v is the velocity component of the current in the direction of the sound path

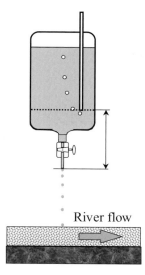

River flow

Figure 10.19. To ensure a constant injection rate of the tracer solution into the river for dilution gauging, a Mariotte bottle can be used: the constant pressure head, shown by the double-arrow pointer, is achieved by having a vent tube to the atmosphere as shown.

and c is the speed of sound (at the temperature and water density at the time). Taking the difference of the flight times,

$$t_u - t_d = (L/c - v) - (L/c + v)$$

This reduces to

$$v = (L/2)(1/t_d - 1/t_u)$$

which is independent of c. Furthermore, $V = v/\cos a$, where V is the velocity of the stream parallel to the banks and a the angle of the sound path relative to the bank.

Taking the sum of the times of flight $t_u + t_d$ gives

$$c = (L/2)(1/t_d + 1/t_u)$$

which is the actual speed of sound at that moment under the particular conditions. This is useful if river depth is also measured by an (upward-looking) ultrasonic sensor on the river bed – for, in addition to measuring the current, the river depth also has to be measured so as to convert the velocity data into volumetric discharge. While it is possible to use an alternative method to measure river level, it can be convenient to use an ultrasonic method since suitable circuits are already to hand; the availability of c from the current measurements is then useful for the adjustment of the depth reading.

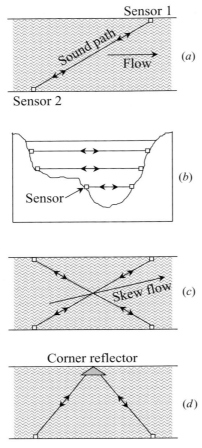

Figure 10.20. Ultrasonic river flow gauging. (*a*) Plan view: the simplest arrange-ment uses one pair of sensors (small white rectangles) at one depth on opposite banks. (*b*) Cross-section: in deeper rivers several levels may have to be measured, with pairs of sensors at each depth. (*c*) Plan view: for better accuracy, particularly where the flow of the river might be skew, on a curved reach, a crossed-path array is used. If having sensors on two banks is not practical, a reflector can be used (*d*).

If the river is deep, it is necessary to install pairs of ultrasonic transducers at several depths (Fig. 10.20(*b*)), each path being measured in succession (Davis 1979). Often the exact angle of flow of the water may not be known, or it may be known to be skew (not parallel to the banks). To deal with this problem, a crossed-path array is used (Fig. 10.20(*c*)), the two readings helping to cancel out any non-parallel flow.

It can sometimes be inconvenient to have transducers on both sides of the river. To get around this difficulty, a reflector can be installed on the far bank, instead of a transducer (Fig. 10.20(*d*)), the sound echoing off it to the second transducer, further along the same bank. A secondary advantage of using a reflector is that it produces an effect similar to a crossed-path array and so helps overcome any skew in the flow. A corner reflector is used because it is not so critical to alignment with the sound beam as a flat plate is. (The reflector has to be kept clean in order to operate reliably.) The angle of the sound beam is a cone of about 10°, although there is spread up to ±25° with some additional side-lobes in the polar diagram (Herschy 1979a).

The sound frequency used ranges from 30 kHz to 1 MHz, depending on path length, the lower frequencies being less affected by sediment suspended in the water and so more suited to wider rivers and those with heavy sediment loads. The radiated audio power ranges from 500 watts for rivers under 50 metres width to 1 kW for anything wider, although these are pulsed, not continuous powers. Nevertheless this makes the technique higher in power consumption than most instruments so far considered and mains power is almost an essential, although isolated examples use large 12 volt batteries, perhaps charged from the mains or by wind power. It is a heavy-weight system in all senses, including cost; this starts at £15 000, rising to well over £50 000. Because cost does not increase with river width, it is best used for wide rivers (rather than the electromagnetic method described below in which cost is very much width-dependent).

It is difficult to be precise about accuracy because it depends on the width and depth of the river, on whether it is rapid or slow moving, turbulent or placid and on how many levels of acoustic path are used (Herschy 1979b). But manufacturers quote around 5% of full scale for the current velocity measurement. Because the method measures velocity and depth continuously (or very frequently) it will inevitably produce better results than the much cheaper and simpler indirect techniques that measure just the level and rely on empirical stage–discharge curves to convert level to volumetric flow. But this higher precision is obtained only at higher cost and so the technique is something of a rarity.

Nor is it suited to all situations (Walker 1979). The growth of plants can affect the passage of the sound, as can the passage of boats, the entrapment of air bubbles or variations in the density of the water owing to temperature gradients or salinity. Nor can it be used in shallow water or, for the same reason, near to the bed or the surface, because of reflections from them. An alternative, free of these problems, but only feasible for the smaller river, is the electromagnetic method.

Electromagnetic flow gauging

A brief history

The Faraday principle (Fig. 10.13) as applied to hand-held current meters (see earlier) can also be used to measure the velocity of the complete river, continuously and automatically. Indeed it was Faraday himself who noted that water flowing in the earth's magnetic field induces a voltage which can be detected by two electrodes (Faraday 1832). The induced voltage in each filament of water is directly proportional to the average velocity of the water as it moves through the vertical component of the Earth's magnetic field.

This principle was used in 1953 to measure the flow through the Straits of Dover (Bowden 1956), using a telephone line to pick up the voltage induced on the French side of the Channel. It was also used two years later to measure the tidal flow in the River Humber (Cox 1956). However, the extremely low potentials that are generated (5 μV per metre width of river) and the fact that the signal is DC prevent the method being used in small rivers, owing to the swamping effects of the much higher levels of interference that occur. These are caused by electrochemical and thermoelectric effects, by mains 'hum', by radio signals and by boats, as well as by noise originating in the amplifier electronics.

A practical system

With the advance of electronics in the 1970s, the development began of a technique that uses artificially induced, much larger, and alternating magnetic fields (Fig. 10.21) to measure river flow electromagnetically (Herschy & Newman 1974). In principle it is identical to that of the electromagnetic hand-held current meter, simply scaled up. For the same reasons, of better zero stability and immunity to noise, the magnetic field is again induced by an AC signal.

However, the scale is hugely different: the coil has to be slightly larger than the width of the river and about as long as it is wide. It is usually laid on the bed of the river, although, if circumstances permit, it can be over the water. While the frequency of the square wave used in the small hand-held current meters is around 130 Hz, the frequency here has to be low enough to avoid any capacitive and inductive pick-up by the electrodes, but high enough to allow differentiation from noise. This turns out to be around 1 Hz. The coils that carry this square wave are from 50 to 300 turns of insulated cable of about 3.5 mm diameter, carrying a pulse of 3 to 5 amps at a voltage of 24–240 volts with a power of 5–1000 watts. Because the coil is an inductive load, the magnetic energy stored in it has to be dissipated at each reversal of pulse polarity. This is done using the conventional capacitor and diode circuits, which results in a back emf voltage spike at the moment of reversal. This is allowed to die away, and the magnetic field to stabilise, before the signal voltage (the induced voltage across the river) is sampled.

The signal is picked up by two stainless steel strips, laid on opposite banks. In this way the full depth of the river is measured, unlike in the ultrasonic method where discrete levels only are sensed. The construction is simpler if the river is channelled through a flume or vertically sided containment.

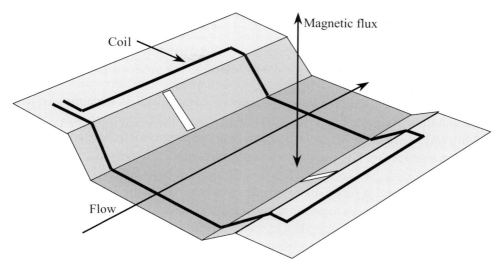

Figure 10.21. Similar in principle to a hand-held electromagnetic current meter, but of greatly increased scale, an electromagnetic river-flow gauging station generates an AC magnetic field spanning the whole river, the flow of water through it generating a voltage from bank to bank which is picked up by the two electrodes (the white rectangles), the amplitude of the signal being proportional to velocity. Knowing the cross-sectional area of the river, through a measure of its depth, the discharge can be computed. It is best suited to smaller streams.

The electrical conductivity of the river bed can be quite high, thereby reducing the signal level by drawing current from the 'source' and so producing a voltage drop across the source's internal resistance. There are two ways to deal with this. Either the bed can be insulated from the river by laying a plastic sheet beneath the coil and under the signal pick-up strips, or the bed and river-water conductivities can both be measured and a correction made. The latter is less desirable than the use of an insulating sheet because if the bed conductivity is high the signal level may be reduced so much as to make its reliable detection from amongst the noise difficult.

Evaluated against several other methods of volumetric flow measurement, the electromagnetic method shows an accuracy of ±2% to ±5% of the actual flow. However, owing to the complexity of the equipment and the civil engineering requirements of installation, costs range from £20 000 upwards. As with the ultrasonic method, the electromagnetic technique is only used where the highest precision is required. Throughout the world, the velocity–area method using a current meter, combined with a stage–discharge curve, remains the simplest and by far the most used technique to measure volumetric flow, and where the rating curve is stable, it is probably as good as any for most practical purposes.

Measuring water quality

The variables

The term *water quality* includes the bacterial content of water, the presence of oil in water and a host of ions, in particular calcium, magnesium, ammonium, nitrite, fluoride, phosphate, sodium, chloride and potassium as well as trace metals such as aluminium, iron, copper and lead. But, in practice, there are just about six water quality variables that are of regular and universal concern, and it is these on which we will concentrate here. They are: water temperature, conductivity (which merges into salinity), pH, dissolved oxygen (DO), biochemical (or biological) oxygen demand (BOD) and suspended solids (merging into turbidity). The many other water quality variables are related more to public health, land-fill monitoring, industrial and waste water and potable water rather than to natural rivers and groundwater.

Method of measurement

The options

There are six principal ways of measuring the quality of natural waters.

- Portable instruments can be submerged in the water and readings taken manually on site.
- Samples can be collected manually and measured immediately *in situ* with portable instruments.
- A sample can be collected manually in a container and taken back to a laboratory for tests.
- Samples can be collected automatically for later laboratory analysis.
- Submerged instruments can be left unattended and their measurements logged or telemetered automatically.
- Water can be pumped from a river on a continuous or intermittent basis into a hut or small self-contained unit containing automatic sensing and logging systems.

An intercomparison of methods

Some variables cannot be measured in the laboratory; temperature, for example, must obviously be measured in real time, preferably directly in the water. But other variables such as pH and DO may also be affected by storage, particularly if it is long term in bottles, owing to interaction with the container, with carbon dioxide in any air in the container and through biological activity, although these can be reduced by the correct choice of materials and, at some expense, by cooling. Filling the container completely in order to exclude air also helps. Pumping can also affect the sample: heavy sediment may not be transported fully, the temperature will change and pressure changes due to pumping may affect dissolved gases such as oxygen.

In some cases, the only practical method is to collect a sample. For such situations, special samplers are available that open at the required instant or take a sample at the appropriate depth. If the samples are measured quickly, the results will generally be acceptable. Automatic samplers, containing perhaps 24 bottles under vacuum, or with a peristaltic pump, are also used; a timer opens the bottles, or starts the pump, at preset intervals, allowing the water to be drawn in. Some designs allow samples to be collected when preset events occur, such as when a given river velocity is exceeded. Pumped systems may have to back-purge the tubes prior to taking a sample, so as to displace the water retained since taking the previous sample. But sample collecting will only suit those variables that do not deteriorate or change under storage.

The submersion of sensors directly in the river, borehole or sea avoids many of these problems, but if the sensors are left to run unattended, many of them quickly lose their calibration (within days) – much more quickly than any of the meteorological or water quantity sensors. If the sensors are deployed out of the river in a hut (or in a more compact container) with the water pumped to them, it is possible to clean and recalibrate them automatically, although this is very complex and expensive. Which method is used will depend on the variables being measured and on the nature and source of the water. For example, a submerged pH sensor may perform perfectly for months in the stable cool water of a deep borehole, away from light and algae growth, whereas it might drift off calibration in a few days in a tropical river.

If samples are returned to a laboratory for test, the same sensors can still be used as in the field, although there is the option of using different techniques which may be more accurate, such as conventional chemical analysis by titration or by the *colorimetric* method, in which reagents are added to the water sample, producing colour changes that are measured manually by eye using a colour chart or automatically, and more accurately, by spectrophotometer. For trace metals, a method known as *voltametric analysis* can be used. These laboratory methods, however, are not discussed further here.

By whatever means water quality sensors are deployed, the same types of sensor are used, and it is the sensors themselves that are the subject of the remainder of this chapter.

Temperature

Water temperature is important not only because it is a crucial climatic variable in the case of sea surface temperatures but also because it affects the metabolism of organisms and the speeds of chemical processes. It also needs to be known in order to correct the readings of some of the other sensors.

It is possible to measure water temperature manually with mercury-in-glass thermometers, and this is indeed one of the methods still used for obtaining sea surface temperatures from ships. But for most water quality applications, as well as most sea-surface temperatures, platinum resistance thermometers or thermistors are the usual choice. Their characteristics have already been covered in Chapter 3.

pH value

Acids and bases; a brief history

Acids and bases were known about long before their chemistry was understood – acids taste sour, bases feel soapy. In the eighteenth century, oxygen was discovered jointly by Joseph Priestley in England and Karl Scheele in Sweden. In 1777, Lavoisier in France interpreted the nature of oxygen as an element. He also suggested that acids were compounds that contained 'oxygen' (the name 'oxygen' means 'acid producer'). But this proved to be wrong when acids that did not contain oxygen were found. In 1816, Sir Humphry Davey proposed in England that all acids contained hydrogen, not oxygen. About 20 years later, the German chemist Justus von Liebig elaborated on Davey's idea by defining an acid as a compound which contained hydrogen that reacted with metals, producing hydrogen gas. But none of them attempted to explain the nature of acids at the atomic level. It was half a century later before Ostwald in Germany and Arrhenius in Sweden suggested that an acid was a compound that produced hydrogen ions in a water (aqueous) solution and a base or alkali was a compound that produced hydroxyl ions in water. That is, these compounds break up (dissociate) either completely or partly into charged particles called *ions* – positive *cations* and negative *anions*. Acids produce the hydrogen ion H^+ (a proton) whereas bases produce hydroxyl ions OH^- and a metal ion. Strong acids and bases are highly dissociated; weak ones are perhaps only dissociated to the extent of a few percent. For example, hydrochloric acid is completely dissociated, $HCl \rightarrow H^+ + Cl^-$. This concept was a great advance, and is still useful in most situations. Its weakness was that it applied only to aqueous solutions. Others, notably Bronsted in Denmark and Lowry in England, independently proposed in 1923 the more general definition that an acid is a species that can give up a proton and a base is a species that can accept a proton. At the same time the American chemist Gilbert Newton Lewis defined acids and bases in terms of electron, rather than proton, transfer. However, for aqueous solutions, the concept that an acid is a compound that produces hydrogen ions and a base is a compound that produces hydroxyl ions in a water solution is the most useful.

Units and terms

The pH of an aqueous solution is a number expressing the relative amounts, or concentration, or activity, of hydrogen ions H^+ (protons) and hydroxide ions OH^-.

Molarity is the most commonly used measurement of the *concentration* of a solution, being the number of moles of solute per litre of solution. A mole of a substance is the amount of that substance whose mass, in grams, is the same as its relative molecular mass. For example, a mole of molecular oxygen, O_2, has a mass of 32 g. A mole of any chemical species, such as an ion, always contains the same number of particles (6.02×10^{23}, Avogadro's constant). To make a solution of, say, 0.5 mol, half a mole of the solute is weighed and dissolved in the solvent (in the present case water), more solvent then being added to make the total volume one litre.

However, to be precise, it is the *activity* (or *effective concentration*) of the solute that is important rather than its actual concentration – that is, its tendency to take part in a given reaction. Because solutions do not always behave as 'ideal', some acting as if their concentration was more, or less, than it actually is, an *activity coefficient* is used to express the difference between activity and concentration. pH is actually a measure of hydrogen ion activity rather than the concentration of hydrogen ions. However, the coefficient approaches unity in dilute solutions and then the difference is more of an academic one from our point of view, but is mentioned because it can lead to confusion. The above terms were introduced in America in 1907 by Gilbert Lewis.

In 1909, Soren Sorensen defined pH as the negative logarithm (to base 10) of the molar concentration (H^+) of hydrogen ions. That is, $pH = -\log_{10}(H^+)$, or $(H^+) = 10^{-pH}$; the symbol pH stands for the power, or puissance, of the H^+ ions. For example, a pH of 3 indicates a molar concentration or activity of 10^{-3} hydrogen ions (10^{-3} mol).

(The logarithm of a number, to base 10, is the power to which 10 has to be raised to give the number. For example, $\log_{10} 100 = 2$, since $100 = 10^2$, and $\log_{10} 1000 = 3$, since $1000 = 10^3$, etc.)

The pH scale ranges from 0 (corresponding to a molar concentration of 1 mol H^+ – a strong acid) to 14 (corresponding to a molar concentration of 10^{-14} mol H^+ – a strong base). For example, a 0.01 mol solution of fully dissociated hydrochloric acid has a hydrogen ion concentration of 10^{-2} and thus a pH of 2. Water self-ionises slightly and this can be expressed as an acid–base equilibrium:

$$H_2O(l) + H_2O(l) \rightleftharpoons H_3O^+(aq) + OH^-(aq)$$

What this equation expresses is that since hydrogen ions consist of single protons, there is evidence to suggest that they attach themselves to water molecules, the resultant species being known as the hydroxonium ion, $H_3O^+(aq)$. pH is, therefore, more correctly, $-\log_{10}(H_3O^+)$. There being equilibrium between H_3O^+ and OH^- ions, the pH of water is 7, defined as neutrality. The dissociation is only slight and the water concentration is very high compared with that of the hydroxyl and hydroxonium ions.

Buffer solutions are solutions that buffer themselves against changes in pH by consuming added H^+ or OH^- ions through chemical association with other ions in the buffer. These solutions are used as references when calibrating sensors (see below) and are made up from capsules, typically for pHs of 4, 7 and 9, that are dissolved in water at the time they are needed. The capsules contain equimolar amounts of a weak acid or base and one of its salts. Although buffers are mostly associated with pH measurement, they are also made for many other ion species, such as calcium.

Ionic strength is a term that takes into account both the concentration or activity of ions in solution and the charge on the ions. The exact computation used is not important here, but it is useful to be aware of the term. (See the discussion later regarding the difficulty of measuring the pH of low-ionic-strength solutions). Conductivity measurements (see below) give a rough estimate of ionic strength.

The pH of unpolluted, natural river waters ranges from about 6.5 to 8.5, while rainwater is naturally slightly acidic (pH 5.6), because the carbon dioxide dissolved in it produces carbonic acid.

Measuring pH by indicators

The simplest way to get a rough indication of the pH of a solution is to use an indicator – a substance, usually a complex organic molecule (a weak acid or base) that changes colour at different pH levels. Litmus is the best known because it changes from red to blue as the pH crosses the neutral threshold of 7 from acid to base. However, many others exist, such as methyl green (with pH 0.2–1.8) and alizarin yellow (with pH 10.1–12.0). While these indicators are useful and cheap they cannot, alone, give a precise measurement, although they are normally used with titration methods that are precise. As titration is a laboratory technique it will not be included here. For field work, *ion-selective electrodes* are widely used to measure pH and, for automatic measurements in the field, they are the only practical option. Ion-selective sensors can also be used in titration, to detect the *equivalence point*, but this, too, is a laboratory procedure.

Measuring pH by ion-selective electrode

pH electrodes fall into the class of sensor known as specific-ion sensors or ion-selective electrodes (Durst 1969, Covington 1979); in this case it is the hydrogen ions that are to be measured. Figure 10.22 shows the electrode construction. Two electrodes are necessary, the ion-sensitive electrode and a reference electrode, each sometimes referred to as half-cells since both are necessary to form a complete cell. Although shown as two separate electrodes in Figure 10.22, in practice they may be brought together into one *combination electrode* (Fig. 10.23).

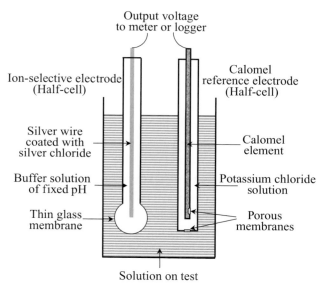

Figure 10.22. A pH electrode (on the left) develops a potential across its thin glass membrane; the magnitude depends on the concentration, or activity, of hydrogen ions in the solution on test. This voltage is measured relative to the fixed voltage generated by the reference electrode (on the right).

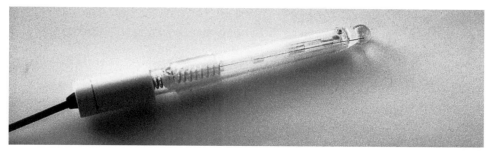

Figure 10.23. About 15 cm in length, this pH combination electrode integrates into one probe both the ion-sensitive electrode (the bulb) and the reference electrode (which makes contact with the solution via the small porous aperture just visible to the left of the bulb).

The ion-selective electrode is made of glass, the sensing part being a thin glass bulb at the end of the tube, this being placed in the water being measured. The bulb is filled with a standard acid buffer solution of known pH. Owing to the different concentrations of hydrogen ions in the fluid filling the bulb and in the water on test, a small voltage builds up across the thin glass membrane (in the hundreds of millivolts range). The composition of the glass determines to which ions the sensor responds. For example, glass electrodes are available that respond to sodium ions.

Electrical contact with the buffer solution in the glass bulb is made via an electrode of silver coated with silver chloride, this acting as one side of the cell. The voltage across the glass membrane has to be measured relative to something and this is why a reference electrode has to be used. This electrode generates a fixed voltage and forms the second half of the cell.

The usual reference electrode for pH measurement is known as a *calomel* electrode, because it contains a central 'calomel element' made of a paste of mercury and calomel (mercuric chloride, Hg_2Cl_2) immersed in a saturated solution of potassium chloride (KCl), the paste being contained in a tube that has capillary holes in it to allow contact to be made with the solution (Ives & Lanz 1961). The outer tube containing the 'filling' solution, KCl, also has small capillary holes or a porous membrane allowing electrical contact with the solution on test. To avoid a build-up of ions inside the reference electrode from the sample under measurement, there has to be a very slow but definite flow of the reference electrode's filling solution into the sample – just sufficient to counter back-diffusion. Flow rates are in the order of 10–200 µl h^{-1}. The voltage at the point where the filling solution and the sample solution meet, the *liquid–liquid junction potential*, is stable, and it is this voltage against which the sensing electrode voltage is measured, giving an indication of the sample's hydrogen ion activity.

In 1889 the Nobel Prize winner, Herman Walther Nernst, introduced the Nernst equation, which relates the potential developed between the two electrodes of any electrical cell. The equation can be most conveniently expressed as:

$$E = E_0 \pm S \log a,$$

where E is the voltage measured between the sensing and reference electrodes, E_0 is a constant dependent largely on the type of reference electrode used, S is the slope of the mV versus pH curve, and a is the activity or concentration of the ion species being measured, in this case hydrogen ions. Because the slope of the curve is temperature dependent, it is necessary to compensate the reading for temperature, this being done manually or automatically.

The potential difference between the two electrodes may be measured manually with a portable meter (see Fig 20.1) displaying the measurement as millivolts or directly in pH units, or the voltage can be measured automatically and logged. By whatever means the voltage is measured, the high membrane resistance of the glass bulb, around 300×10^6 ohms, means that the amplifier circuit must have a very high input impedance so as not to load the cell, draw current and cause a drop in the detected voltage. Amplifiers, whether in a portable meter or a logging station, must have input impedances of at least 10^{12} ohms.

In the relatively pure waters of the natural environment, unlike those of industry and laboratories, the amounts of acid, bases and salts can be very low – as in normal rainwater for example. Measuring these is more difficult (Davison & Harbinson 1988) since the low ionic strength, and thus the low conductivity, of these solutions allows electromagnetic noise to be picked up rather easily. The use of lower-resistance glass membranes ($<50 \times 10^6$ ohms) and low-ionic-strength buffers as electrode fillers, along with more complex (double-junction) reference electrodes, helps improve measurements. Natural waters also have a much lower buffering capacity, allowing for example the slow take-up of carbon dioxide from the atmosphere (if not already saturated with the gas) to appear as a drift in the pH reading.

Those planning to measure the pH of natural waters should, therefore, think twice before buying standard laboratory types of sensor and using them without precaution and awareness. Errors as high as 0.5 pH units can easily occur (Neal & Thomas 1985) if care is not taken. Some 'acid rain' work may well have ignored this factor. The accuracy of sensors used in laboratory conditions and in solutions of moderate ionic strength is in the order of ± 0.1 pH. The accuracy of measurements made in the field in low ionic strength water will depend very much on whether the above precautions are taken. Errors could be in the order of 1 pH unit.

Conductivity

The dissociation of compounds into charged ions in aqueous solutions (see the above discussion of pH) makes solutions conductive of electricity. If two electrodes in the solution have a potential difference applied between them, cations (+) migrate to the cathode (−), accepting electrons, while anions (−) move to the anode (+), giving up electrons. This transfer of electrons through the solution completes the electrical circuit. The conductive solution (*electrolyte*) obeys Ohm's law just as any other conductor does. As noted earlier, the conductivity of an electrolyte gives a fair indication of the ionic strength of the solution.

Units

The conductivity of a solution is defined as the reciprocal of the resistance in ohms of a one centimetre cube of the solution at 25 °C across two opposite faces of the cube. The unit of conductivity is the reciprocal ohm per centimetre (mho cm^{-1}); mhos are also now known as Siemens (S). As this unit is too large in practice for water conductivity, it is more usual to use the reciprocal megohm per centimetre, that is, the microsiemen per centimetre (μS cm^{-1}), or the millisiemen per centimetre (mS cm^{-1}), which, using the older names, are micromhos per centimetre or millimhos per centimetre (μmhos cm^{-1} or millimhos cm^{-1}). The microsiemen cm^{-1} is now the most common unit, the 'cm^{-1}' often being omitted leaving just μS.

However, a more basic definition of the conductivity of a substance is the reciprocal of the resistance of a cube of the material with sides of one metre, across two opposite faces, this being expressed as $\mu S\ m^{-1}$. As the resistance across the faces of a 1 cm cube of material is 100 times that across a 1 m cube, its conductivity is 0.01 of that of a 1 m cube. Thus, if a value of conductivity is, for example, $2\ S\ m^{-1}$, this would be equivalent to $0.02\ S\ cm^{-1}$. Converting to microsiemens (by multiplying by 10^6) gives $2 \times 10^4\ \mu S\ cm^{-1}$.

Very pure water has a conductivity of around $0.05\ \mu S\ cm^{-1}$, distilled water around $3.0\ \mu S\ cm^{-1}$, with natural waters ranging from 35 to $500\ \mu S\ cm^{-1}$. In comparison, a 2% (by weight) solution of sulphuric acid has a conductivity of $93\,000\ \mu S\ cm^{-1}$.

For saline water, the term *salinity* is used in place of conductivity, although it is still the conductivity that is being measured and by the same type of sensor. The units are also different, with parts per thousand (ppt) or parts per million (ppm), by weight, being used instead of units of conductivity. Seawater salinity averages about 34 grams per kilogram (or 34 ppt). This is equivalent to 34 000 ppm. As 1000 ppm of salt solution is equivalent to a conductivity of $2000\ \mu S$, the conductivity of seawater is in the order of $64000\ \mu S$.

Conductivity sensors

Conductivity sensors consist of an electrode assembly designed to measure the resistance of a precise volume of water between two electrodes of precise area. Since, by definition, the conductivity is the reciprocal resistance of $1\ cm^3$ of the water, a sensor that has two electrodes of $1\ cm^2$ spaced 1 cm apart (or the equivalent), is said to have a *cell constant* of 1.0. Variations in the area and separation of the electrodes give different cell constants, for example 0.1, 1.0 and 10. Instruments are usually switched to cover different ranges, depending on application, the most common being from 0 to 10, 100, 1000, 10 000 or 100 000 μS. A cell constant of 1.0 will cover most of these ranges, with perhaps 0.1 being used for the lower end and 10 for the top. Accuracy ranges from $\pm 0.5\%$ to $\pm 2\%$ of full scale, although these values are for well-maintained, clean sensors.

Most conductivity sensors consist of an epoxy resin tube containing electrodes in the form of rings, flush-fitting on the inside of the tube. To minimise corrosion, the electrodes are made of stainless steel, tungsten or carbon or have a platinised finish. To ensure that a precise volume is measured, the electrodes are arranged as shown in Fig. 10.24. As when measuring soil moisture with gypsum blocks and humidity with resistive sensors, AC excitation has to be used to avoid polarising the sensor, a constant AC square-wave current being made to flow through the water between the two outer rings. This generates a voltage along the length of the water path, the voltage being related, by Ohm's law, to the resistance of the water and the current flowing. The two inner electrodes, precisely positioned to encompass the

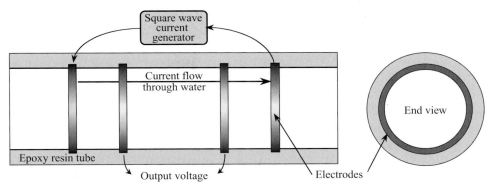

Figure 10.24. This section through a conventional conductivity sensor shows the four electrode rings, the outer two of which introduce a current through the water, the voltage drop across the resistance of the water being picked off by the two inner rings, which define a precise volume of water.

Figure 10.25. By measuring the magnetic coupling between two adjacent coils wound on toroidal ferrite cores (in the circular head in the figure), the conductivity of the water can be measured without the need for any exposed electrodes.

required volume of water, pick off the voltage, which is converted to resistance and thus to conductivity.

Because the electrodes can become dirty, an alternative design has been developed based on the magnetic coupling between two toroidal ferrite cores, parallel to each other and both wound with coils, as shown in Fig. 10.25. One coil is energised with an AC current, the resultant magnetic field inducing a voltage in the second coil. Water occupies the space through the centre of the two coils, the degree of coupling between them being proportional to the conductivity of the water; any increase in ion mobility causes a corresponding increase in the output from the detector coil. The response, however, is non-linear and temperature dependent, these being corrected by an on-board microprocessor. Accuracies of around ±1% of full scale are claimed, with the same ranges of conductivity as are available with the conventional electrode type of sensor.

Oxygen

Since animals living in water depend on the *dissolved oxygen* (DO), it is one of the most important water quality variables to be measured. The amount present depends on the balance between *inputs* of oxygen from the atmosphere and from photosynthesis by aquatic plants and *extraction* of oxygen by respiration and as a consequence of the bacterial decay of organic matter (the *biochemical oxygen demand*, BOD), the latter sometimes resulting in a marked decrease in the amount of oxygen in water.

Units

Oxygen makes up 20.95% of the volume of dry air (that is, air containing no water vapour). It is only slightly soluble in water and the amount is dependent on the pressure of the gas and the temperature of the water. At one standard atmosphere (1013.2 hPa, or 760 mm of mercury) and a water temperature of $0\,°C$, the maximum solubility of oxygen is 14.6 mg per litre (or 14.6 ppm). This falls to 13.5 ppm if the pressure is reduced to 700 mm (at $0\,°C$) and to 7.7 ppm at one atmosphere if the temperature rises to $30\,°C$. So under the normal range of pressure and temperature, there will be a maximum of between 7.5 and 15 ppm of oxygen dissolved in natural open water; there may be much less if the BOD is high. The concentration of DO is sometimes expressed as a percentage, 10 ppm of oxygen being the maximum concentration at standard pressure and temperature, and this is taken to be 100% concentration. Thus if the water contains 7.5 ppm (as at $31\,°C$) the concentration is 75%, while at $0\,°C$ it is 146%. Meters are usually calibrated in ppm (mg litre^{-1}) or in per cent (0–20 ppm or 0–200%).

Oxygen sensors

The most commonly used oxygen sensor, or variations on it, is still the type designed by Mackereth (1964). In this design, oxygen diffuses through a permeable membrane of polypropylene of thickness 0.05 mm into an electrochemical cell (Fig. 10.26) made up of a perforated cylindrical silver cathode, contained within the membrane and insulated by a nylon-weave insulator from a porous lead anode. The whole is bathed in a saturated solution of potassium carbonate (there are variations on this). The cell generates a current proportional to the partial pressure of the oxygen – about 10 μA at 0.21 atmospheres and $25\,°C$.

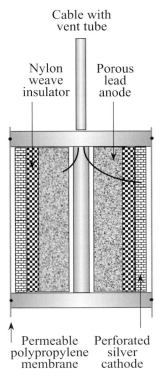

Figure 10.26. The cylindrical lead and silver electrodes of a Mackereth oxygen sensor are bathed in an electrolyte of potassium carbonate, the whole assembly being sheathed in an oxygen-permeable polypropylene tube 0.05 mm thick. One wire from the cable goes to the porous lead anode, the other to the perforated silver cathode.

The cell is calibrated by placing it in distilled water, stirred to ensure air saturation. When the current it produces is stable, this corresponds to the maximum concentration of O_2 that is possible at the prevailing barometric pressure and water temperature. By referring to a table relating temperature and barometric pressure to the maximum solubility of oxygen in water saturated with air, in ppm, a graph can be drawn relating sensor current to the maximum possible amount of DO in ppm. The graph is a straight line going almost through the origin. There is in fact a slight current of about 1% of the air-saturated value when no oxygen is present, and for precise work this current can be measured by immersing the sensor in de-aerated water (obtained by bubbling nitrogen through the water for five to ten minutes) and a more precise graph plotted.

Not only is the amount of oxygen that can be dissolved in water very temperature-dependent, but the current produced by the sensor also changes considerably with temperature, increasing by 2.5% per degree rise in temperature. To convert sensor current into ppm of oxygen, therefore, requires that the temperature also be measured, with the necessary adjustments being made either in the sensor electronics or in the computer software if the data are logged.

The current is usually converted to a voltage (by passing it through a resistor), so that it can be displayed on a millivoltmeter, such as a pH meter, for manual readings, or the voltage readings can be logged. Accuracy ranges from ±1% to ±5% of full scale. But the accuracy also relies, as does the speed of response of the sensor, on the water being circulated over the sensor membrane effectively and continuously and on the membrane being clean. When these conditions are ensured, a sensor has a response time of about 30 s to reach 90% of a step change.

Biochemical oxygen demand

BOD is of more concern to the operators of sewage treatment plants than to those working with the natural environment, and its measurement also involves a laboratory procedure. It is included here because it could, on occasion, be of use in natural waters. BOD is a measure of the use by bacteria of the oxygen dissolved in the water.

The measurement of BOD involves placing a sample of the water in an airtight bottle, the DO level having first been measured, typically with a Mackereth sensor. The sample is then incubated at $20\,°C \pm 1\,°C$ for several days. The DO content is then measured again and the consumption of oxygen expressed as ppm per litre over the period. Demand may vary over the period and this can be followed by measuring the DO level throughout, giving a measure of the rate of consumption.

Turbidity

Particles suspended in water range from small to large; the smaller are generally said to cause *turbidity*, while larger particles such as sand are described as *suspended sediments*. Generically, both smaller and larger suspended particles are often referred to as turbidity. Turbidity affects light transmission, fish eggs and life on the bottom of rivers; it is present in lakes and reservoirs and may have bacteria and pesticides attached to it. Land use and vegetation cover are the main determining factors of turbidity levels. How turbidity is measured depends on the amount of matter in suspension.

Measurement of turbidity

The turbidity of water is measured by the way it affects the passage of light through it, either by its attenuating effect on a direct beam or by its sideways or backward

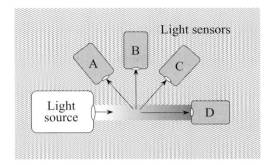

Figure 10.27. By detecting the degree of attenuation or the amount of scattering of a beam of light, the concentration of suspended particles can be measured. A: for very high turbidity, the back-scatter is measured. B, C: for low levels, 90 degree or forward-scatter is sensed. D: for intermediate levels, attenuation of the direct beam is used.

scattering of the beam (see also Chapter 13 on atmospheric visibility). Methods using scattering are often referred to as using the 'nephelometric' principle, nephology being the study of clouds.

For low levels of turbidity, the amount of light scattered at 90° to a light beam is measured. For higher levels, the attenuation of the direct beam is sensed – usually based on a one metre or quarter metre path length, sometimes folded back on itself, using a prism, to shorten the instrument. For yet higher turbidities and for suspended sediments (such as mud and sand), the back-scatter from angles ranging from 180° to 140° is measured. Figure 10.27 illustrates these options, while Fig. 10.28 shows an actual sensor.

In the past, filament lamps had to be used as the light source, and some designs still use them, although they take a large amount of power. In place of filament lights, many present-day instruments use light emitting diodes, which are more stable, longer lived and less power demanding; some use xenon-gas-filled tubes. These latter two sources can also be pulsed, and this reduces power consumption further as well as providing an AC signal in place of a DC one, which improves electronic stability. The pulse rates used are in the 5 and 10 Hz range. The optical windows of these instruments can become dirty, but this, and changes in the light-source intensity as well as in detector sensitivity, can be compensated by measuring absorption over two paths of different lengths in the water, cancelling out the variations. This method can also be used with filament lights, which slowly age. Some instruments include automatic cleaning, either ultrasonic or by a mechanical pump-like piston which periodically moves past the optical windows. Some designs combine 90° scatter and direct transmission in one instrument, thereby allowing a distinction to be made between light losses due to dissolved matter in the water

Figure 10.28. In this particular turbidity sensor, two LEDs (lower window) emit IR light into the water, the IR detector (upper window) measuring the light scattered back from the suspended particles.

(giving the water its colour) and particles in suspension; both attenuate the light beam but only the particles scatter it.

The wavelength of the light used ranges from the visible to both the infrared and ultraviolet. (UV from a xenon light source is used for the measurement of fluorescence in detecting, for example, chlorophyll, although this is not considered here.) IR has the advantage that longer wavelengths do not penetrate far below the water surface (being absorbed), thereby reducing the interference from natural light. IR transmission is sufficient, however, over the short distances used in instruments, to measure sideways-scatter or back-scatter over a few centimetres. Filtering the light, so as to sense just the wavelengths used, further reduces the interference effects of natural light. In deeper water, there is little penetration of light from the surface and so no effect on the signal from natural light.

Units and calibration

Calibration is normally done using formazin – the turbid precipitate formed by mixing a solution of hydrazine sulphate with a solution of hexamethalinetetramine $(CH_2(6)N_4)$, the combination producing a precipitate with a mix of particle sizes similar to those found in naturally turbid water. It is prepared as a stock solution equal to 4000 FTU (*formazine turbidity units*, also known as NTUs – nephelometric turbidity units). In use, the stock solution is diluted to the required levels, from very low values, 1–10 FTU, up through the hundreds to one or two thousand. (Once diluted, the solution has a useful life of only 24 hours; the stock solution has a life of one year.) Because hydrazine sulphate is a carcinogen, safer, long-life alternative calibration solutions using copolymer particles have been developed, but they are more expensive and do not have such a wide range of particle sizes.

The absorption and scattering of light in water depends very much on the size distribution, shape and colour of the suspended particles, which vary from white chalk to dark mud, and it would be impossible to calibrate an instrument to encompass all of these. Indeed, one of the problems with formazin is that it is a white precipitate and there can be a 5 : 1 ratio, or more, between the reflective and scattering properties of formazin and those of natural dark particles. To be certain of good turbidity measurements, therefore, it is necessary to calibrate sensors in a range of suspensions that match the water being measured. Over-reliance on factory calibrations or formazin standards could lead to large errors.

Low turbidities, measured by $90°$ scatter, cover the range 0 to 100 FTU (NTU). Higher values, measured by direct-path absorption, fall in the 100 to 500 FTU range. Although there is considerable overlap, anything much greater than 500 FTU is usually considered as suspended sediment rather than turbidity and is expressed in ppm or mg l^{-1}, values ranging from 5 to 5000 ppm for mud and 100 to 100 000 ppm for sand.

References

Ackers, P., White, W. R., Perkins, J. A. & Harrison, A. J. M. (1978) *Weirs and Flumes for Flow Measurement*. John Wiley & Sons, Chicester.

Barnes, H. H. (Jr) & Davidian, J. (1978) in *Hydrometry, Principles and Practices*, ed. Herschy, R. W., pp. 149–203. John Wiley & Sons.

Benson, M. A. (1968) Measurement of peak discharge by indirect methods. WMO No. 225. TP 119, Technical Note No. 90.

Biswas, A. K. (1970) *History of Hydrology*. North-Holland, Amsterdam and London, ISBN 7204 8018 3.

Bowden, K. F. (1956) The flow of water through the Straits of Dover related to wind and differences in sea level. *Philos. Trans. R. Soc.*, **248**, 517.

Corbett, D. M. *et al.* (1945) Stream-gauging procedures. US Geological Survey (USGS), Water Supply Paper 888.

Covington, A. K., ed. (1979) *Ion-selective Electrode Methodology*, Vols. I and II. CRC Press, Boca Raton, Florida.

Cox, R. A. (1956) *Measuring the Tidal Flow in the River Humber*. The Dock and Harbour Authority, July 1956.

Dalrymple, T. & Benson, M. A. (1967) Measurement of peak discharge by the slope-area method. Chapter A2 in *Techniques of Water Resources Investigations*, Book 3. USS.

Davis, J. M. (1979) Calibration of multi-path ultrasonic stations. In *Proceedings of the Ultrasonic River Gauging Seminar*, pp. 27–38. Water Research Centre and Department of Environment Water Data Unit.

Davison, W. & Harbinson, T. R. (1988) Performance testing of pH electrodes suitable for low ionic strength solutions. *Analyst*, **113**, 709–13.

De Pitot, H. (1732) Description d'une machine pour mesurer la vitesse des eaux courantes et le sillage des vaisseaux. *Mémoires de l'Académie Royale des Sciences*, 363–76.

Durst, R. A., ed. (1969) *Ion Selective Electrodes*. Special Publication 314, National Bureau of Standards.

Faraday, M. (1832) *Philos. Trans. R. Soc.*, 1832, 175.

Frazier, A. H. (1969) Dr. Santorio's water current meter, circa 1610. *J. Hydraul. Div., ASCE*, **95**, 249–54.

Gilman, K. (1977) Dilution gauging on the recession limb: (1) constant rate injection method. *Hydrol. Sci. Bull.*, **22** (3).

Herschy, R. W. (1979a) Site selection and specification of configuration. In *Proceedings of the Ultrasonic River Gauging Seminar*, pp. 6–15. Water Research Centre and Department of the Environment Water Data Unit.

Herschy, R. W. (1979b) Uncertainties in ultrasonic flow measurement. In *Proceedings of the Ultrasonic River Gauging Seminar*, pp. 103–5. Water Research Centre and Department of the Environment Water Data Unit.

Herschy, R. W. (1995) *Streamflow Measurement*. Chapman & Hall, London.

Herschy, R. W. & Fairbridge, R. W. (1998) *Encyclopedia of Hydrology and Water Resources*, p. 800. Kluiver Press, Dordrecht.

Herschy, R. W. & Newman, J. D. (1974) Electromagnetic river gauging. In *Proceedings of the Symposium on River Gauging by Ultrasonic and Electromagnetic Methods*, UK Water Research Centre and Department of the Environment Water Data Unit.

Horton, R. E. (1907) Weir experiments, coefficients and formulas. US Geological Survey, Water Supply Paper 200.

Ives, D. J. G. & Lanz, G.J., ed. (1961) *Reference Electrodes, Theory and Practice*. Academic Press, New York.

Leliavsky, S. (1951) Historic development of the theory of the flow of water in canals and rivers. *Engineer*, **1**, 466–8.

Leliavsky, S. (1965) *River and Canal Hydraulics*. Chapman & Hall, London.

Limerinos, J. T. (1970) Determination of the Manning coefficient from measured bed roughness in natural channels. US Geological Survey, Water Supply Paper 1898-B.

Mackereth, F. J. H. (1964) A sensor for dissolved oxygen. *J. Sci. Inst.*, **B**, 38.

Neal, C. & Thomas, A. G. (1985) Field and laboratory measurements of pH in low conductivity natural waters. *J. Hydrol.*, **79**, 319–22.

Riggs, H. C. (1976) A simplified slope-area method for estimating flood discharges in natural channels. *US Geological Survey J. Res.*, **4**, 285–91.

Walker, S. T. (1979) Where is ultrasonic gauging appropriate? In *Proceedings of the Ultrasonic River Gauging Seminar*, pp. 3–5. Water Research Centre and Department of the Environment Water Data Unit.

Water Research Association (1970) River flow measurement by dilution gauging. WRA Bulletin T. P. 74.

11

Data logging

He who first shortened the labour of copyists by the device of *Movable Types* was disbanding hired armies, and cashiering most Kings and Senates, and creating a whole new democratic world: he had invented the art of printing.

Thomas Carlyle *Sartor Resartus*.

Before the development of modern data loggers in the 1960s, the only means of automatically recording measurements of the environment was on paper charts, either mechanically or on electrical strip-chart recorders with electrical sensors. It was the arrival of solid-state electronics, in particular its ability to operate digitally, that enabled computers and data loggers to be developed. Both have greatly enhanced the way in which the natural environment can be measured, indeed they have revolutionised it.

The construction of a data logger

The schematic of Fig. 11.1 shows each main section of a data logger. With the development of large-scale integration on one integrated circuit (IC) chip, and of the microprocessor, many of these functions are now carried out on a single IC, supported by a range of peripheral chips such as serial data communicators, memory access controllers, counters and clocks (Fig. 11.2), although even many of these are now on one single chip. However, to explain the functioning of a logger, it is useful to keep the boxes separate. Indeed they were, in reality, physically separate until the development of the larger ICs in the 1980s, the first loggers using individual transistors, resistors and capacitors with wires interconnecting them. Today, a small number of ICs, mounted on printed circuit boards, perform all of the functions required – in a reduced space, at reduced cost, with increased reliability and with very low power requirements.

Measuring the Natural Environment, second edition, Ian Strangeways. Published by Cambridge University Press. © Ian Strangeways 2003.

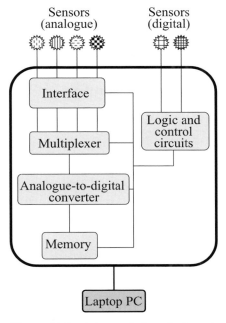

Figure 11.1. A logger's interfacing circuits first standardise the wide variety of analogue signals that sensors produce, the processed signals then being switched sequentially by a multiplexer to an analogue-to-digital converter. Pulse signals, such as those from raingauges and anemometers, bypass these processes, being already digital. As a final step, all the readings are stored in memory. A separate memory is programmed to hold user instructions, controlling how the logger processes the sensor signals and stores them, this process, and the subsequent downloading of the logged data, usually being via a laptop PC.

Interfacing

Most sensors are analogue, producing electrical signals that vary continuously in response to the variable being measured. Few analogue signals are immediately useable in their raw state and it is the role of the logger *interface* to process them into a standard form, usually to one volt (sometimes five volts) full scale. There are a number of kinds of conversion that need to be done to meet this requirement, but the majority are covered by a few simple processes.

Potentiometer sensors

Potentiometer sensors can be made to generate the required voltage signal by applying a precision reference voltage across them; the 'wiper' picking-off a proportion of this as the signal, depending on its angular position, which is related to the amplitude of the variable. The reference voltage need only be applied for the few milliseconds required for the logger to take a reading, and so the power consumed

Figure 11.2. A present-day data logger will typically have all the integrated circuit chips, and the few necessary discrete components such as resistors, on one printed circuit board with tracks on both sides, through-hole plated. Connections to the sensor inputs, to the PC (for programming and downloading of data) and to the power supply are made via the plugs and sockets seen along the lower and upper edges of the board. The two chips with the round windows are EPROM memories, the chip components being visible through the windows for erasure by UV light. This is a complete logger, able to handle eight analogue and two pulse circuits. It consumes only a few microamps in its quiescent state.

is minimal. An error in the reference voltage will cause an extra error in the reading, additional to those generated by the sensor itself.

Low-voltage sensors

Where sensors generate a voltage directly, but not of sufficient amplitude, as in the case of radiation sensors, for example, which have outputs in the millivolt region, an amplifier must be used to increase the signal level to the required one volt full scale. Like so much electronics following the development of the IC, amplifiers today are contained on one chip, indeed one chip may contain several. There are those that handle AC signals, such as for audio and video applications, and those that amplify DC signals, such as those generated by most environmental sensors. Before ICs were available, it was difficult to amplify DC voltages accurately, the signals first having to be 'chopped', to convert them to AC for amplification; they were then converted back to DC signals afterwards. While this is still the most precise method, for most environmental applications there are now many low-cost, low-power, IC operational and instrumentation amplifiers which perform extremely well, and it is generally these that are used in data loggers.

Any error in the zero or gain (amplification) settings of the amplifier will be added to those of the sensor, and this can be affected by the voltage rails which power the amplifier and by the means through which the gain and zero are adjusted. Adjustment may be by fixed or trimmable resistors external to the IC chip, although loggers now tend to use programmable-gain amplifiers, in which the gain is set by logger software, ensuring good accuracy. Amplifier temperature coefficients are typically ± 10 ppm K^{-1}.

DC amplifiers are either single- or double-ended (*differential*). Most loggers today use the latter, which can be operated in either mode to suit the particular application, the differential mode being used when the sensor output is electrically 'floating', neither side being connected to earth (or common or ground).

Most of the water quality specific-ion sensors, typified by pH, have very high internal resistances and the amplifiers used with them must have a much higher input resistance than normal (at least 10^{12} ohms), so as not to load the sensor and reduce the signal.

Resistive sensors

If a sensor, such as a platinum resistance thermometer (PRT), does not produce a voltage signal directly, it must be derived. In the case of a PRT, the usual method is first to generate a millivolt signal by making the sensor one arm of a bridge circuit, this being followed by amplification. While manually operated bridges are balanced by hand to give a null output on a meter, in an automatic system it is usual to allow the bridge to go out of balance and to log the unbalanced output voltage, after amplification.

When two or more stages of signal conversion are necessary, as here, there are more potential sources of error, including errors in bridge-excitation voltage as well as in the amplifier, these being in addition to the sensor errors and sensor exposure errors. The overall total error will depend on how the individual errors combine, some adding and some subtracting depending on the circumstances, which may change with time.

Some resistive sensors cannot have a DC excitation voltage applied across them because they become polarised. In such cases, for example soil-moisture gypsum blocks and some resistive humidity sensors, an AC excitation voltage must be used. Otherwise the principle is the same, although the final output from the interface must be converted to a DC voltage in the required 0–1 volt range. In such cases, the logger usually supplies the required AC excitation internally.

Capacitive sensors

A few sensors, such as those for humidity and pressure, vary in capacitance as the variable changes. The value of the capacitance is usually very small (picofarads) and so it is not possible to connect them to the logger over a long cable, which will have its own (much larger) capacitance. Such sensors have the interfacing electronics within a few centimetres of them, usually supplied by the manufacturer of the sensor in the actual sensor housing, although the logger may have to supply the power to operate them.

Pulse-producing and digital sensors

Pulse signals, such as those from rain-gauges and wind-speed sensors, are dealt with differently, there being no need for analogue-to-digital conversion. Instead, they go directly to the logic circuits of the logger, as described in the subsection below on logger programming.

A few sensors, such as shaft encoders, produce serial-binary or parallel-binary signals (such as the Gray code). Loggers must include a special interface card to handle these, which enable most of the normal logger processing operations to be bypassed.

Multiplexing

Some loggers accept just one input, such as for a stand-alone raingauge, but most are multichannel with provision for 2, 4, 8, 16 or more analogue inputs, and one or more pulse-counting sensors, as well as serial or parallel digital signals. When there are two or more analogue inputs, the logger samples them sequentially, switching (*multiplexing*) rapidly from one to the next. While this was once done electromechanically by reed switches, today the switching is done electronically: each of the outputs from the interface circuits are switched, one after the other, to the next stage of the logging process, which is the *analogue-to-digital* converter (ADC). In fact the multiplexer and ADC are now usually on the same chip, probably combined with the processor.

Analogue-to-digital conversion (ADC)

Digital terms

It is useful to clarify the meaning of three words used frequently in discussing digital data. A *bit* means a binary digit (1 or 0). A *byte* is a group of bits treated as a unit, nearly always involving 8 bits, and is usually, but not always, a subdivision of a *word*, which is a string of bits, usually upwards of 16. But 'word' is also used more loosely in place of 'byte' (Illingworth 1998).

It is also useful to look at the various *number codes* in common use. The numerical examples below are taken from Horowitz & Hill (1984), from which more detail can be obtained.

Decimal numbers

A decimal number (base 10) is the familiar form of numbering used in everyday life. It needs ten symbols (0–9) to express a number. An example follows; the subscript indicates that base 10 is being used:

$$137_{10} = 1 \times 10^2 + 3 \times 10^1 + 7 \times 10^0$$
$$= 100 \quad + 30 \quad + 7$$

Straight binary numbers

A binary number (base 2) needs just two symbols to express a number (0 and 1). This is useful when working with computers and loggers because 0 and 1 are easily represented by two positions of a switch (on or off) or by two ends of a voltage range (for example, $+5$ and -5 or $+5$ and 0). To see the relation between decimal and binary numbers, consider the following:

$$1101_2 = 1 \times 2^3 + 1 \times 2^2 + 0 \times 2^1 + 1 \times 2^0$$
$$= 8 \quad + 4 \quad + 0 \quad + 1$$
$$= 13_{10}$$

Octal and hexadecimal codes

Binary numbers tend to be long, and so it is convenient to break them up into shorter lengths. Octal numbers (base 8) are binary numbers broken up into groups of three binary digits, each group requiring eight symbols (0 to 7) to express a number. Consider the number

$$835_{10} = 1101000011_2$$

Broken up into groups of three, starting from the least significant bit, this number becomes

$$1\ 101\ 000\ 011$$
$$= 1 \quad 5 \quad 0 \quad 3$$
$$= 1503_8$$

Hexadecimal numbers (base 16) are obtained by breaking a binary number into groups of four digits, each group having values from 0 to 15 and requiring 16 symbols (0–9 together with A–F) to express a number. Thus one counts 0123456789ABCDEF. As an example,

$$707_{10} = 1011000011_2$$

Broken up into groups of four binary digits, this number becomes

$$10\ 1100\ 0011$$
$$=\ 2\quad C\quad 3$$
$$=2C3_{16}$$

Hexadecimal is useful because of the convention of organising numbers into bytes (8-bit groups), usually as 16- or 32-bit words; a word becomes 2 or 4 bytes in hexadecimal format. The problem with this code is getting used to strange numbers and doing arithmetic with them.

12- and 36-bit words were common in earlier computers and this suited the octal format better, with its divisions into groups of three binary digits. But with 16- and 32-bit words hexadecimal is more useful.

Binary-coded decimal (BCD)

This number code converts each decimal digit into binary form – hence the term BCD – using a 4-bit group for each decimal digit. Using a previous example,

$$137_{10} = 0001\ 0011\ 0111\ (\text{BCD})$$

This is not the same as the binary version of this number, which would be 10001001_2.

BCD is wasteful of bits since each group of four binary digits could express a number 0 to 15, while BCD never represents a number greater than 9. However, BCD is useful if a number is to be displayed in decimal, since all that is necessary is to convert each BCD group of four binary digits to the equivalent decimal number. Many electronic ICs are available to do this – BCD decoders, drivers and displays. So the BCD format is the one used most commonly for the input and output of numerical information.

The Gray code

In Chapter 5, in the subsection on automatic wind direction sensors, and in Chapter 10, in the subsection on sensing water level by float, the concept of a shaft encoder was introduced, in which the sensor turns a shaft that rotates a disc encoded digitally. It would be a simple matter to produce such a disc with a pattern on it, coded in straight binary, that could be read optically, producing a parallel output indicating the state of each of the bits of the number representing the angular position of the shaft. But, rather than use a straight binary code on the disc, it is preferable to use what is known as the Gray code. This code has the property that, unlike in binary, only one of the bits changes at any one time in going from one state to the next. The reason for preferring this is that in straight binary format several bits can change when going from one state to the next. In practice it is very difficult

Table 11.1 *The Gray code*

Decimal	0	1	2	3	4	5	6	7	8	9
Gray	0000	0001	0011	0010	0110	0111	0101	0100	1100	1101

to ensure that the mechanical alignment is good enough to guarantee that all the bits that should change, do change together and at exactly the same instant. If they do not, large errors in the output can occur. The Gray code avoids this by changing the single least significant bit that gets it to the new state. Table 11.1 shows the decimal numbers 0–9 in the Gray code.

Gray codes can have any number of bits; just four are shown here. It is possible to convert from Gray to binary and back using suitable electronic gate circuits (Horowitz & Hill 1984). In the case of shaft encoders used with float-operated water level sensors, their outputs can be chosen to be Gray coded or can be converted to BCD (for ease of display) or straight binary. But apart from special cases such as BCD and Gray codes, computing and logging circuits work in straight binary.

Resolution

In Chapter 1, the term 'resolution' was defined as the smallest change in a variable that could be measured by a sensor (Fig. 1.2). The same considerations also apply to the logging of the measurements: the larger the number of bits used in ADC, the finer the resolution obtained. Table 11.2 shows the resolutions obtained for different word lengths. The larger the number of bits, the finer the resolution obtained. Thus, if an 8-bit ADC is used, the resolution obtained is 0.4% of full scale; this might be adequate for, say, humidity, which cannot be sensed more accurately than this. However, 12-bit word lengths will give a resolution of 0.02%, which can be of value in measuring, say, river level, where this can be matched by sensors and where such a resolution can be of undoubted value in hydrological applications.

A justification sometimes made by users regarding high-resolution loggers is that they want to look at small changes, not at absolute values, and that the extra resolution provided by 12 bits helps them do this. Although they may have a case, care is still needed so as to be certain that the small changes revealed are real and not due to random noise in the system or to sensitivity or zero drifts within the sensor or power supply. Also, when the data get into the public domain a warning about the fact that high resolution might not mean high accuracy is rarely printed alongside the data. All data should bear an indication of their origin, of the type of instrument used to collect them and a guide as to how accurate they are.

Table 11.2 *Bit numbers and resolution*

Bit number	Bit value	Total value	Resolution (%)
1	1	1	100.00
2	2	3	33.33
3	4	7	14.29
4	8	15	6.66
5	16	31	3.23
6	32	63	1.59
7	64	127	0.79
8	128	255	0.40
9	256	511	0.20
10	512	1023	0.10
11	1024	2047	0.05
12	2048	4095	0.02

The electronics of analogue-to-digital conversion

How electronic circuits operate is not the concern of this book; however, it is useful to look at the principle of analogue-to-digital conversion. There are several means of achieving it:

In *successive approximation*, the conversion is performed by a trial-and-error process. Using a *voltage comparator* circuit, the voltage equivalent to the most significant bit (MSB) is compared with that of the signal (Fig. 11.3). If the signal is the greater, that bit is left on (1), if less it is turned off (0). This process is repeated for each next-most-significant bit in turn until the least significant bit (LSB) is reached, each step progressively homing-in on the correct reading. It is a quick and accurate method and probably the most widely used.

In a similar technique, the voltage to the comparator increases, in steps equivalent to the LSB, until it is the same as the input voltage, when the process stops; the number of steps is counted and represented as the binary value of the voltage. Because it has to step through each possible voltage from zero upwards, this can be a (relatively) slow process.

Lesser-used circuits are the *dual slope* (or *charge balancing*) method, and the fastest, and most expensive, known as the *parallel conversion* technique.

Memory

Brief history

Once converted to digital form, the sensor measurements are stored in memory. In the early to mid 1960s, quarter inch magnetic tape was used as the storage medium, both reel-to-reel and cassettes. After a few years, a change was made to the then

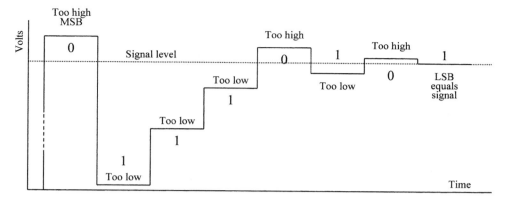

Figure 11.3. In the successive approximation method of analogue-to-digital conversion each bit, starting with the most significant bit (MSB), is compared with the signal and if greater is set as 0, if less as 1. The 8-bit word illustrated, therefore, represents the binary number 01110101 (or 117 in decimal).

new *compact cassette* (Fig. 11.4) (Strangeways & Templeman 1974). A few loggers used half inch computer tape in large cassettes. All these tapes were played back on special readers into the new mainframe computers of the day, such as Digital's PDP8. Tape had many good characteristics – compact cassettes were available worldwide in every town, they were cheap (£1), could store between 50 000 and 100 000 8-bit bytes, could be posted, the data on them were secure, the tapes could be reused and no power was required to maintain the memory. Their main disadvantage was that they needed mechanical tape-drives and these could be a problem in the field, particularly at low temperatures. Downloading of data was also time-consuming, requiring the tapes to be played back at a rather low speed. Nor was tape-replay possible in the field. (It is useful to be able to look at the logged data while still at a field site as a check on the correct functioning of the station.) During the 1980s, solid state memory became available, ending three decades of tape dominance.

Types of solid state memory

Memory is used in a logger both to store the programmes for operating it and to store the measurements made by the sensors, the former requiring memories that do not need frequent erasure and reprogramming, while the latter must be cleared and reused many times over.

EPROM, or electrically programmable read only memory, can be programmed and then erased for reprogramming. Erasure is by UV light through a quartz window in the chip, the UV acting directly on the components of the IC chip (Fig. 11.2). It was designed for occasional erasure, and today is widely used to hold the programmes that operate the logger. At one point in the 1980s, however, before

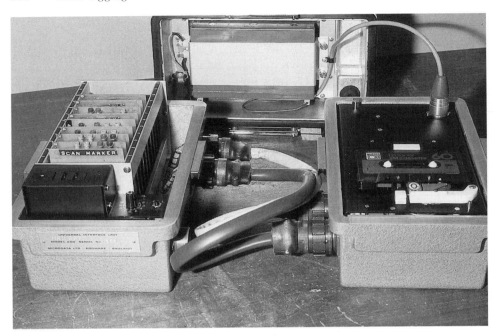

Figure 11.4. Prior to the availability of low-cost, high-capacity, low-power-consumption RAM, magnetic tape, usually in the form of compact cassettes as in the figure (right front), was widely used and served well from the late 1960s up until the late 1980s, although now replaced entirely by solid state memory. This particular logging system has a separate interfacing unit (left), each card containing a circuit for the conversion of a sensor signal into a 5 volt standard for multiplexing and analogue-to-digital conversion; these processes are carried out in the right-hand case. The rechargeable battery is seen in the lid.

RAM (below) became cheap, some loggers did use EPROMs to store the data but the UV erasure was a disadvantage and very frequent reuse over a period could eventually lead to some data corruption.

EEPROM (E^2PROM) or electrically erasable and programmable read only memory, is similar to EPROM but erasure is done electrically, which is more convenient than by UV light. They too are meant for occasional reprogramming.

RAM, or random access memory, is intended for the temporary storage of data and for frequent reuse, making it ideally suited to the storage of data in a logger. RAM is usually of complementary metal-oxide semiconductor (CMOS) fabrication. MOS transistor construction allows very high packing densities, while CMOS logic circuits use complementary MOS transistors, that is a pair of opposite type (n–p–n and p–n–p), which has the advantage of low power consumption (Illingworth 1998). Many loggers now use removable (RAM) memory cards with capacities up to 4 Mbyte and increasing.

Logger programming

Early loggers were not programmable. The number of channels and their functions were selected by the addition or removal of interface cards from a rack (Fig. 11.4), sensor ranges were adjusted with a screwdriver by turning a trimpot, the analogue clock was set by hand and the interval at which readings were taken was adjusted by switches. But with the ability to incorporate a microprocessor, logger function became much more flexible and under the control of software.

Today, the user can specify the following: how many channels are to be used; the type of sensors that will be connected; the ranges to be covered; the *scan interval* (5, 15, 30, 60 seconds – the frequency at which readings are taken and stored temporarily); the *logging interval* (the frequency, for example hourly, at which values are stored in memory for later collection, which can be different for each channel); whether any processing is required, such as the averaging of the readings taken at the scan intervals or the logging of extremes measured during the logging interval, spot readings at the logging interval or the units to work in (metres, %, °C); how the data are to be output to the PC and plotted on-screen; the date and time at which logging is to start. Details of the field site (name, or latitude and longitude) can also be included. Whether data should continue to be stored in memory when the memory is full (new data overwriting the old), or whether logging should stop, can also be specified. Details of communication between the logger and the PC via their serial ports also need to be set up, specifying such things as the rate at which data are transmitted, in bits per second, parity bits (for error checking) and byte length. Although most loggers are programmable in the above way via a PC (temporarily attached to the logger's serial port), some have a built-in liquid-crystal display and simple keyboard that allow interaction with the logger.

There are two options for handling pulses from sensors, which is especially useful in the case of a tipping bucket raingauge. The total number of pulses during the logging interval can be logged, giving the rain total for the period, or the time and date of each pulse can be logged – in the *event mode* – the logger coming out of its quiescent state momentarily to log each event as it occurs. This gives an indication of rain intensity, as well as allowing totals to be calculated. Which mode is used depends on requirements. For example, in an arid environment it is wasteful of memory to log six months of hourly zeros, so the event mode may be preferable. However, if there is frequent high rainfall the event mode may take up too much memory, since there will be a great many events to log. It is also possible to use a *delta function*, in which the logger logs only when a variable has changed by a preset amount (for example in river level). Logging might also be initiated only when an event such as the tip of a raingauge bucket occurs, the logger remaining inactive until rain starts, or when a preset threshold, such as river level, is reached.

Often all these instructions to the logger can be set up on a PC in advance of use, stored on the PC's disc and finally loaded into the logger with just a few key-strokes, cutting the amount of programming work needed in the field, which often has to be done under difficult conditions and under the pressure of time.

Power supplies

CMOS logic circuits consume little power, and with loggers in a quiescent state most of the time (when only microamps are drawn) overall power consumption is very low indeed. For these reasons, field stations can be operated unattended for long periods at remote sites where there is no mains power. There are four common ways of powering a logger, as follows.

Replaceable cells, normally of the alkaline type, can operate a station for up to a year and can be bought worldwide. *Rechargeable nickel–cadmium batteries* are an alternative, but *lithium batteries* are increasingly being used as they can power a logger for up to five years. *Solar panels* (Fig. 11.5), charging *lead–acid* (sealed-gel) batteries, are also now commonplace and can power a station indefinitely, even in higher latitudes. Except under special circumstances the logger battery also powers all the sensors that need power and all the interface circuits. Just occasionally, however, special external interface circuits may be necessary, for example with water quality sensors, and these may need a dedicated power supply of their own.

Figure 11.5. To avoid the need to change batteries and to allow long-term operation, many environmental instruments are now powered by lead–acid-gel batteries charged by a solar panel, the one illustrated being large enough to operate most forms of remote instrument, even in higher latitudes.

Collecting the logged data

After a day, a week, a month or a year, the logging station will be visited and the recorded data retrieved. Although removable memories are a feature of some designs, retrieval is most usually done by downloading the data to a PC's floppy or hard disc, via the serial port of the PC and logger (Fig. 11.6). This is a fairly rapid process taking but a few minutes, and the data can then be returned to base. An advantage of on-site downloading to a PC is that it is then possible, while still at the field site, to check the performance of the logger – perhaps by plotting the data on-screen. But this requires some training of the operator and mistakes can be made. Straight memory swapping is simpler and requires less skill.

It is usually possible to make the choice whether to reprogram the logger at each visit after downloading the data or to let it run on. If the former is selected, all the old data are cleared from memory; in the latter case the old data remain, the new data following on after them. An advantage of reprogramming at each visit is that the logger clock is automatically resynchronised with the PC's clock, while if the logger is left to run on then any drifts in its clock setting will accumulate with time. An advantage of letting the logger run on is that the old data are still there if the downloading process has failed or the data are subsequently lost.

Figure 11.6. The most usual way of collecting data from a logger at a field site is to connect a laptop PC and to download the data to disc. Here data are being collected from a tipping bucket raingauge; the logger, normally housed in the box, is standing next to, and connected to, the PC on top of the box. The raingauge cover has been removed for testing, making the tipping bucket visible.

Protection from the environment

Loggers are typically housed in hermetically sealed boxes of plastic or metal, with sealed plugs and sockets for the connection of the sensors and the PC. A bag of silica-gel crystals is often placed inside the box to keep it dry. It is also usual to protect the logger further, by housing it in a hut at the top of a stilling well for river level measurement or in a small extra enclosure at an AWS.

Loggers can thus be fully protected from dampness and most physical damage – but they are not so easily protected from temperature extremes. The inside of an unventilated box or hut exposed to the sun can become hotter than the surrounding air, even if the box is white, and in situations where air temperatures can reach $+45\,°C$ or more some additional shielding from the sun may be advisable to prevent too high a rise in temperature. At the other extreme, in polar regions, temperatures can fall to $-45\,°C$, or lower. The temperature specifications of ICs, batteries and other components vary considerably, typical ranges for ICs being 0 to $+70\,°C$, -20 to $+85\,°C$ and -45 to $+85\,°C$. Nickel–cadmium and lithium batteries can work over temperatures from -40 to $+60\,°C$, although the former may not retain all their capacity at the extremes and can suffer from a high self-discharge at raised

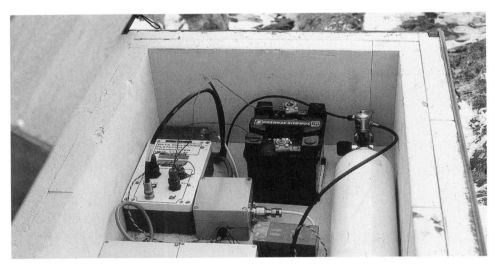

Figure 11.7. Here, two layers of 5 cm thick expanded-polystyrene insulation, contained in a wooden box, protect the logging and satellite telemetry units contained within it. This was used at the BAS Faraday base in Antarctica in connection with experiments carried out by the Institute of Hydrology, on experimental cold-regions sensors (Fig. 18.5). While anything other than extremely thick insulation cannot prevent the equipment from experiencing the mean temperature, insulation does protect against the extremes (including very high temperatures in hot climates).

temperatures. Logger manufacturers quote temperature specifications ranging from −20 °C to +50 or +60 °C, with just a few covering −40 to +70 °C. By use of military-specification chips (at increased cost) the range can be extended, but often normal chips will also operate well beyond their specified limits, although tests are necessary to confirm this. By burying the logger in the ground or in a snowpack, or by using well-insulated boxes (Fig. 11.7) temperature extremes can be reduced, but usually the logger must simply be able to work over the full range.

References

Horowitz, P. & Hill, W. (1984) *The Art of Electronics*. Cambridge University Press, Cambridge UK. ISBN 0 521 29837 7.
Illingworth, V. ed. (1998) *The Penguin Dictionary of Electronics*. Penguin books. ISBN 0 14 051402 3.
Strangeways, I. C. & Templeman, R. F. (1974) Logging river level on magnetic tape. *Water and Water Eng.*, **178**, 57–60.

12

Telemetry

I'll put a girdle round the earth in forty minutes.
                    Shakespeare, Puck from *A Midsummer Night's Dream.*

Reasons for telemetering data

Telemetry is the transmission of data from one point to another. If data are needed in real time they must be telemetered, for example for weather forecasting and flood warning. Telemetry also has two significant advantages over in situ data logging, even if the measurements are not required in real time: the cost of visiting field sites to collect data is saved and the failure of field stations can be detected – months of data could be lost if a logging station failed soon after a visit. Logging is best suited to applications where stations are within relatively easy access or where the loss of some data is not a serious problem.

The general process of telemetering data is sometimes referred to generically as *system control and data acquisition* (SCADA), although the term applies more strictly to management applications – where not only are data acquired from a remote location but remote control is also exercised back. A dam managed from a distant control-room, for example, is a more appropriate use of the term SCADA than is the one-way collection of environmental data.

The structure of a telemetry system

Figure 12.1 is a schematic of a telemetry system, showing its main subdivisions into sensors, logger, modem, communications link and a PC at the base station. This basic arrangement is similar for all telemetry systems although it will differ in detail, mostly depending on the communications link used. A telemetry system is

Measuring the Natural Environment, second edition, Ian Strangeways. Published by Cambridge University Press. © Ian Strangeways 2003.

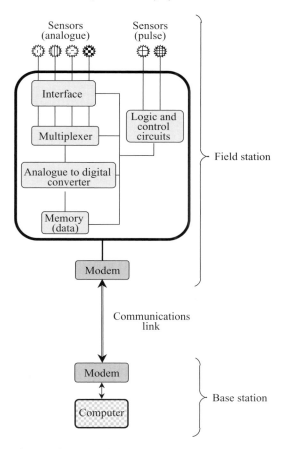

Figure 12.1. A telemetry system is a logging station with a communications link added.

in effect a logging station with a communications link appended and with a remote base station to receive the transmitted data. The front end of such a field station is composed of sensors, identical to those used at a logging station, and a unit that performs the same functions as a logger, even if it is not called such – interfacing, multiplexing, analogue-to-digital conversion and memory. Previous chapters have been concerned with the sensors and with logging; all that needs to be addressed in this chapter is the communications link.

Modems

Loggers and computers express their digital '1's and '0's as DC voltage pulses; typically '1' corresponds to +5 volts and '0' to −5 (or 0) volts. DC pulses cannot, however, be transmitted over communications links, and must be converted into *voice frequency tones*, which can modulate a communications link. This is done using a modem (MOdulate–DEModulate) – for modems also perform the reverse function of converting tones back to DC pulses when they are received at the far end of the link. The process by which a modem converts digital 1s and 0s into tones is known as *frequency-switch keying* (FSK), there being one frequency for 1s and another for 0s.

The base station

The base station is generally at a central office where the data are required, either for immediate and automatic input to a computer model for a real-time application or for archiving for later use. The hardware of the base station comprises the communications link and a PC, or perhaps a larger computer. The PC performs several functions: calling up (*polling*) the field stations, receiving their returned data and processing them.

The communications link

Certain basic principles apply to the way in which communication is carried out, whatever the communications link. If the link is able to send data both ways at the same time, such as voice over a telephone line, it is termed *full duplex*. With two-way communication the base station can poll the field stations at any time and instruct them to send their data. If the link can send data both ways, but not at the same time, it is known as *half duplex* and this is more than adequate for telemetry. Some links, however, can only carry information in one direction, and these are known as *simplex* channels. Polling is not then possible, and the base station has to wait for the field station to send its data, usually but not necessarily, at set times.

There is much benefit to be had from being able to poll field stations from base, since they can then be interrogated as often or as little as necessary and, in the case of a network of stations, each station can be called at different times and at different intervals. A failed transmission can also be requested again. With simplex channels there is no option but for the field station to send its data and for the base station to wait for the transmissions. This, however, may be perfectly adequate in most cases.

Telephone, ground-based radio or satellites can all be used as telemetry links. Which is most suitable will depend on circumstances:

Terrestrial communication links

Telephone lines

The *public switched telephone network* (PSTN), as used for normal telephone voice communication, also now carries digital signals from faxes and PCs. *Baud rates*, more usually now referred to as bits-per-second, as high as 56 000 are now commonplace. Before the introduction of electronic exchanges and fibre-optic links, the telephone network was not as reliable as it now is. Telephone lines are now very well suited to telemetry and, because communication is duplex, field stations can be polled; nevertheless, there remains some possible truth in the view that they are unreliable, since the last stretch of a telephone link to a field site may be vulnerable to damage by wind or flooding, and so particular care is needed in this final step of the link. The cost of installing lines to the remoter sites may also be high, or impractical, in which case an alternative will need to be found. But where telephone lines can access a site and where the country's network is modern and reliable, telephone links are very attractive and are widely used for telemetry. Where a mobile telephone service is available these can also be used to telemeter data, but unfortunately the signal level is often too low in many remote areas, which is where it would be most useful.

The only hardware that is required to convert a logging station into a telephone-telemetering station is a modem, the cost of which is small. The cost of calls will depend on the usual factors: the number of transmissions made, their duration and distance and the time of day. Where real-time data are not required, measurements can be stored and sent at night or at the weekend, in one batch, but where this is not possible, low-cost scheduling may not be an option. Indeed, the ongoing cost of calls can be a deterrent to the use of telephone links, in which case radio may be preferred.

VHF and UHF radio

Where telephone lines are not practicable, technically or because of cost, but where distances are not great, terrestrial line-of-sight *very high frequency* (VHF) and *ultra-high frequency* (UHF) radio is the usual alternative. However, satellite links are increasingly being used even for these local situations. The VHF band extends from 30 MHz (10 m wavelength) to 300 MHz (1 m), merging at the upper frequency into the UHF frequencies, 300 MHz to 3 GHz (10 cm). UHF, in particular, behaves like light, the waves not bending significantly beyond the horizon, restricting their range to about 30 km (Fig. 12.2). There may be some slight refraction beyond the line of sight, especially with the lower VHF frequencies and in certain weather conditions, but such communication is not reliable. High antenna masts (Fig. 12.3) can extend the range, as can high ground, but high ground is only useful if it is advantageously positioned; hills more often restrict the range by obstructing the path of the beam. Repeaters can be used (Fig. 12.4), but they add to the complexity of a network, increase cost and reduce reliability.

Because these radio bands can be focused into narrow beams using a Yagi antenna, the power required is small and 5 watts is generally sufficient; power levels are also restricted by law to about this level. Interference from other transmitters

Figure 12.2. Although there can be refraction beyond the horizon, UHF and VHF radio offer essentially only line-of-sight links, even if nothing more than the curvature of the Earth intervenes. Hills and buildings often obscure the view and make links shorter, although, conveniently placed, they can be used to extend the range, by elevating a base or field station antenna.

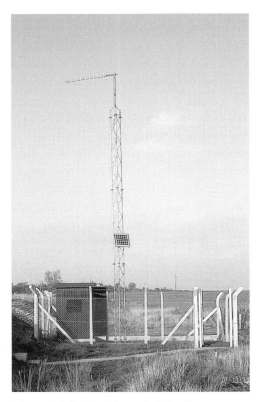

Figure 12.3. The line-of-sight distance covered by this river-gauging station's UHF radio link has been increased by installing the antenna on a high mast.

is reduced by the directional nature of the beam. These bands are very crowded, however, and it can be difficult to get a licence to use them. Despite line-of-sight limitations and a crowded spectrum, VHF and UHF provide good communication and are the only ground-based alternative to telephone lines and in consequence are widely used. They also have the advantage that the calls are free. In the long term, the cost of operating a transmission link (the cost of telephone calls, or of using a satellite) can be the greatest overall cost, exceeding even the initial capital cost of the equipment. The UHF frequencies are the more commonly used because their

Figure 12.4. If a UHF link cannot cover the required distance in one span, a repeater can be used, usually sited on a hilltop as at this water tower; many organisations make use of this facility.

higher frequency gives a greater bandwidth and thus a greater number of channels are available.

Recently a UHF method using low-power transmitters (half a watt) have become available; these can cover distances up to 15 km. Their main, and great, advantage is that they do not require a licence for operation (in the UK at least).

As with telephone lines, two-way communication (half-duplex in this case) is possible with UHF and VHF links but, because the radio receiver must be in operation waiting for a polling call from base, too much power may be consumed if the site is reliant on solar power or replaceable batteries. A similar problem also occurs with those satellite links that allow duplex communication. For this reason, networks operating in the simplex mode have some attraction, since they are simpler and do not need to have a receiver at the field stations. Such systems transmit the data from the field stations either at set intervals or when an event occurs, such as the tip of a raingauge bucket.

However, in the simplex mode transmitting at random, even though the small amounts of data usually involved can be transmitted in under a second, there remains the possibility that two stations will transmit at once, thereby resulting in the loss of both messages. Provided that the number of field stations is not too large, nor the transmissions too frequent, the chance of coincident transmissions is usually acceptably small. A technique, also used by the Argos satellite (see later), can, however, be employed to minimise loss through coincident transmissions. Each field station can be arranged to transmit its message several times, each station at slightly different repetition rates, such that if two transmissions clash, the next two will not.

But if for any reason data are not received, owing perhaps to a fault at the base station or at a repeater or at the field station, in such a simplex system these data will be lost in real time. It may be possible to retrieve them later by visiting the field stations affected, provided that the stations include a data logger for in situ recording, but if the data are needed in real time, they will not be available. Polling can minimise such losses, since it allows the stations to be called a second time, but if the fault is permanent, even this will not retrieve the lost data accessed.

In a polling network, as a compromise to save power, it is possible to call the field stations just at set times, say during a 15-min window every six hours, to reprogramme the stations if necessary or to call up data missed in the intervening six hours; between times the stations simply transmit at set times or when an event occurs. To achieve this, the field station is programmed to switch on its receiver at prearranged times to await any calls from base.

HF radio

Before satellite telemetry was available, *high-frequency* (HF) radio was the only channel available for telemetering data over a long distance where there was no telephone network. It is included here more for completeness than in any likelihood of its being used, now that satellites are available.

More commonly known as *short-wave* radio, HF radio frequencies extend from 3 MHz (100 metres wavelength) to 30 MHz (10 metres). Radio waves in this band

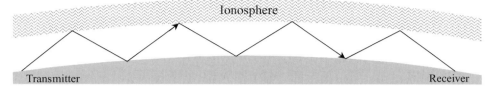

Figure 12.5. The ionosphere reflects HF radio waves (3–30 MHz), permitting multiple reflections between it and the ground and thus enabling communication over great distances. The *C*-layer of the ionosphere extends from 40 to 60 km above the Earth's surface, the *D*-layer from 60 to 90 km, reflecting the lower part of the band, the *E*-layer from 90 to 150 km, reflecting the medium frequencies, and the *F*-layer from 150 to 1000 km, reflecting the high-frequency end of the band. However, fading and interference make it a less than ideal communications link and it is little used for telemetry today.

are propagated by successive reflections between the ionosphere and the ground (Fig. 12.5), thereby spanning great distances with a power of only 50 watts. But the ionosphere moves up and down during the day, as the seasons change and as a result of solar activity, so the signals suffer from fading and from interference from other transmitters. The waves can also travel by several slightly different paths (multipath propagation) and this distorts the received waveform because they arrive at slightly different times. While these problems can be reduced by using multiple-antenna and multiple-frequency techniques, these are very expensive. However, before satellites, HF was successfully used for telemetering flood warning data in Brazil (Strangeways & Lisoni 1973; Fig. 10.4 illustrates a stilling well from which river level data were telemetered by HF radio for flood warning). Today, however, satellites do the job more reliably, and about the only thing to be said for HF now is that, like UHF, calls are free; perhaps also it has the advantage that negotiations do not have to be made with any agency, although a frequency has to be negotiated.

Meteor-burst communication

Micrometeorites, the size of particles of sand, enter Earth's atmosphere continuously, leaving small ionised trails that last a few seconds. These trails will reflect radio transmissions in the 30–200 MHz range and this property has been used as a means of long-distance communication. The most effective band is between 30 and 50 MHz (the low end of the VHF band).

Field stations using this technique direct their (rather large) Yagi antenna at that part of the sky where an ionised trail, when it occurs, will reflect the signal to the distant base station (Figs. 12.6 & 12.7). The base station keeps calling the field station until an ionised trail occurs at the right place and the right time, reflecting

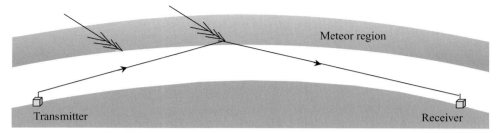

Figure 12.6. The short-lived ionisation trails produced by micrometeorites can be used to reflect radio waves in the 30 to 200 MHz band over distances up to 2000 km. Such meteor-burst communication is used for a number of environmental projects, but not as widely as satellites, in part because of the relatively high cost of the base station.

Figure 12.7. The large size of the antenna necessary for meteor-burst communication (because of the low VHF frequency used) is apparent from this substantial tower at a base station. Four antennas are necessary to achieve 360° coverage. Despite the initial capital cost, the technique has the advantage that the calls cost nothing.

the call to the field station. Averaged over a year, the waiting time is typically about two minutes. When the field station receives the call it replies, and if the ionisation is still active, the base receives the message; if not, the base keeps calling until it gets a reply without errors. The main problem is the very high cost of the base station (£100 000+), which is complex. But it also has practical problems. Signal levels reflected off meteor trails are extremely low and a compromise is necessary between transmitter power (100 watts being the minimum) and the gain (or sensitivity) of the antenna. If the antenna gain is high, that is it has many dipole elements, it directs its beam more narrowly and so with more concentrated power, but it then reaches fewer ionised trails. If, at either end of the link, the receiver gain is increased beyond a certain threshold to detect weaker signals, it may respond to man-made or natural electromagnetic noise.

The maximum distances that can be covered are up to about 2000 km from base to field station, although this is adequate for all but the largest networks. While meteor-burst communication has occasionally been used for environmental data telemetry, this use is restricted because of cost and complexity.

However, recently, cheaper field stations have been developed and meteor communications are now being used quite extensively in the Western USA to telemeter snowpack data (Schafer & Verner 1996) by the US Department of Agriculture, Natural Resources Conservation Service (http://www.wcc.nrcs.usda.gov click on 'Data Collection Techniques').

Satellite communications

Since the 1970s, satellites have been launched for a variety of environmental applications, the foremost of which are meteorological, notably for the collection of images, particularly of clouds but also for the measurement of variables such as temperature, rain, humidity and evaporation. While in situ surface measurements are the main topic of this book, remotely sensed observations are closely related and Chapter 19 illustrates what can be measured remotely and how the two sets of data complement each other. However, in addition, the same satellites have in many cases also been equipped with the capability of relaying data telemetered from remote field stations to a distant base. It is this capability that is the topic of the remainder of the present chapter.

Because environmental weather satellites have the dual capability of image-generation and telemetry, it is convenient to look at the characteristics that affect both functions first, and one of the important distinguishing features of satellites is their orbits. There are two types of orbit into which satellites with telemetering, remote-sensing and communications capabilities are normally placed – geostationary and polar.

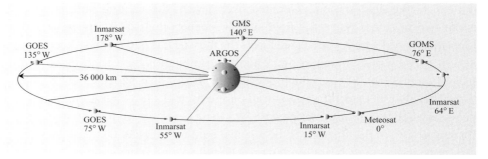

Figure 12.8. The principal satellites used for telemetering environmental data include Meteosat, the two US GOES satellites (that at 75° west being GOES East and that at 135° west being GOES West), the GMS satellite of Japan, the Russian GOMS and the three ARGOS satellites in polar orbit. In addition the four Inmarsat satellites are used for telemetering environmental data, although at present to a much lesser extent

Orbits

Geostationary orbits

Satellites with an orbital distance of 35 900 km from the Earth's surface circle the Earth exactly once a day. If placed in such an orbit around the equator and rotating in the same direction as the Earth, the satellite appears stationary in the sky, and is known as *geostationary-equatorial*. More precisely they should be termed *geosynchronous* (Fig. 12.8).

Polar orbits

Satellites that orbit the Earth over the poles can have any altitude required because they do not have to keep step with the Earth's rotation. Meteorological satellites in polar orbits fly at an altitude of about 850 km, giving an orbital period of around 100 min. The plane of the orbits of polar satellites used for imaging and telemetry is kept facing the Sun, the Earth rotating beneath it, and these satellites are thus known as sun-synchronous; a satellite passes over a different swath of ground at each orbit, gradually scanning the whole of the globe, with the Sun always at the same angle at each pass (Fig. 12.8) (See also Chapter 19 for more details). Polar orbits that are not truly polar are also used, in which the angle of the orbit is not exactly vertically north–south. Some communication satellites, such as the new Orbcomm network, use a mixture of orbits, equatorial, polar and skewed, to achieve full global communication at all times.

Orbits compared

Geostationary orbits present the satellite with the same view of the Earth (*footprint*) all the time, the whole circle of the globe being visible (Fig. 12.9). Immediately

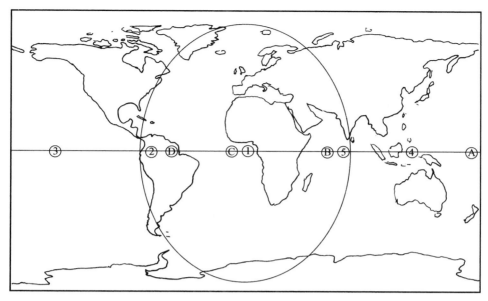

Figure 12.9. All the geostationary satellites have a footprint of similar area and shape. In this figure, for clarity, only that of Meteosat is shown, centred on point 1. The others are centred on the points indicated: 2, GOES (East); 3, GOES (West); 4, GMS; 5, GOMS. The Inmarsat satellites are centred at *A, B, C* and *D*.

beneath the satellite, the Earth's surface is seen square-on, with a resolution of up to 2.5 km × 2.5 km, but in all other directions away from the nadir point, the view is increasingly oblique, to a point where there is no view at all. In particular, the high polar regions can never be seen from geostationary satellites.

Polar orbits, in contrast, give an ever-changing view of the surface, the satellite sweeping out a path up and down the Earth and over the poles; each part of the Earth is viewed more or less vertically downwards and to much higher resolution, pixels representing areas as small as 30 m square or even smaller. However, repeat images may be many hours apart, even days or weeks.

From a telemetering field station's viewpoint on the ground, geostationary satellites appear at a fixed point in the sky, their angle above the horizon becoming progressively lower the higher the latitude of the observer and the further away the observer is longitudinally from vertically beneath the satellite. Higher than about 75° latitude, the satellite's angle above the horizon is too low for reliable telemetry, although this depends partly on the nature of the terrain. In valleys, for example, the cut-off latitude can be lower, perhaps no higher than Scotland. Because the satellite's position is fixed, telemetering field stations can use directional antennas to concentrate the radio wave into a narrow beam towards the satellite, thus making the best use of limited power. By this means, the 35 900 km to the satellite can be bridged with as little as 5–10 watts.

In contrast, a polar-orbiting satellite passes over a telemetering field station quickly and disappears from view until the next pass. So while a field station can telemeter via a geostationary satellite at any time, a polar satellite is only available for about ten minutes or less during each orbit. Further, while polar satellites pass over the poles at every orbit, at any point on the equator a polar satellite is only in range for telemetry about four times a day – even though it crosses the equator the same number of times as it crosses the poles. With an orbital time of 100 min there are around 14 passes a day at the poles. Thus telemetry via a polar satellite is possible only during the approximately 10-min overpasses, at a repeat interval dependent on latitude. Because the satellite passes across the sky, field-station antennas cannot be directed in any one direction and instead have to be omni-directional, directing the radio emission evenly in all directions. This reduces the signal level received at the satellite but, because polar satellites are around 25 times nearer than those in geostationary orbit, the signal level is adequate, although a higher radiated power of 20 watts has to be used.

As well as being a disadvantage, satellite movement relative to the ground can also be used to advantage since it results in a Doppler shift in the radio frequency received at the satellite from the telemetering field station which has some advantages (see the discussion, later in this chapter, of the Argos system).

The family of satellites

At the present time, most of the satellites used for telemetering environmental data are the weather and environmental satellites, put into orbit primarily as collectors of remotely sensed images. Those in geostationary orbit are Meteosat (Europe), GOES East and GOES West (USA), GMS (Japan) and GOMS (Russia). The Inmarsat network of four commercially operated geostationary satellites encircles the globe, offering an alternative to the weather satellites. Polar-orbiting satellites operated by the US National Oceanic and Atmospheric Administration (NOAA) are used for telemetering data through the French Argos system. All these various satellites are shown in Fig. 12.8. The Russian Meteor weather satellites are another set in polar orbit and EUMETSAT (the organisation managing the European Meteorological Satellite) plans to have a polar orbiting network of three satellites in operation by 2003, which will produce high-definition images (EUMETSAT 1997).

In all cases there are spare satellites in orbit as standbys, for use if an operational satellite should fail, and there is an on-going programme of launches to replace satellites nearing the end of their life and also to improve performance through new developments. Failures can occur and the need to fine-tune the orbits of the satellites from time to time requires on-board fuel, which gradually gets used up over the years; this gives satellites a finite lifetime.

There are also many national satellites, such as India's geostationary Insat, America's Domsat and Brazil's SCD-1 satellite (in equatorial orbit with an inclination of 25°, giving coverage of all Brazil and of all the world's tropical and subtropical countries). The last was designed specifically for environmental data collection. None of these can be dealt with in this short treatment. The situation will change considerably over the next decade, with new, commercial, systems being launched (as described in the last two sections of this chapter – Orbcomn and Iridium). Indeed the present weather satellites should be viewed as just the beginnings of environmental satellite telemetry.

But Meteosat is a convenient satellite to use to explain the general principles of satellite environmental telemetry, for it incorporates many of the features found in the other weather satellites, although telemetry via the commercial communications satellites is different, as will become clear later.

Meteosat

Meteosat is operated by the European Space Agency (ESA) and managed by EUMETSAT, its mission control centre being situated in Germany in Darmstadt and its primary ground station at Fucino in Italy (EUMETSAT 1996).

Data collection platforms

Field stations telemetering via satellite are usually referred to as data collection platforms (DCPs), but the term is used somewhat loosely and may refer to the whole field station or to just the electronics package containing the transmitter. Whichever satellite is used, the field stations are similar to a terrestrial radio telemetering station, being made up of sensors, logger and communications link, the only difference being that in this case the transmitter is directed at a satellite. A DCP performs identical functions to any other telemetering field station, making the measurements, multiplexing through them, converting them to digital form and storing them ready for transmission, as Fig. 12.1 shows (Strangeways 1990).

Transmission time-slots

Telemetry via the geostationary weather satellites is (generally) one-way only (simplex), the DCP transmitting its measurements at fixed times, these being allocated by the satellite operator. An address, a string of (usually eight) numbers to identify the DCP, is also allocated to each station. Time-slots are usually spaced at three-hourly intervals, to keep in line with the main use of all the weather satellites (weather forecasting by the NWSs), data being collected every three hours. More frequent time-slots are technically feasible (since the satellite is in constant view), but this needs special approval. Each time-slot has a duration of one minute, during which the station's measurements are transmitted to the satellite at a slow baud rate, typically 100 bits per second, along with housekeeping data such as battery voltage, DCP address and perhaps indications of equipment performance. This slow data rate restricts the amount of information that can be telemetered, but it is usually adequate for most environmental applications. Two consecutive time-slots can be arranged when data amounts require it.

Although transmissions are at set time intervals, the field station can nevertheless make measurements at whatever interval the user cares to choose, such as every 15 min, these being stored ready for transmission at the prescribed time. For many uses three-hourly transmissions are adequate, but if events are liable to unfold rapidly, such as in the case of flash-floods, widely spaced intervals may not be sufficient. It can also happen that occasional transmissions fail to get through, for one reason or another, and these data are lost in real time (although they may be retained in on-site memory for later retrieval). This can mean that no data are received for six hours, which is perhaps unacceptable for many real-time applications.

As DCPs transmit at set times, their clocks have to be precise, to avoid transmissions overlapping adjacent time-slots. To guard further against such interference between two stations, there is a buffer of one minute between each time-slot, allowing a certain amount of clock-drift. Transmissions out of time-slots are viewed with considerable disapproval by the satellite operators and immediate remedies requested. As DCPs are often sited at remote locations, which are usually difficult to reach, high reliability and good clock stability are important so that visits to readjust clocks or to reprogramme a system that has malfunctioned do not have to be made too often.

There is an exception to out-of-time transmissions in the form of *alert* or *random* messages, which can be sent at any time – on a special channel free of fixed-time-reporting DCPs. Because two DCPs might send an alert message at the same time, they send them several times, each DCP having a different repeat time. As noted earlier, this technique is widely used.

Polling DCPs

It is possible to poll DCPs, but not directly by the user from the base station. Instead, calls are initiated by the satellite operator, from the *ground station*. The call message contains the DCP's address and all DCPs (that have receivers) receive the call, only the one with that particular address responding. Such DCPs must be equipped, however, with a receiver, and these have to be powered-up all the time to await a call, which requires more power and makes the station much more complex and expensive. The great majority of DCPs are self-timed, transmitting at set time intervals.

Meteosat currently has 66 normal channels for DCP telemetry, but the situation is always in flux (the EUMETSAT web site is http://www.eumetsat.de). Channels are divided into *international* and *regional* categories, the former for DCPs that move from area to area, such as those on ships or planes, and the latter for fixed stations.

DCP hardware

A field station may take the form of a data logger connected to a DCP, the DCP containing just the transmitter, some logic and a small memory (Fig. 12.10), the logger and DCP communicating through their serial ports. In the simplest of cases the logger may be completely omitted, a sensor being connected directly to the DCP; an example of this is a telemetering rain station (Fig. 12.11) (Strangeways 1985, 1994, 1998). DCPs of this simpler type need to be programmed by the operator only to the extent of setting the clock: the time is input by the operator via a laptop PC or by a simpler dedicated synchroniser unit (Fig. 12.12). The time at which transmissions are to be made and the DCP's address are programmed into the DCP at manufacture, in these simpler DCPs, and these can only be changed by changing an EPROM memory chip.

The DCP and logger combination consume little power and most of the time are in a quiescent mode, coming into action only briefly to log the sensor readings, perhaps every 15 min, and to transmit the accumulated data, generally every few hours. Because the average power consumed is small, stations can operate using solar power – even in higher latitudes in winter.

Other DCPs are more sophisticated, containing all the elements, logger included, in one integrated package. Assembled in modular form with individual plug-in cards or rack-mounted units, the design becomes multipurpose, it then being possible to select a combination of cards to suit needs. Thus it is possible to construct a station able to be operated as a telephone or a UHF ground-based telemetry system or as one able to telemeter via satellite; or it can perform simply as a logger. A DCP of

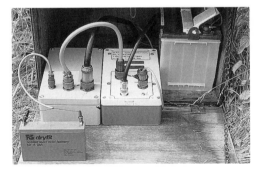

Figure 12.10. A data collection platform (DCP) of the simpler type (central) is here connected to a data logger (left) that records measurements from an AWS of the type shown in Figure 7.5. The car battery (right) powers the DCP, the small battery (left) the logger.

Figure 12.11. Although the simpler DCPs of the type illustrated in the previous figure do not have a logging capability, they can accept direct inputs from sensors producing suitable analogue voltages or pulses, in this example from a tipping bucket raingauge in the pit (centre). The Yagi antenna is directed at Meteosat from this Institute of Hydrology site in central Wales.

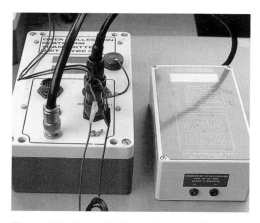

Figure 12.12. A DCP of the type illustrated in Fig. 12.10 (and in this figure to the left) requires little setting up. What programming is required is done using a 'synchroniser' (right), the main function of which is to set the DCP's internal clock precisely, to the second, to avoid transmitting out of the allotted time-slot. In addition, the synchroniser can perform diagnostic checks on the DCP and on the sensors and initiate test transmissions into a dummy load. A laptop PC could perform the same functions as the synchroniser.

this type is programmed by the user in much the same way as a logger, using a PC (Fig. 12.13). The precise time is input automatically from the PC's clock, but the times of the transmission slots, the DCP's address, the format the data should take (such as the SYNOP or CREX codes – see below) and the units in which the data

Figure 12.13. Although the basic DCPs illustrated in the previous figures can be connected to a data logger to allow a greater variety and number of sensors to be operated, DCPs of the type illustrated here perform the dual role of DCP and logger. They are also capable of more complex programming, such as being set up to telemeter the data in the SYNOP or CREX codes and to input their address number and details of the time-slots. Programming is achieved using a laptop PC (seen on top of the DCP case) in the same way that a logger may be programmed (Fig. 11.6).

are to be expressed must all be prescribed by the operator via the PC. This makes programming a much more complex matter than with the simpler DCPs. Some such DCPs have built-in keyboards and displays that can be used, rather than a separate laptop PC (Fig. 12.14), and some incorporate a *global positioning system* (GPS) that automatically resets the clock regularly.

From DCP to ground station

The DCP radio transmitter operates in one of the bands between the frequencies 401 and 402 MHz; these are spaced at 3.0 kHz intervals, with a power of between 5 and 40 watts, the actual power depending on the type of antennas used. In the case of fixed stations the signal can be beamed at the satellite using a Yagi antenna (Fig. 12.11), in which case 5 watts is sufficient, but with a moving station the

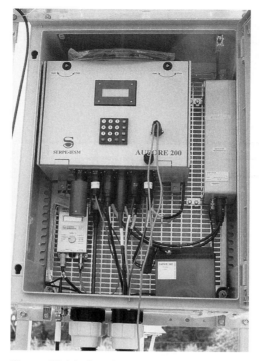

Figure 12.14. In a further variation on DCP design, illustrated here, the laptop PC is replaced by a built-in display and simple keyboard, although a PC can also be connected for more complex requirements, allowing less specialised staff to attend to the DCPs. This particular DCP also incorporates a GPS receiver that allows it to synchronise its clock regularly, and to confirm its position. It also has a removable back-up memory module (seen beneath the case at the left).

antenna must be omnidirectional and the radiated power must be at the higher level to achieve an adequate signal strength at the satellite. Upon arrival at the satellite, Meteosat retransmits the signal, at a frequency of around 1675 MHz, to the *ground segment* (Fig. 12.15). The ground segment is focused on the mission control centre at EUMETSAT's HQ in Darmstadt in Germany, the primary ground station being in Fucino near Rome; the two are connected by high-speed links via a commercial satellite, with a ground-based link as a back-up. At Darmstadt the data are quality controlled and then disseminated by two routes – over the *global telecommunications system* (GTS) and back to the satellite for retransmission to the user (EUMETSAT 1995a, b).

The global telecommunication system

The GTS is a communications channel interconnecting all the world's NWSs for the interchange and dissemination of meteorological data for weather forecasting.

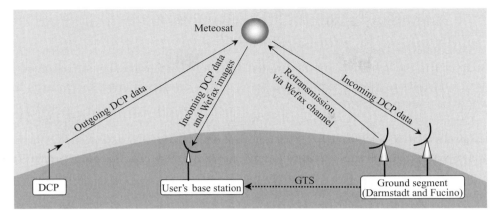

Figure 12.15. Typically every three hours, the field-station's DCP transmits its data to Meteosat during a one-minute time-slot, the data being received in real time at the ground segment. Here the data are disseminated either over the global telecommunication system (GTS) or are retransmitted over the Wefax channel for direct reception by the user's base station.

Data put onto the network must be presented in internationally agreed WMO codes, such as SYNOP. Whereas SYNOP is intended primarily for reporting meteorological measurements from land stations, there are other codes, such as SHIP for observations from sea, HYDRA for hydrological stations and the newer CREX code, which has provision for encoding both meteorological and hydrological measurements. Details of the codes are not relevant here, but if used they have to be followed to the letter or they will fail (for more information see WMO 1988). However, all of these codes will soon be replaced by a new one called the *Buffer* code. Programmable DCPs (Fig. 12.13) allow these codes to be set up by the user via a PC to suit the variables being transmitted, but this is quite a specialised matter. The simpler DCPs and loggers may not be capable of this and may have to be set up at manufacture.

Data sent over the GTS can only be received if access to the GTS is available, and this is generally only possible in co-operation with an NWS. This can be something of a disadvantage to all except operational meteorologists and is a drawback to the GTS when it is used for anything other than meteorology.

The DCP retransmission system, and Wefax images

In the case of Meteosat (only), however, there is an alternative to the GTS, for as well as being circulated over the GTS the data can be retransmitted from Meteosat for direct reception by the DCP operator. To understand how this is done, it is necessary to digress into how Meteosat's remotely sensed images are collected and disseminated, for they are closely linked with data transmission. This digression, however, also provides some necessary background for Chapter 19, on remote sensing.

Meteosat's images are generated by an on-board multispectral radiometer (see Chapter 19), the images being transmitted digitally to the ground station in Fucino, Italy, and then relayed to the European Space Operations Centre in Darmstadt, where they are processed.

For immediate use, the images then follow two paths back via the satellite for retransmission to users. The first channel carries high-quality digital data for reception by what are known as *primary data user stations* (PDUSs). These need receiver dishes typically two to three metres in diameter. It is the second path, which carries the images in the form of an analogue signal known as Wefax (Weather facsimile), that is of concern here, for this channel also carries the retransmitted DCP data (Fig. 12.15). Image definition is lower than in the case of PDUSs, but the signal is simpler to handle (EUMETSAT 1998). These transmissions are received on *secondary data user stations* (SDUSs), which not only receive the Wefax analogue images but also, interspersed with them, the retransmitted DCP data. Transmissions to SDUSs can be received on small dishes, typically 1.5 m diameter (Fig. 12.16), because the signal level is high, allowing low-cost reception of the images (Harris 1996). The signal is down-converted from its received frequency of 1694.5 MHz to 137 MHz at the dish, allowing cables of up to 100 m length between the dish and the receiver. The format taken by the Wefax transmissions is relevant to the reception of DCP data and so needs explanation, as follows.

Wefax images build up line by line, as in a normal TV picture but much more slowly, taking $3\frac{1}{2}$ minutes rather than $\frac{1}{25}$th of a second, the incoming data being handled by a PC and displayed on its monitor or stored on disc. The images cover different sections of the globe and different spectral bands, scanned by Meteosat during the previous 25 min (see Chapter 19). An image is sent every 4 min, taking $3\frac{1}{2}$ min to build, followed by a half-minute gap before the next image starts to arrive. During this 30-s gap, the DCP data are retransmitted from Meteosat, all DCP data received at the Meteosat ground station during the previous 4 min being disseminated (Fig. 12.17). A low-cost SDUS receiver is thus all that is required to obtain the data from DCPs in near-real time (with 4 min delay at most). The data rate is much faster than the 100 bits per second of the up-link used to send the data to the satellite.

The data from all DCPs can be viewed, printed out or stored on disc, or just those with the required addresses can be selected.

While at the Institute of Hydrology in the 1980s, I explored the use of Meteosat for use in environmental telemetry by operating 10 DCP in various parts of the UK from Southern England to the Scottish Highlands as well as for a year in Antarctica. These various stations transmitted rainfall, river level and AWS data, a few examples being seen in Figs. 12.10, 12.11, 12.12, and 7.5. A Wefax receiver (Fig. 12.16)

Figure 12.16. A small dish antenna is sufficient to receive the retransmission of DCP data via the Wefax channel from Meteosat. Unfortunately this very useful facility is not available from any of the other geostationary weather satellites, although direct reception is possible from the polar-orbiting Argos satellites under certain circumstances.

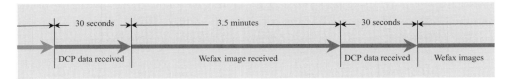

Figure 12.17. DCP data transmitted from field stations during the previous four minutes can be received via the Wefax channel in the half-minute gap between picture transmissions, which take 3.5 minutes to build, line-by-line.

at Wallingford performed well apart from one problem with the down converter that was easily put right. The complete system (DCPs, satellite, retransmission and receiver) functioned extremely well and the great value of satellite telemetry was made very clear to me.

However, the Meteosat analogue transmissions are due to be phased out over the next few years, possibly by 2003, to be replaced by a new digital service called *low-rate information transmission* (LRIT) from the geostationary satellites (and *low-rate picture transmission* (LRPT) from the polar-orbiting satellites). Present receiving equipment can probably be updated (information on this, on dates and on costs can be obtained from EUMETSAT's web site).

Cost of using Meteosat

There is no simple answer to the question of whether the use of Meteosat to telemeter data has to be paid for, and if so how much it costs. It depends partly on the nature of the data: purely meteorological data of synoptic value are handled free if the data are coded for SYNOP transmission on the GTS. A charge may be made for other environmental data, such as hydrological or oceanographic data. If the project is sponsored by the WMO the use of Meteosat will probably be free. It also depends which country is involved, those making a financial contribution to the European Space Agency (ESA) budget (the 17 member states of EUMETSAT) being more likely to get free use. If a charge is made, it will be in the order of £1000 per year, per field station making three-hourly transmissions. This amounts to 2920 transmissions a year, making the cost of each transmission about 34 pence (50 cents). But charges change with time and EUMETSAT should be contacted; the above is a rough guide only.

GOES

The *geostationary operational environmental satellite* (GOES) is the American equivalent to Meteosat and is operated by the National Oceanographic and Atmospheric Administration's (NOAA's) national environmental satellite, data and information service (NESDIS), the ground station being at Wallops in Virginia. There are in fact two satellites (Fig. 12.8), GOES East and GOES West, covering half the Earth's surface including the entire Pacific region. Both collect images and support DCP telemetry, just as does Meteosat – but not in quite the same way.

Cost of use

An important difference between GOES and the other three geostationary satellites is that it is not intended exclusively for NWS use and weather forecasting. As its name indicates it has a broader role, encompassing environmental applications generally, and can normally be used free of charge for telemetering most types of environmental data. However, the three-hourly synoptic time-slot seems to be adhered to. Any country within the range of the two GOES satellites can use them.

Routes taken by data

While data suitably encoded in the SYNOP format are distributed over the GTS, there is no retransmission of DCP data from the GOES satellites on the Wefax channel (although there is a Wefax image transmission service). There are, however, a number of alternatives: DCP data can be received by users in the US through a DOMSAT spacecraft while those outside its range can use the GOES dial-in service, which allows users (in any country) to download their data by telephone by calling a number at the Wallops Command and Data Acquisition Station, the only cost of using the GOES satellites being the cost of the telephone calls. Web access through a client/server setup is also possible. Currently the system is being upgraded and will offer some automated e-mail access. Many details can be seen at http://noaa.wff.nasa.gov.

For users outside the USA who are operating a large network of DCPs, it is possible to install a replica of the NOAA ground station and thereby to receive the DCP data directly from the satellite. However, this type of receiver is very much more expensive and complex and requires a five-metre receiver dish. It is not, therefore, an option that many users can afford. These various options are shown in Fig. 12.18.

I was involved in planning a river and rainfall telemetry system for Honduras a few years ago, for the Danish Hydraulic Institute, using the GOES East satellite, to provide data on the three rivers that feed the large El Cajon dam that supplies most of the country's electricity. During field excursions to existing river gauging stations we visited a site where a man, living with his young son in a one-roomed

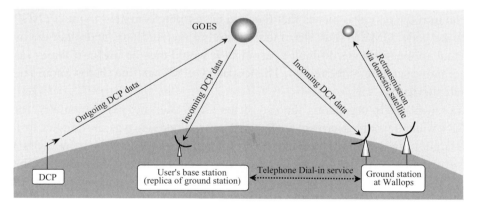

Figure 12.18. DCP transmissions can be received via three routes from the GOES satellites. In the US, they can be received directly by retransmission via a domestic satellite. Elsewhere there is a telephone dial-in service to the ground station. For larger organisations, it is possible to install a replica of the ground-station receiving equipment at Wallops, although this is an expensive option.

small hut by the river, took visual level readings every few hours. In a corner of his windowless shelter, in which detail was lost in the surprisingly intense darkness, one bright red LED glowed in the corner from a VHF radio that he used to send his readings to base. On another visit that day, the gauge-reader had to be got from his house in the afternoon, somewhat unsteady on his feet and in danger of falling into the river quite a way below, which he skilfully avoided doing. I often feel that this sort of telemetry might be best not modernised. It works well. I have felt this many times. I certainly feel it should be retained as back-up. I felt the same over two projects in India. In one, river levels and rainfalls were telephoned- or radioed-in and typed on an old typewriter in a sparsely furnished office with heavy curtains. The list was then reproduced on an aged cyclostyle machine with a handle, to be carried away and distributed by hand. It had worked well for 25 years. There were no computers in sight. I felt a half way stage to modern telemetry would have been better and more likely to succeed than the proposed jump to full automation. An upgrade of an existing system may well be more successful than starting from scratch.

GMS

The *geostationary meteorological satellite* (GMS) is operated by the Japan Meteorological Agency (JMA), and, as its name spells out, it is intended for synoptic and weather forecasting use. Indeed, other environmental applications are not allowed unless the DCP also takes readings appropriate to forecasting, such as barometric pressure or temperature. Nor is there any retransmission system or dial-in service, data having to be collected via the GTS. In addition to the organisational difficulties this can introduce, it also means that the data have to be formatted in the SYNOP or similar code. GMS is thus very much geared to meteorology, and other environmental applications are difficult to arrange. It would thus be useful if there was an alternative satellite in the region. The Russian GOMS satellite (below) could be such an alternative.

I was involved in organising the use of the GMS satellite, experimentally, for a WMO water resources project in Papua New Guinea. (It was an adjunct to a much larger project to modernise the river level and rainfall data logging network.) Once everything is set up with the JMA it works well, but it took a long time. The project also demonstrated to me just how important the reliability of equipment is when installed in remote places. Getting to the test site involved a commercial flight from Port Moresby to the Highlands followed by a helicopter flight into very remote rainforest that occupied the whole day. Failed equipment at such remote sites is an expensive matter to correct and is ill-afforded by a relatively poor country. Back at the NWS in the capital I waited in trepidation by the telex machine for the first incoming GTS transmissions from the station – and then, each day, for the transmissions to continue. As intriguing as these missions are, there is a large element of uneasiness, complemented by jet lag.

GOMS

The Russian *geostationary operational meteorological satellite* (GOMS) was launched in 1994 although it was not fully operational until more recently. However, it would seem that DCP data may only be accessible over the GTS, as with the GMS satellite, which again would limit its usefulness to broader environmental applications. I have not been able to get firm confirmation on this yet. Time will tell, however, exactly what the satellite can offer, but it does seem that, with GMS and possibly GOMS not offering a retransmission facility nor a dial-up service for DCP data, that this half of the globe is poorly served for environmental telemetry by the geostationary weather satellites, although well served for meteorology.

Indeed, with Meteosat's restrictions over non-meteorological use (from the payment point of view), only the area covered by the two GOES satellites is well served, and even here it is necessary to collect DCP data by telephone (which could prove expensive for a large network, for an extended period, in a poor country). Something needs to be done to open up the use of all these satellites for hydrology, oceanography and other environmental applications. Or perhaps the Inmarsat system, or the Argos system of polar-orbiting satellites, is the solution, although cost is again a problem for the poorer countries.

Data could now be distributed via the Internet by e-mail at very low cost and this would appear to be starting. However, it depends entirely on whether satellite operators are prepared to provide this option. It would certainly help. It is difficult to get up-to-the-minute information on such matters and it is not on their web pages.

Inmarsat

The international marine satellite system (Inmarsat) started operation in 1982. It is a commercially operated communications network based on four geostationary satellites (Fig. 12.8) giving worldwide coverage (except in the high-latitude polar regions, which are out of range of all geostationary satellites). It was initially intended for marine communication, and this is still one of its more important applications, but it now supports the transmission of voice, fax, e-mail, video and data for a wide variety of uses, including the telemetry of environmental data. Inmarsat is operated in several modes.

Inmarsat-*A* is the original service (1982) offering direct-dial telephone, fax and e-mail. It requires a dish antenna and is full duplex. Thousands of ships are equipped with it. Inmarsat-*B* was introduced in 1993 and is the same as *A* but uses digital techniques rather than the analogue mode of *A*, allowing more efficient use of the satellite and so a reduction in cost. Inmarsat-*E* is for global maritime distress alerts. Inmarsat-*M* is a more compact, portable system, used, for example, by reporters in remote areas. The sound quality is lower than that of *A*, more akin to telephone quality. It can also be used for low-speed fax transmission. Inmarsat-*D* is the latest introduction, using pocket-sized terminals to transmit messages of up to 128 characters. Inmarsat-*C* is a compact system used for the transmission of low-speed digital data or text messages. This is done on a two-way *store-and-forward* basis, not in real time. (Inmarsat-*A*, -*B* and -*M* offer full-duplex communication.) Inmarsat-*C* is the mode most suited to environmental telemetry and is our only concern here, although *M* and *D* might later find a place in such applications.

Routes taken by data

Figure 12.19 illustrates the route taken by data from a field station to and from the base station using Inmarsat-*C*. The data rate is 600 bits per second. There are two ways of operating such a system: self-timed or polled. In the former, field stations send their data at fixed times to the satellite, which retransmits them to one of its *land earth stations* (LESs). There are many LESs around the world – at a recent count, about 36. The received message can then either be held in an e-mail-like box at the LES until the base station collects it by telephone, much as in the case of the GOES dial-in service, or it can be forwarded to the base station as soon as it is received, provided that the base station is in readiness to receive it. Field stations can send their data as often or as infrequently as wished, there being no need to

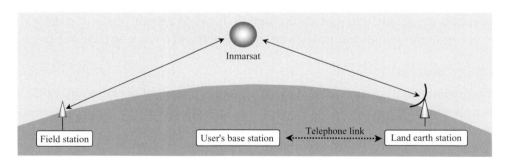

Figure 12.19. Unlike the weather satellites, the Inmarsat satellites are duplex in operation, allowing two-way communication. The field station transmits its measurements via the satellite to a *land earth station* (LES), where they are stored and then forwarded to, or collected by, the user's base station, by telephone. The base station can also send information to the field station by a similar but reverse route. Although not illustrated here, the base station can also communicate with the LES via the satellite, using an Inmarsat field station in place of a telephone link, by a so-called *mobile to mobile link*.

fit in with three-hourly transmissions as there is with the geostationary weather satellites, and therein lies one of Inmarsat's advantages.

Messages can also be sent to the field station from base via the LES through a reverse route, and so a polling network can be operated. But there is a problem with this, for, in order to receive a call from base, the field station receiver must be switched on, and this may take more power than is available at many field stations. To get around this difficulty, it is possible to programme the field stations to call the LES at preset intervals, say daily, to collect any new instructions sent to the box number from base. The receiver then has to be powered up only for a time long enough to receive the new instructions, which might concern such matters as the time interval at which they should report. While this is not as versatile as being able to poll field stations in real time, it does give considerable more flexibility than is provided by fixed three-hourly transmissions to the weather satellites. This compromise can also be used with UHF ground-based telemetering field stations, which must also conserve power for the same reasons.

Although the Inmarsat satellites are geostationary, and so stay at a fixed point in the sky, Inmarsat terminals are normally mobile (for use on trucks and ships) and so an omnidirectional antenna is used. The antenna is small and unobtrusive, and so is less subject to vandalism than the much larger, higher-profile DCP Yagi antennas. However, for fixed operation, it is possible to use a helical antenna, directed at the satellite, reducing the transmitter power required and thus the power consumption of the field station.

Cost

The cost of communicating via an LES varies from country to country, and so it is not possible to give a general estimate. The service provider in each country must be contacted. The cost also depends on the amount of data sent. If this can be limited to 32 bytes, the cost can be kept low (for latest details see http://www.inmar.sat.com.).

Argos

The Argos system differs from the other satellites, being in a polar orbit. The Russian Meteor system is similar to it. Argos is run on a commercial basis, by the French space agency (CNES) through the company CLS (Collecte Localisation Satellites) based in Toulouse, in co-operation with NASA using the NOAA satellites.

Method of operation

Currently there are two Argos satellites in general operational use (although a third is in orbit and can be used if necessary) and, with orbital times of about 100 min, each satellite passes over the poles 14 times a day, giving a total of 28 passes. The total number of passes by the two satellites combined, at any point near the equator, is six to eight. Because the satellites are not visible all the time, transmissions from field stations cannot be made at fixed intervals, but instead are repeated continuously every 100–200 s. When a satellite comes into view it will receive one or several transmissions, depending on the exact path it takes relative to the field station, before passing beyond the horizon again (it may not pass immediately overhead).

Because the satellite is moving, an omnidirectional transmitter must be used. This means that the transmitted power cannot be concentrated by a Yagi antenna towards the satellite. However, because the satellite is in a low orbit and passes close to the field stations (compared with the distance to a geostationary satellite), less radiated power is required and this is why an omnidirectional antenna is practicable. Because of the low orbit, the satellite is only in radio contact up to a distance of about 2500 km. Twice this range has been reported, and may at times be feasible, but a distance greater than 2500 km cannot be wholly relied upon in planning a network. Field stations in valleys, for example, will have a more restricted view of the satellite (this is also true for geostationary satellites), which reduces the range of communication and also the time during which communication is possible; in flat terrain, the satellite is in view for a maximum of about 10 min at each pass, but in valleys it could be less. Since the data are transmitted every 100–200 s, they will be received several times during each pass. But because of this rapid repetition rate, the message length is limited to 64 bytes, which may not be enough for some applications, especially near the equator where passes are less frequent.

Routes taken by data

Figure 12.20 shows the path taken by data via Argos. Upon reception at the satellite, the message processed in two ways: it is immediately retransmitted and is also stored on board the spacecraft. With suitable equipment, users can receive the retransmissions directly in real time, with no delay, provided that both the field station and the receiver have a view of the satellite at the time of transmission. If not, then the data must be collected later, after the satellite has passed over one of the Argos ground stations – one being in France and two in the US. As it passes over a ground station, the satellite downloads all of the data collected since its last overpass and these are either disseminated on the GTS, sent by post, or collected by telephone via a dial-in service similar to that for GOES. But with ground stations only in France and the US, telephone calls could be expensive for some countries, although e-mail is now a cheaper alternative being made available.

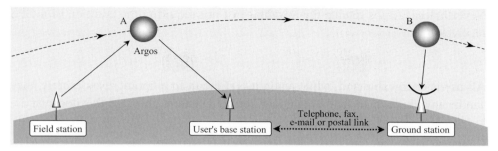

Figure 12.20. Argos field stations transmit their data every 100–200 s (on a frequency of around 401 MHz). When the satellite passes into view, the data are received and immediately retransmitted (on 136–138 MHz). This retransmission can be received by a base station (provided that both the field and base stations are in view of the satellite at the same time – position A). This enables direct reception of data up to a radius of about 2500 km centred on the base station. The transmissions are also stored on the satellite and retransmitted to one of the three ground stations in the US and France (position B) as the satellite passes over. This provides an alternative to bases that are out of range of direct reception or in situations when real-time data are not required.

Avoidance of interference between DCPs

To reduce the probability that two field stations will transmit at the same time and thus interfere with each other, slightly different repetition intervals are allocated to each field station in any one area (in the range from 100 to 200 s). Should, despite this, two stations transmit at the same time, there will be a subsequent second or third opportunity to send the data while the satellite is still in view, since the stations, having different repeat times, will not interfere again for some time. If there are many stations then the chance of simultaneous transmissions increases, but there is a second strategy to overcome interference, as follows.

Position location

The Doppler shift of the received radio signals that occurs because the satellite is moving relative to the field stations results in each DCP's transmissions being received at slightly different frequencies, since each has a different geographical position, and this is sometimes sufficient, on its own, to separate two simultaneous transmissions. The shift in frequency has an additional benefit, for it can locate the position of field stations, so that not only can data be collected from the field stations but the station's location can also be determined and transmitted. This is one of the strengths of the Argos system where field stations are mobile, such as drifting buoys at sea (Chapter 17) and in animal experiments. No other satellite with a DCP relay facility has this capability, although with GPS now available, which can be used as one of the inputs to a DCP, this advantage is lessened.

Nevertheless, Argos retains the advantage of not requiring the addition of a GPS receiver.

Cost

All use is chargeable and, while again it is difficult to pinpoint costs exactly, they can be anywhere up to €3640 per year per field station, although this does vary.

Polar versus equatorial cover

Argos and Meteor are the only weather satellites that cover the polar and high-latitude regions at present (although EUMETSAT is to introduce three polar satellites in 2003). Indeed, they cover these high latitudes better than the equator. But therein also lies one of their disadvantages for, at the equator, passes are limited to six or seven times a day, with occasional gaps of up to $6^1/_2$ hours. But in those large regions of the world covered by the GMS and GOMS satellites, which lack a retransmission facility or a dial-in service and which require the dissemination of data in the SYNOP format over the GTS, Argos may be seen as the better choice, even if it has to be paid for, despite its reduced time coverage at lower latitudes.

Orbcomm

A new network of satellites is now in operation in low Earth orbit, known as Orbcomm, one satellite always being in range from anywhere on Earth. They are not weather satellites but commercial communication satellites intended to carry very short rapid bursts of digital data. There is already action to use this network for environmental telemetry, but it is too new to comment upon further at the moment. However, it would seem that the system will be most economical if used to send very small amounts of data (6 to 250, typically 20 bytes), not the larger volumes often sent through the weather satellites. See http://www.orbcomm.com.

Iridium

The Iridium network of satellites became operational in March 2001 and is now a viable system for environmental telemetry. It is composed of 66 low Earth orbit (LEO) satellites with six spares covering the whole globe, operated by Boeing. It allows hand-held phones to communicate directly with the satellite from anywhere on the globe and to interface to the ground telephone networks. This allows, in theory at least, telemetry via Iridium as if it were a land-based mobile phone network. Latest details can be found at http://www.iridium.com.

My investigations into using satellites other than the weather satellites for telemetry is that the devil is in the detail – of getting the various units to communicate with each other, for example baud rates and compatibility of the ASCII codes used by the satellite equipment and the user's logger/DCP, or signal levels and power supply requirements. But sooner or later, probably sooner, the commercial communication satellites will be the route by which much environmental data will be telemetered (NWS apart, who will probably continue to use the weather satellites). They offer in effect a mobile telephone link from anywhere on the globe. What might be more of a problem is cost, especially if used by a country with a limited budget. Those who can get the use of the weather satellites free will obviously opt for them if they are otherwise appropriate.

Which of all the above satellites is best suited for environmental telemetry will, therefore, and for some time, depend on very many factors, all of which have to be weighed against each other. And as the situation is changing rapidly, the above is just a snapshot of the situation in mid 2003. I suggest you watch the web sites, from where you can e-mail questions.

References

EUMETSAT (1995a) Meteorological data distribution, user guide. EUM UG 01.
EUMETSAT (1995b) Data collection system, user guide. EUM UG 02.
EUMETSAT (1996) The Meteosat system. EUM TD 05.
EUMETSAT (1997) The EUMETSAT polar system. EUM BR 06.
EUMETSAT (1998) Meteosat high resolution and Wefax imagery. EUM UG 03.
Harris, L. (1996) *Satellite Projects Handbook.* Newnes (Reed Elsevier), Oxford ISBN 0 7506 2406 X.
Schafer, G. L. & Verner, J. G. (1996) SMOTEL, into the year 2000. In: *Proceedings of the American Meteorological Society's 12th Conference on Biometeorology and Aerobiology*, January 1996, Atlanta.
Strangeways, I. C. (1985) Automatic weather station and river level measurements telemetered by data collection platform via Meteosat. In: *Proceedings of an International Workshop on Hydrologic Applications of Space Technology.* Cocoa Beach, pp. 194–204.
Strangeways, I. C. (1990) The telemetry of hydrological data by satellite. *Inst. Hydrol. Rep. No. 112*, p. 60, ISBN 0 948540 222.
Strangeways, I. C. (1994) Satellite transmission of water resources data, Technical Report 42. In: *Proceedings of the WMO Regional Workshop on Advances in Water Quality Monitoring*, Vienna, pp. 292–301. WMO/TD No. 612.
Strangeways, I. C. (1998) Transmission of hydrometric data by satellite. In: *Hydrometry: Principles and Practice*, second edition, ed. Herschy, R. John Wiley & Sons, London, pp. 245–64. ISBN 0 471 97350 5.
Strangeways, I. C. & Lisoni, L. (1973) Long-distance telemetry of data for flood forecasting. UNESCO. *Nature Res.*, **IX**, 18–21.
WMO (1988) *Manual on Codes*, Volumes I and II. WMO No. 306.

13

Visibility

When the sun rose there was a white fog, very warm and clammy, and more
blinding than the night, it did not shift or drive; it was just there, standing all
round you like something solid.

Joseph Conrad *Heart of Darkness* (from a boat far up the Congo river).

Fog and mist figure deeply in the human psyche, casting an air of mystery and
danger. Ships and boats still rely on reports of visibility, as now do aircraft, despite
the arrival of radar; and lighthouses remain with us, albeit automated. But as a
subject of scientific scrutiny, visibility is relatively new. It is surprisingly complex.

The variable and its history

Bouguer (1760), in his classic work on photometry, has a chapter on the transparency of the atmosphere. Thirty years later de Saussure (1789) (also inventor of
the hair hygrometer) described a 'diaphanometer' for 'measuring the clearness of
air', which although not very successful showed that he was aware of 'air-light'
(below). Wild (1868) was the first to use photometric methods to measure the luminance of distant objects, but his theory was wrong, being based on the idea of
absorption rather than scattering. Shortly after, however, the British mathematician
Lord Rayleigh (1871a, 1871b, 1871c, 1899) wrote his classic papers on the scattering of light by air molecules and by small spherical particles, including the first
correct explanation of why the sky is blue. Following this, and based largely on the
work of Mie (1908) and Wiener (1907, 1910), many other papers followed. Weber
(1916) made the first measurements of luminance, but did not go the next step to
relate it to visual range. This important step was taken by Koschmieder (1924) and
there then followed a series of investigations to test the validity of his equations in
practice. At first, experimental work was done chiefly in Germany, but later in

Measuring the Natural Environment, second edition, Ian Strangeways. Published by
Cambridge University Press. © Ian Strangeways 2003.

France, the US and the UK. The Second World War stimulated research on the subject particularly in the US and the UK, during which time the whole subject changed dramatically.

The first instrument to measure visual range came out of the First World War, designed by Wigand (1919) and Jones (1920), but they were of low accuracy. After Koschmieder's work, many *telephotometers* (measuring by eye or instrument) were devised, but not widely used in meteorology. Indeed Middleton says, in his detailed study of the subject (1952), that no instrument was in widespread use at meteorological stations, not all the fault lying with the designers, a barely veiled criticism of the conservatism of the meteorological community.

Bearing this in mind it is not surprising that visibility was originally defined in terms of an intuitive quantity to be estimated by eye, and visual estimation is still widely used. But such observations are partly subjective, and to put measurements on a more quantifiable footing, the WMO introduced, in 1971, a new measure of visibility – the *Meteorological Optical Range* (MOR) (WMO 1990a). This is the same range as is judged by eye, but expressed in terms that can be quantified mathematically and measured objectively by instruments.

If the atmosphere contained no water vapour, visibility would extend to 350 km (given a suitable vantage point). But containing water vapour, air becomes an *aerosol* (or aerial colloid – a suspension with air as the medium: 'aerosol' refers to the whole system not just to the suspended particles). In addition to water, the particles include *lithometeors* (non-aqueous particles) such as smoke from industry and burning forests, minute living organisms, dust from sandstorms, volcanic eruptions and from space and salt particles from the sea. But the most important particles are water droplets and ice crystals, the water droplets ranging in size from 1 mm to 0.01 μm. There is no mechanism for the conversion of water vapour to liquid or solid form in the atmosphere that does not involve minute nuclei in the form of groups of several water molecules, sulphuric acid, combustion products and sea salt, around which the drops or crystals form (Simpson 1939).

Units and terms

Some radiation units are described in Chapter 2. However, in the case of visibility, different units are used. There are two types of unit that it would be possible to use, *photometric* units, being related to the eye, and the radiant energy units of physics. They are in fact the same thing, but with different units. Since visibility is essentially a matter of vision and of light, photometric units are the most appropriate to use, although the physical units might be marginally more applicable to instruments. To avoid confusion, through mixing them, the photometric will be used exclusively throughout. These units are:

Luminous flux (*F* or *Φ*): Power emitted or received as light (lumen).

Luminous intensity (*I*): Luminous flux emitted per unit solid angle (candela).

Luminance (*L*): Luminous intensity per unit surface area (candela m^{-2}).

Illuminance (*E*): Luminous flux received on unit surface area (lux or lumen m^{-2}).

Note: *Luminance* is also known as *brightness*.

Intensity is also known as *candlepower*.

Illuminance may be called *illumination* when the light falls on a surface, or *flux density* when there is no surface.

Other terms include:

Luminance contrast (*C*): Ratio of difference between the luminance of an object and of its background to the luminance of the background.

Contrast threshold (*ε*): Smallest level of contrast at which an object can just be distinguished by eye from its background (nominally taken to be 0.05).

Illuminance threshold (*E$_t$*): Smallest illuminance an eye can detect as a point source of light against a background of specific luminance. (So varies according to lighting conditions.)

Extinction coefficient (*σ*): Fraction of initial luminous flux lost while travelling along unit length of atmosphere. (Attenuation is due to both absorption and scattering, but absorption is negligible compared with scattering.)

Transmission coefficient (or *factor* or *Transmissivity*) (*T*): Fraction of initial luminous flux remaining after travelling along a path of *unit length* of atmosphere. The term *Transmittance* (or *Transmittancy*) (also symbol *T*) is the same, but the term is used when the length of the path is defined, for example in the case of a transmissometer. Transmittance is usually expressed as a percentage.

Visibility by day: The greatest distance at which a black or dark object of suitable dimensions, located near to the ground, can be just seen *and recognised* against a scattering background of fog or sky (recognition is critical).

Meteorological Optical Range (*MOR*, symbol *M*): The length of a path through the atmosphere required to reduce the luminous flux in a collimated beam from an incandescent lamp at colour temperature 2700 K to 0.05 (5%) of its original value.

Airlight is light from the sky and Sun scattered by particles suspended in the atmosphere, usually in the form of water droplets, but also as dust, long-known by artists as *aerial perspective*, or *atmosphere*. Air molecules also scatter the light, but only to a slight extent. Airlight is the main factor limiting daytime horizontal visibility.

The above units are related as follows:

Basic to visibility measurement is the *Bouguer–Lambert law*:

$$F = F_0 \, e^{-\sigma x}$$

where

F = flux remaining after travelling over distance, x, through the atmosphere

F_o = the initial flux at source

σ = the extinction coefficient

An alternative way to express light attenuation during its passage through the atmosphere is by the *Transmission Coefficient*, defined as:

$$T = F/F_o$$

where F and F_o are as above. So:

$$T = e^{-\sigma x} \tag{1}$$

MOR (M), by definition, is when $T = 0.05$, so:

$$0.05 = e^{-\sigma M} \text{ (where } x \text{ is replaced by } M) \tag{2}$$

This reduces to:

$$M = 3/\sigma \tag{3}$$

But Koschmieder's law (1924) states that

$$C_x = C_0 e^{-\sigma x}$$

where C_x = the apparent contrast of an object at distance x, as estimated by eye.

C_0 = Contrast close to the object.

But visibility by eye is defined as when $C_x = 0.05$, so

$$0.05 = e^{-\sigma x}$$

Comparing this with eq. (2), above, shows that when the apparent contrast by eye is 0.05, the object is at distance M, the MOR. Thus the visibility range, as estimated by eye, and the MOR, as measured by an instrument, are the same distance.

(A complete mathematical treatment, with full derivations, can be found in Middleton 1952, with a shorter analysis in Met. Office 1982 and WMO 1996.)

Estimating visibility by eye in daylight

Visibility in daylight is assessed by observing suitable objects at known distances. The objects must be black or dark, should stand above the horizon and subtend an angle of at least 0.5°, in both width and height, but must not be more than 5° in width. (A hole of 7.5 mm diameter in a card, held at arm's length, subtends an angle of 0.5°, as does an object 10 m wide at 1 km.) Beyond 1 km, it is usually necessary to use topographic features such as a hill top, but as these can be covered by low cloud or change colour with the seasons, they are not ideal. If sky is not possible as the background, the distance between the object and the terrestrial background should be at least half the distance between the object and the observer.

An object remains visible in daylight so long as there is sufficient contrast between it and the background. If the luminance of the object is L_o and that of the background horizon is L_h, the contrast is defined as:

$$C = (L_o - L_h)/L_h$$

which can range widely but which rarely exceeds 10. As L_o approaches L_h, a value of C is reached when the eye can no longer distinguish object from background, a condition known as the *contrast threshold*, ε. This value for the average eye is 0.05 (or 5%). This is also the definition of MOR (see above). To obtain a good estimate of visibility it is clearly necessary to have suitable objects up to a distance of many kilometres at regular intervals. This may not always be possible.

On open plains and at sea, or when the horizon is very close as in valleys, or when there are no suitable objects to view, it is not possible to make visual esti- mates of visibility except when visibility is very low. In such situations, values of MOR (greater than those for which close objects can be used) must be estimated by less precise methods. Such expedients as observing, in a general way, the de- gree of transparency of the atmosphere have to be used by looking at the most distant object available and judging the distinctness of its features and colours. But this gives only a crude estimate. In such situations, instrumental methods are necessary.

Accuracy and calibration

To minimise errors, the observer must have normal vision and be properly trained; even then, interpretation will vary from person to person. If the observer's contrast threshold is close to 0.05 (using the criterion of *recognition* of the object – see 'Visi- bility by day', above), the estimate of visibility will be in reasonably good agreement with the MOR as measured by instruments (see below). Middleton (1952) found that the mean threshold of 10 trained airmen was 0.03, with a range (for individual observations) of from 0.01 to 0.2. WMO tests found that manual visibility estimates were on average 15% higher than instrument measurements. If visibility varies with direction, the lowest value is used (WMO 1990b).

Estimating visibility by eye at night

The change from daylight to night-time does not alter visibility and so observations made at night should report the same visibility as they would if it were daylight.

The most suitable object for determining night-time visibility is an unfocused point-source of light at a known distance. Lenses and mirrors should not be used to direct the light towards the observer. Flashing lights can be used so long as the flash is at least one second long.

Since the atmosphere transmits colours to different extents, coloured lights should not be used (beyond about 100 m) as considerable errors can result. Sensitivity of the eyes also varies with colour, in bright light peaking at 0.56 μm with cut-offs at 0.42 and 0.73 μm. In low level illumination, sensitivity peaks at 0.51 μm with cut-offs at 0.35 and 0.64 μm. There is also the matter of eye adaptation to the dark. In daylight, the eye operates in the *photopic* mode (in which vision is by means of the fovea), concentrated in the most sensitive, central part of the retina, sensitive to colour and detail. When going from a bright room outside into the night, the central, direct vision of the eye adjusts fully in a few minutes. However, when looking slightly to the side of an object in the dark, sensitivity is greater than in the central part of the retina (using the rods of the retina – peripheral vision), allowing objects to be seen which would not be seen by looking directly at them. Furthermore this mode of vision becomes increasingly more sensitive with time, full adaptation taking up to two hours. For these reasons, indirect vision should not be used when estimating whether or not a light is visible; the light must be looked at directly, after adapting to the dark for about five minutes.

The distance a light can be seen at night is not the same distance as that at which an object can just be recognised in daylight. The distance depends not only on the intensity of the light, its distance from the observer and on visibility, but also, and crucially, on the background illumination from all sources, natural and artificial, even if they are not directly in the field of view of the observer. Observations through windows also affect the observation. A good observation can only be made at an outside, and dark, location.

In 1876, Allard proposed a law quantifying the attenuation of light from a point source:

$$E_t = (I/d^2)e^{-\sigma d}$$

where

I = the luminous intensity of the light

d = the distance in metres at which the light can just be seen

E_t = The threshold of illumination at which the eye can see a point source.

But it has been shown (eq. 3) that MOR $(M) = 3/\sigma$, allowing graphs or tables to be derived for lights of different intensities. The graph of Fig. 13.1 illustrates the relationships between the distance a light is visible and the MOR, for a lamp of 100 watts. Other graphs are required for other lamp brightnesses.

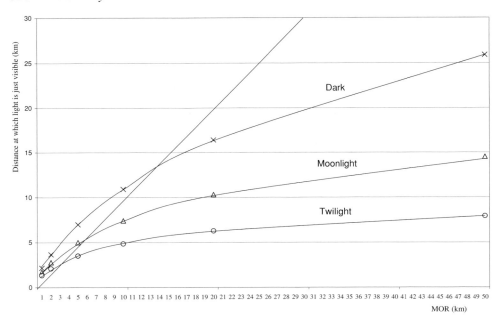

Figure 13.1. The distance a light can be just seen in the dark is shown here plotted against the equivalent MOR. The distances are not the same as that at which an object can be just recognised in daylight because of the considerable effect the background ambient illumination has on the visibility of a light. This is made apparent by the straight line which is a one-to-one linear relationship between the distance and the MOR.

The WMO has recommended that E_t should have values of 10^{-6} lux for twilight and dawn, $10^{-6.7}$ for moonlight or when it is not quite dark, and $10^{-7.5}$ in complete darkness or starlight.

Accuracy and calibration

Variations in ambient light conditions can introduce large errors, as the graph of Fig. 13.1 illustrates, and there is a range of possible interpretations between the three shown. The light sources may not be well sited, or stable, or a point source. The colour of the lights may not be the required colour temperature of 2700 degrees. The observer must also take the necessary five minutes to allow adaptation of the eyes to the dark. WMO (1990b) found that estimates of night-time visibility were 30% higher, on average, than that measured by instruments.

Visual extinction meters

To estimate visibility at night by eye, without instrumental aid, requires many lights extending as far as possible into the distance, spaced at frequent intervals, so that

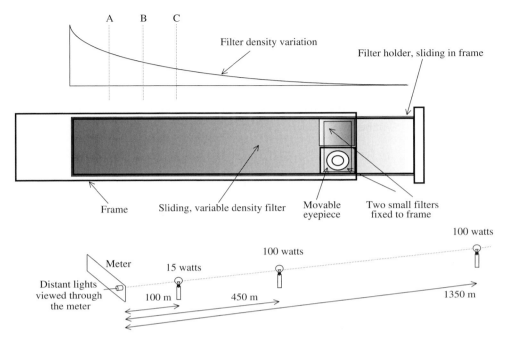

Figure 13.2. An extinction, or visibility, meter is a simple photometer that allows visibility to be measured at night using just three lights instead of many, by enabling extrapolation between them. The filter is of neutral density, absorbing all colours equally, graduated so that the change in density between any two equidistant points along its length (e.g. A to B and C to D) is the same.

one is always close to being just visible. Extinction meters avoid the necessity of having so many lights, and allow visibility to be measured using, typically, three lights only, by extrapolating observation between and beyond them.

Extinction meters are simple photometers, the prototype having been developed by Gold (1939). They consist of a neutral density variable filter in a frame (Fig. 13.2), the filter being moved along the frame until the incoming light is only just visible through the eyepiece. The filter, which transmits light of all wavelengths equally, is almost transparent at one end, while at the other it transmits only 2.5×10^{-4} (1/4000) of the incoming light, varying in density along its length in a uniform way. There are also two smaller, square, fixed filters at the open end of the frame by the eyepiece, having the same gradation of density as the sliding filter, but in the opposite direction. One or the other of the smaller filters is used in conjunction with the sliding filter, producing a uniform density across the field of view, the denser or lighter filter being selected to obtain the required level of attenuation depending on conditions. The lighter filter gives a range of attenuation of from 0.4 to 2.5×10^{-4} (0.4 to 1/4000), the darker from 4×10^{-4} to 2.5×10^{-7} (1/2500 to 1/4 000 000). The frame is marked with two scales, one for each fixed filter, in units called *nebules* (introduced by Gold 1939). According to Middleton (1952), this unit is limited mostly to the UK, but some explanation of it is useful.

If an optical filter with transmittance T is inserted in the path of a beam with luminous flux F, the flux is reduced to $F\,T$. If n identical filters are interposed, the flux reduces to $F\,T^n$. If T is chosen so that when there are 100 filters, $T^n = 0.001$, one filter is defined as having an opacity of 1 nebule, and so 100 will have an opacity of 100 nebules.

So, $T^{100} = 0.001$, or $T = 0.933$.

But any length of atmosphere behaves in the same way as several filters, and so can be said to have an opacity of a certain number of nebules. If a beam travels through a path length of d, and the atmosphere has an opacity of n nebules per unit distance, the transmittance of the sample of atmosphere is T^{nd}.

If expressed in terms of the extinction coefficient $T^{nd} = e^{-\sigma d}$, so:

$$T^n = e^{-\sigma}$$

Taking logs gives $\sigma = 0.069n$.

So one nebule (per unit length) is equivalent to an extinction coefficient of 0.069 (per unit length).

One nebule is also chosen to be equivalent to the smallest *change* in illumination that the eye can detect.

At least three lights, of constant brightness, should be used. Typically these are at 100, 450 and 1350 m from the observing position. A 15-watt light is used at the nearer position, with 100-watt lamps at the remoter sites.

To use the meter, after a few minutes' adjustment to the dark, each light is viewed in turn through the filter, the slide being adjusted until the light just disappears, the filter then being slid slowly backwards until the light just reappears. This is done first with the clearer small filter, changing to the darker if necessary. A measurement is read off (in nebules) from the graduated frame at this setting. Tables are then used for the conversion of the reading in nebules into visibility in metres.

Accuracy and calibration

Because the nebule is equivalent to the smallest change in illuminance that can be detected by a human eye, an error of ± 1 nebule is inevitable. The sensitivity of the eye may also vary from day to day and this can introduce an error up to ± 5 nebules (Bibby 1947). There may also be random variations in the brightness of the lamps, which can cause a further error of ± 3 nebules. These errors amount to an overall value of ± 6 nebules. As the visibility increases, errors increase proportionally, such that for each doubling of the visibility the error rises by about 15%, up to as high as 70%. It follows that it is best, therefore, to use the most distant lamp visible.

To minimise these errors, each observer derives their own individual *calibration figure* for each meter and for each lamp, the process being repeated periodically to allow for changes. The value of the calibration figure is derived by taking readings when the atmosphere is very clear, thereby allowing for all the human and instrumental variations, while excluding any attenuation due to atmospheric scattering. When routine measurements are subsequently made, the meter reading is first subtracted from the calibration figure to give the difference, and it is this value that is then converted to visibility, in meters, by means of a conversion table. Manufacturers supply a skeleton card for this, with full instructions for use. Because the meter measures the amount of light lost in transmission, the higher the instrument reading the lower the visibility. Typically for a light at 100 m, 10 nebules represents a visibility of 400 m, while 45 nebules represents 100 m.

Installation and maintenance

Lights should be installed at a height of two to three metres above the ground. They need not be switched on from the observation position: it is usually cheaper to have a switch on the lights and to leave them on all night, or alternatively to have a light- or time-activated switch. The voltage supply should be stable to avoid variations in light output. A tungsten-filament light has a life of about 1000 hours and will radiate a fairly steady level of illumination after being aged for about 5% of its life (50 hours). New lights should, therefore, be aged before use by operating them continuously for two days. The same type of lamp should be used all the time, and at every change of bulb a new calibration figure needs to be obtained (as above).

If practical, there is advantage in viewing the lights through tubes about a metre long and 35 mm in diameter, matt black on the inside. If the tubes can be fitted through the wall of a hut so much the better, for such an arrangement ensures that the correct light is being viewed, allows the observations to be made in complete darkness, and the observer, being more comfortable, is less likely to rush and make errors. It is also difficult to use the meter in the rain, and the outdoor level of illumination can vary and introduce error. The distance of each light from the point of observation must be measured carefully.

Runway visual range (RVR)

For the special case of airfields, visibility is defined as 'The maximum distance in the direction of take-off or landing at which the runway, or specified lights or markers delineating it, can be seen from a position above a specified point on its centre line, at a height corresponding to the average eye-level height of pilots at touch-down', typically about five metres above the ground. Any lights used should only be visible from the observing point and not along the runway. Otherwise the methods of observation are the same as for general use as described above.

Single-path transmissometers

Estimates of visibility by eye in daytime are for most purposes adequate, provided suitable visibility objects exist. However, at night instruments can be useful, as well as being essential at unattended stations or when continuous observations are required, such as for warning purposes.

Automatic instruments are of two types: *transmissometers*, which measure the transmittance along a column of atmosphere, and those instruments that measure the scattering of light in a small volume of air. The former measure light attenuation due to both absorption and scattering. There is not complete agreement as to the relative proportions of attenuation due to absorption and scattering: Twitty and Weinman (1971) suggesting they can be of equal amounts, while the measurements of Foot (1979) suggest that absorption is comparatively small.

According to definition, a MOR measurement would, strictly speaking, require the transmitter and receiver units of a transmissometer to be moved apart, on say rails, until the transmittance was reduced to 5%. But this would be impractical due to the large distances involved and so shorter-path, fixed instruments are used, on the assumption that the extinction coefficient is independent of distance.

Transmissometers are of two types, those with a transmitter and receiver in separate units, at a fixed distance from each other, and those in which both are in the same unit, the light being reflected back by a remote retroreflector. The total distance the light travels is referred to as the *baseline* of the instrument, and this can range from just a few metres up to as many as 300, depending on the ranges to be measured, it being accepted that the MOR range is between about 1 and 25 times the baseline. *Double baseline* instruments use two receivers (or retroreflectors) at different distances, extending both the upper and lower ranges that can be measured.

In a typical instrument, such as the UK Met. Office Mk 4 transmissometer, the light source is a 200-watt tungsten-halogen lamp at the focus of a parabolic mirror housed in a box to which is attached a tube 75 cm long by 15 cm diameter directed along the baseline at the distant receiver unit (Fig. 13.3). The receiver houses a lens which focuses the incoming light beam onto a photocell via an iris and several diaphragms to minimise the entry of stray ambient light. Because of drifts in DC circuits, later designs use a modulated light beam to increase the signal-to-noise ratio (as in cloud measurement by searchlight – see Chapter 14, 'Measuring cloud height automatically by searchlight').

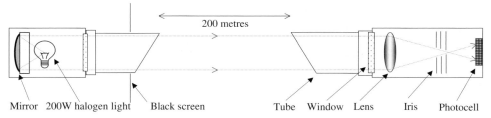

Mirror 200W halogen light Black screen Tube Window Lens Iris Photocell

Figure 13.3. Transmissometers measure the atmospheric attenuation of light, which is due to both absorption and scatter. (It is not entirely certain what the relative proportions are, but normally it is assumed that scatter is by far the greater.) Typically the transmitter and receiver are separated at opposite ends of a long optical path. The output of the photocell is zero for 0% transmission and 15 mV for 100%, with a linear response between. Transmittance is a simple ratio relating the actual reading to that at 0 and 100% from which the MOR can be calculated (see text).

The voltage output of the diode detector circuit is linear, with zero volts output for 0% transmittance and 15 mV for 100%. Transmittance is given by:

$$T = (V_T - V_0)/(V^1 - V_0) \text{ (expressed as a percentage)},$$

where

V_0 = the voltage output from the detector when visibility is zero

V^1 = the output for full visibility and

V_T = the output during a routine measurement.

From eq. (1) transmittance (T) is shown to be related to the extinction coefficient by:

$$T = e^{-\sigma x} \text{ from which:}$$
$$\sigma = 1/x . \log_e 1/T$$

where x = the baseline of the transmissometer.

But it was shown earlier that the visual range (MOR) = $3/\sigma$. So:

$$\text{MOR} = 3x/\log_e 1/T$$

Thus from the measurement of T by the instrument, the visible range or MOR can be calculated.

Rapid changes in signal due to natural variations in transmittance are smoothed out by circuitry that introduces a time constant of about one minute. The natural time constant will depend on the flow of wind and the degree of scintillation along the optical baseline. The two time constants will be summed. The output of the receiver (telemetered to a suitable office location) can either be recorded continuously or sampled every 10–20 s.

Accuracy and calibration

During low visibility, the light received from the transmitter is very low, making its measurement difficult. This results in a rapidly increasing error below about 5% transmittance, reaching 100% error at the lower limits of visibility. Under high transmittance conditions, the problem becomes one of error due to small variations in lamp brightness and slight misalignments of the beam. This results in a rapidly increasing error as transmittance increases beyond 90%, such that when nearing perfect visibility the error approaches 100% +.

Errors also arise due to inaccuracies or drift in calibration, in the value of V^1 (the instrument's full scale setting). This can arise due to a change in the intensity of the light source that has not been corrected, although the filaments of halogen lamps age less quickly and some designs monitor the brightness of the source, thereby minimising these errors. But contamination of the optics of the system, in particular the windows of the transmitter and receiver units, can be the main problem. This error increases as the visibility increases, and can reach 20% to 30% error at a visibility of 10 km.

Similarly, an error in the value of V_0 (the instrument's zero setting), through drift in the electronics or interference due to high ambient stray background radiation, causes error in the estimate of visibility. These can be minimised by using chopped light rather than a fixed DC level, and by baffles in front of the light detector. Errors range from 5–10% at low visibility (below 200 m), reaching a minimum of just a few percent between 300 m and 2 km, thereafter increasing again to 10–20% when visibility is 10 km. It is obvious, therefore, that even automatic instruments can be in considerable error at the extremes of visibility.

Installation and maintenance

The two units must be mounted on very stable towers to ensure alignment does not drift and they should be deployed at about three metres above the ground. The site should be away from any local disturbance that may affect conditions, such as aircraft run-up areas. A north–south baseline is preferable with the receiver to the north as this minimises the effect of sunlight reflected off the background in the field of view of the instrument. A terrestrial background, as viewed from the receiver, is preferred to one of the sky, without any brightly reflecting object within five degrees. Regular window cleaning and checks on lamp brightness are the principle maintenance requirements.

Folded-path transmissometers

Where sites cannot provide the long baseline required, or when there is advantage in having all the instrumentation together in one unit, the light path can be artificially

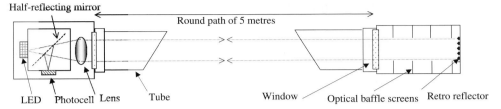

Figure 13.4. In a folded-path transmissometer, the transmitter and receiver are in the same housing, with a reflector at the remote end, thereby doubling the path length, allowing installation in more limited space. More complex designs of reflector allow multiple reflections and thus an even shorter path and a more compact installation.

extended by repeated passes of the light beam back and forth through the same small sample of atmosphere. In its simplest form, the remote receiver is replaced with a single-corner cube retroreflector, the receiver being housed in the same unit as the transmitter (Fig. 13.4), the required path length being halved. To shorten the required baseline further, more complex designs incorporate a single-corner cube retroreflector at the remote end with a multi-corner reflector in the active end of the instrument. In such a design, the actual baseline can be just two or three metres long, but with repeated folding of the beam the distance traversed can be increased to about 20 m thereby improving sensitivity and discrimination.

Such systems generally chop the light beam mechanically to reduce the effects of extraneous light and electronic drift through the use of phase-sensitive detection (see also above and Chapter 14, 'Measuring cloud height automatically by searchlight'). They may also include the capability of automatically checking the light source periodically, by introducing a mirror into the optical path to deflect the light into a second, local detector. This allows automatic correction for any change. A second mirror is interposed once an hour, reflecting the light beam through the optical system without any significant passage through the atmosphere, the resultant signal being used to correct for contamination of the lens and mirror surfaces. These surfaces are also heated to prevent condensation.

In some very short baseline (5 m) single folded-path transmissometers, an LED, emitting in the visible red spectrum, may be used as the light source. This has the advantage that it can be electronically modulated (rather than with a rotating disc). A mirror is periodically inserted into the light beam, reflecting it directly back to the detector without any passage through the atmosphere; the resultant signal, being affected only by the ageing of the LED and of the detector and by contamination of the optics, allows automatic corrections to be made. But WMO (1996) recommends that polychromatic light sources in the visible spectrum are best to ensure a representative measurement is obtained; monochromatic LEDs do not fulfil this advice.

Because shorter path instruments measure smaller lengths of atmosphere, their measurements may not be representative of the general situation. However, with averaging of readings this can be minimised.

Forward-scatter visibility meters

In most situations, the attenuation of a light beam is due largely to scattering. As an alternative, therefore, to measuring the extinction of a light beam as it passes through the atmosphere, it is possible to measure the atmosphere's light scattering properties. The exception to this might be in an industrial area, where there could be enough absorption due to industrial haze to render scattering unsuitable for estimating visual range.

If a small volume of atmosphere is illuminated with a light beam, and viewed from a distance, it may be considered as a point source of scattered light. It can be shown (Met. Office 1982) that:

$$b = 2\pi \sin \varphi \ I_\varphi / \Phi_v$$

Where

b = scatter coefficient (= extinction coefficient σ, assuming no absorption)

φ = angle of scatter (forward, backward, etc.)

Φ_v = intensity of incident flux on the volume of atmosphere

I_φ = intensity of the scattered radiation, at angle φ, relative to incident radiation.

From this, visibility can be obtained from the earlier-derived relationship V (or M) = $3/\sigma$, assuming that $b = \sigma$. This is the basis of visibility measurement using scatter rather than extinction as the variable. All that scatter instruments must measure, therefore, is the intensity of the scattered radiation, at a known angle, with a known flux input.

Various studies were made of the relative intensity of scattering at different angles relative to the incident beam (Foitzik & Zschaeck 1953) and it was shown that it is most intense in the forward direction, particularly at low visibility, reaching a minimum at around 120°, rising again, but less so, when approaching 180° due to back-scatter (Fig. 13.5) (Deirmendjan 1964, Winstanly & Adams 1975). The differences are greater at low visibility, but follow the same pattern under all visibility conditions. Scatter falls by a factor of about five as visibility increases from 1 km to 10 km and by a factor of 100 when it increases to 50 km.

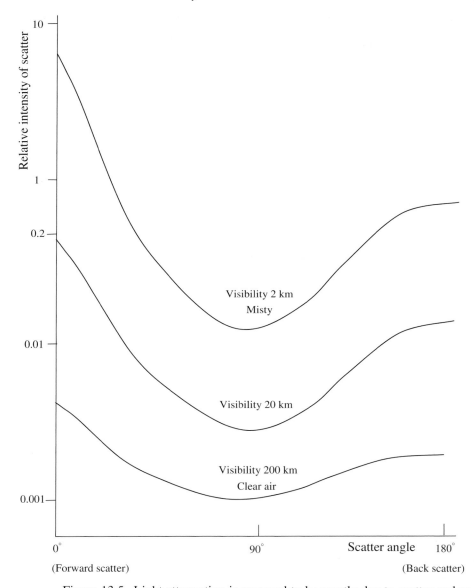

Figure 13.5. Light attenuation is assumed to be mostly due to scatter and so, by measuring how much light is scattered, visibility and the MOR can be inferred. But scatter is not the same for all angles, being greatest in the forward direction (the direction in which the light is being directed). Scatter is least at 90°, rising again as the angle increases further up to 180°, when it is known as backscatter. It does not, however, rise to the same level as forward scatter and it also has a less sound correlation with visibility.

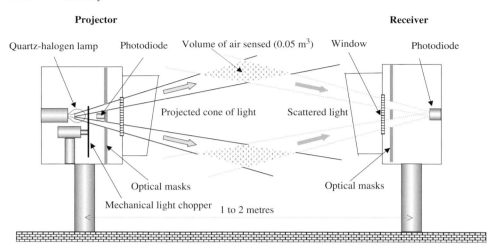

Figure 13.6. Since forward-scatter gives the strongest signal and best correlation with visibility, instruments based on this are the most accurate and most common.

There are two classic designs of forward scatter visibility meter. Figure 13.6 illustrates a typical in-line model, in which the receiver directly faces the transmitter, separated by about a metre, with masks to prevent the direct transmission of light between them. A quartz-halogen lamp – with optics to produce a cone of light between 20° and 50°, chopped by means of a rotating shutter – illuminates a volume of air of 0.05 m^3. The receiver contains a photodiode with optics to collect the light from the illuminated volume of atmosphere. A second photodiode in the transmitter picks off a fraction of the chopped light output to provide timing information for the phase-sensitive circuits and also a measure of the level of the light, so that corrections can be made to maintain a constant level. The output is generally 0–5 volts in two ranges, covering visibility from about 60 m to 6 km (equivalent to an extinction coefficient of 44×10^{-3} to 44×10^{-5}) and from 6 km upwards. (See also 'Present weather sensors' and Fig. 13.9 b.)

In another forward-scatter design, a much smaller volume of air is sampled in a compact instrument (Fig. 13.7), the original having been designed by Waldram (1945). In a modern version, the light source is an electronically modulated near-IR LED housed in a tube with a collimating lens. The beam is sensed directly opposite by a light-sensitive diode, allowing light levels to be maintained at a constant level, and by a second diode in a tube at about 35°, sensing the forward-scattered portion of the transmitted beam from the 2 cm^3 volume of air at the centre of the instrument. To reduce interference, the diode is directed at a light trap (black hole) so as to provide a black background.

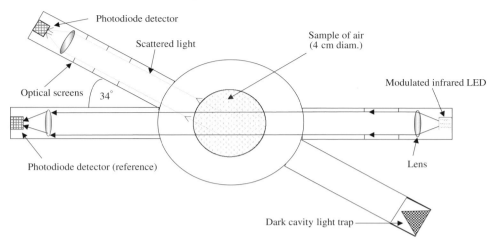

Figure 13.7. In this forward-scatter instrument, only a very small volume of air (about 2 cm²) is measured and the level of scattered light is consequently low. To minimise the interference from stray light, the sensor (top left) looks into a black light trap. To further increase discrimination against natural background radiation, the LED IR source is electronically modulated to produce an AC signal (a useful, widely used procedure).

While these two designs illustrate well the basic principles of forward-scatter instruments, there are variations, the commonest being a system in which the transmitter and receiver do not face each other directly in-line, but are slightly offset, in either the vertical or horizontal or both (Fig. 13.9 b). This type of instrument uses a modulated IR LED as the light source. (See also the last section in this chapter, Present weather sensors'.)

Back-scatter and wide-angle-scatter instruments

Although many attempts have been made to derive theoretical relationships between back-scatter and visibility (Dietze 1957, Curcio and Knestrick 1958, Twomey & Howell 1965, Fenn 1966, Vogt 1968), no simple connection has been found. It is now accepted that reliable visibility measurements cannot be made in this way, although some instruments have been used, based on similar technology and construction to forward-scatter instruments.

However, instruments that sense scatter over a wide angle, known as *integrating nephelometers* (see also 'Turbidity' in Chapter 10), are based on a sound theoretical relationship and it can be shown (Met. Office 1982) that the scatter coefficient is:

$$b = 2\,\pi h\ \Phi/I_e\,w\,s$$

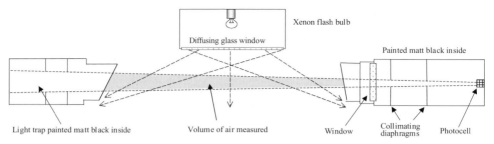

Figure 13.8. By placing the light source to the side of the detection path, the air sample is illuminated by light incident from 7° to 170° rather than from just a narrow angle, covering most of the range of angles shown in Fig. 13.5. There is a reliable correlation between scatter from such illumination and visibility, even though it includes some backscatter.

where

h = the perpendicular distance between the light source and the viewing axis

Φ = the total luminous flux received by the detector as scattered light

I_e = the intensity of the light source

w = the solid angle of the cone over which scattered light is received

s = the area of the detector

As with the forward-scatter instrument, visibility can be derived from this through the relationship V (or M) $= 3/\sigma$ (assuming that $b = \sigma$, and that there is no absorption, just scatter) (see also 'Turbidity' under 'Measuring water quality', Chapter 10).

The first instrument using this characteristic was developed by Beuttel and Brewer (1949) and later adapted by Crosby and Koerber (1963), and by Ruppersberg (1964). A generic example of a wide-angle scatter instrument is shown in Fig. 13.8, in which a xenon flash bulb emits light through a diffusing glass, producing radiation in all directions so that the instrument measures light scattered over the range 7° to 170°. The receiver views the scattered light through two collimating diaphragms. Positioned directly opposite the receiver is a light-trap, which, along with the interior of the receiver, is painted matt black. As with most automatic visibility instruments, an additional sensor measures the light source intensity directly to provide constant light output. A shutter periodically closes the sensor input diaphragms to allow a zero adjustment to be made (simulating clear air). To replicate a predetermined level of scattering, diffuse white surfaces direct a portion of the light from the flash into the receiver. However, WMO suggest that results may not be repeatable.

Comparison of instrument types

Scatter instruments share many of the errors of transmissometers, although the WMO consider the latter to be marginally more accurate, and they are often used as standards for comparisons. The long baseline of transmissometers reduces the uncertainty caused by spatial and temporal variations, which introduce error into the readings of shorter baseline and scatter instruments, but they need a large area of land and solid foundations to prevent movement. Folded path transmissometers avoid the need for space, but they are less sensitive as a result, as well as suffering more from the effect of calibration error. All types can be adjusted to give good discrimination at low visibility but as visibility increases discrimination falls. The calibration of scatter instruments often depends (Chisholme & Jacobs 1975) on the weather at the time. Transmissometers, however, are less susceptible to weather and can work with a single calibration for most conditions. Because scatter and short baseline transmissometers operate on a small volume of atmosphere, heat from the instrument, being in close proximity to the sample, may affect it. Most instruments suffer to varying degrees from the accumulation of ice and snow, often despite heaters. A comparative study of visibility instruments was made by Bond *et al.* (1981).

The effect of rain on visibility

Rain reduces visibility by scattering light just as do the smaller droplets of mist and fog. Middleton (1952) draws attention to an almost forgotten paper by Preston (1920) which demonstrates that the extinction coefficient, or obscuring power as Preston calls it, of falling rain is proportional only to the *number* of drops falling on unit area of ground per second. From this it follows, for example, that a heavy thunderstorm may be much less opaque than drizzle. (See also sub-subsection 'Optical gauges' in Chapter 8.)

Present weather sensors

Instruments with this name (Fig. 13.9*a*) are a combination of visibility meters (Fig 13.9*b*) and optical raingauges (Fig 8.13). To the visibility sensor and receiver is added a third sensor, at right angles to their path, to detect precipitation particles. By measuring the velocity and size of the precipitation particles, it is possible to differentiate between rain, drizzle, hail and snow and from this the intensity can be estimated. Developments are underway which combine the outputs of present weather sensors with AWS data, *arbiter* software merging the two to provide an improved nowcast.

(*a*)

(*b*)

Figure 13.9. *Present weather sensors* (a) are derivatives of forward-scatter visibility meters (b) (and of optical raingauges, Fig 8.13), seen here on test at the Met. Office Laboratories at Beaufort Park, Bracknell. To the lower right in (a) can also be seen a remotely controlled TV camera. As noted in Chapter 14, CCTV is being explored as a means of observing clouds remotely, but it also provides an overall view of weather conditions at remote sites. IR cameras are expensive, but much can be achieved with cameras working in the visual range.

References

Beuttel, R. G. & Brewer, A. W. (1949) Instruments for the measurement of the visual range. *J. Sci. Instrum.*, **26**, 357–9.
Bibby, J. R. (1947) Gold visibility meter Mk II. *Meteorol. Mag.*, **76**, 130–3.

Bond, F. S., Foot, J. S. & Pettifer, R. E. W. (1981) A comparative study of some single-pole visibility sensors, the Meteorological Office Mk 4 transmissometer and estimates of visibility made by observers. Scientific Paper of the Meteorology Office, No. 39.

Bouguer, P. (1760) *Traité d'optique sur la gradation de la lumière.* H. L. Guerin and L. F. Delatour, Paris.

Chisholme, D. A. & Jacobs, L. P. (1975) An evaluation of scattering-type visibility instruments. Air Force Cambridge Research Laboratories, Bedford, Mass., AFCLR-TR-75-0411. Instrumentation Paper No. 237.

Crosby, P. & Koerber, B. W. (1963) Scattering of light in the lower atmosphere. *J. Opt. Soc. Am.*, **53**, 358–61.

Curcio, J. A. & Knestrick, G. L. (1958) Correlation of atmospheric transmission with backscattering. *J. Opt. Soc. Am.*, **48**, 686–9.

Deirmendjan, D. (1964) Scattering and polarisation properties of water clouds and hazes in the visible and infra-red. *Appl. Opt.*, **3**, 187–96.

Dietze, G. (1957) Einführung in die Optik der Atmosphäre. Leipzig, Akademische Veriagsgesellschaft, Geest & Portig.

Fenn, R. W. (1966) Correlation between atmospheric backscattering and meteorological optical range. *Appl. Opt.*, **5**, 293–5.

Foitzik, L. & Zschaeck, H. (1953) Messungen der spektralen Zerstreuungsfunktion bodennaher Luft bei guter Sicht, Wunst und Nebel. *Z. Meteorol.*, **7**, 1–19.

Foot, J. S. (1979) Spectrophone measurements of the absorption of solar radiation by aerosol. *Q. J. R. Meteorol. Soc.*, **105**, 275–83.

Gold, E. (1939) A practical method of obtaining the visibility number V at night. *Q. J. R. Meteorol. Soc.*, **65**, 139–59.

Jones, L. A. (1920) A method and an instrument for the measurement of the visibility of objects. *Phil. Mag.*, **39**, 96–134.

Koschmieder, H. (1924) Theorie der horizontalen Sichtweite. *Beitr. Phys. Freien. Atmos.*, **12**, 33–55.

Met. Office (1982) *Handbook of Meteorological Instruments.* HMSO, London.

Middleton, W. E. K. (1952) *Vision Through the Atmosphere.* University of Toronto Press, Toronto.

Mie, G. (1908) Beitrage zur Optik trüber Medien, speziell kolloidaler Metallösungen. *Ann. Phys.*, **25**, 377–445.

Preston, F. W. (1920) Visibility of landscape during rain. *Nature*, **106**, 343–4.

Rayleigh (Lord) (1871a) On the light from the sky, its polarisation and colour. *Phil. Mag.*, **41**, 107–102; 274–9 (Collected Works **1**: 87–103).

Rayleigh (Lord) (1871b) On the scattering of light by small particles. *Phil. Mag.*, **41**, 447–54 (Collected Works **1**: 104–10).

Rayleigh (Lord) (1871c) The incidence of light upon a transparent sphere of dimensions comparable with the wave-length. *Proc. R. Soc. Lond.*, **A84**:25–46 (Collected Works **5**: 547–68).

Rayleigh (Lord) (1899) On the transmission of light through an atmosphere containing small particles in suspension, etc. *Phil. Mag.*, **47**, 375–384 (Collected Works **4**: 397–405).

Ruppersberg, G. H. (1964) Registrierung der Sichtweite mit dem Streulichtschreiber. *Beitr. Phys. Atmos.*, **37**, 252–63.

Saussure, H. B. de (1789) Description d'un diaphanometer ou d'un appareil propre à measurer la transparence de l'air. *Mem. Acad. Turin*, **4**, 425–40.

Simpson, G. C. (1939) Sea salt and condensation nuclei. *Q. J. R. Meteorol. Soc.*, **65**, 553–4.

Twitty, J. T. & Weinman, J. A. (1971) Radiative properties of carbonaceous aerosols. *J. Appl. Meteorol.*, **10**, 725–31.

Twomey, S. & Howell, H. B. (1965) The relative merits of white and monochromatic light for the determination of visibility by backscattering measurements. *Appl. Opt.*, **4**, 501–6.

Vogt, H. (1968) Visibility measurement using backscattered light. *J. Atmos. Sci.*, **25**, 912–18.

Waldram, J. M. (1945) Measurement of the photometric properties of the upper atmosphere. *Q. J. R. Meteorol. Soc.*, **71**, 319–36.

Weber, L. (1916) Die Albedo des Luftplanktons. *Ann. Phys.*, **51**, 427–49.

Wiener, C. (1907) Die Helligkeit des klaren Himmels und die Beleuchtung durch Sonne, Himmel and Rückstrahlung. *Nova Acta K.-L. Deutsc. Akad. Naturf. Halle*, **73**, iii–xii: 3–239.

Wiener, C. (1910) Die Helligkeit des klaren Himmels und die Beleuchtung durch Sonne, Himmel and Rückstrahlung. (Fortsetzung und Schluss.) *Nova Acta K.-L. Deutsc. Akad. Naturf. Halle*, **91**, iii–vi: 81–292.

Wigand, A. (1919) Eine Methode zur Messung der Sicht. *Phys. Z.*, **20**, 151–60.

Wild, H. (1868) Über die Lichtabsorption der Luft. *Ann. Phys.*, **134**, 568–583; **135**, 99–114.

Winstanly, J. V. & Adams, M. J. (1975) Point visibility meter: a forward scatter instrument for the measurement of aerosol extinction coefficient. *Appl. Opt.*, **14**, 2151–2157.

WMO (1990a) *Guide on Meteorological Observation and Information Systems at Aerodromes.* WMO No. 731.

WMO (1990b) *The First WMO Intercomparison of Visibility Measurements: Final Report.* Instruments and Observing Methods Report No 41, WMO/TD-No. 401.

WMO (1996) *Guide to Meteorological Instruments and Methods of Observation*, 6th ed. WMO, No. 8.

14

Clouds

The sun was setting, and a gentle southerly breeze, striking against the southern side of the rock, mingled its current with the colder air above: the vapour was thus condensed: but as the light wreaths of cloud passed over the ridge, and came within the influence of the warmer atmosphere of the northern sloping bank, they were immediately redissolved.

Charles Darwin *Voyage of the Beagle* (observing the Corcovado Mountain in Rio de Janeiro).

The variable and its history

Observations of the extent, height and type of cloud-cover are important for many purposes, including meteorology and aviation, and also now for climatology, since clouds have a considerable influence on the energy budget of the Earth. At present, climate modellers have a problem in predicting change in part because of the difficulty of representing clouds in the models. Low, liquid-water clouds tend to cool the climate, the higher ice clouds to warm it, but there is the complication of supercooled cloud when the drops remain liquid well below the freezing point, for they reflect almost half the incoming solar radiation and so lead to cooling while the models might assume from their temperature that they are made of ice. At present all the effects appear to be nearly in balance, but with a slight net cooling. To improve the models it is important to compare what they foretell with what actually happens. The measurement of clouds thus takes on a new important role. It is in the nature of the subject of this book that many topics interleave and overlap and this is particularly so with clouds, which are also discussed in the chapters on the upper atmosphere (Chapter 16), lightning (Chapter 15) and visibility (Chapter 13). Rather than give copious cross references, the reader is asked to look at relevant chapters.

Measuring the Natural Environment, second edition, Ian Strangeways. Published by Cambridge University Press. © Ian Strangeways 2003.

The work at Chilbolton in the UK is covered in Chapter 16 and is particularly relevant to clouds (Illingworth & Hogan 2002).

In 1803 Luke Howard gave clouds the names *cirrus, cumulus, stratus* and *nimbus* in a first attempt to differentiate between the main types (Hamblyn 2002, *Weather* 2003). Gradually a more detailed classification evolved culminating in 1891 at the International Meteorological Conference in Munich with a new classification, published in 1896 as the *International Cloud Atlas*. In 1921 an international commission was set up to study clouds and by 1932 the *International Atlas of Clouds and of Types of Skies* was published. An updated version of the atlas was published in 1956 by the WMO, with a further revised edition in 1975 (WMO 1975, 1987).

Clouds are composed of water droplets, ice crystals, or a mix of both, smaller than about 200 μm. Even as low as $-25\,°C$, however, droplets can remain unfrozen in a supercooled state. With few exceptions (nacreous, noctilucent and some cirrus) clouds occur entirely in the troposphere, that is up to about 18 km altitude in the tropics, 14 km in middle latitudes and 8 km in polar regions (Fig. 14.1). (The troposphere contains about 90% of the atmosphere and 99% of atmospheric water vapour.) Clouds form mostly due to air rising – through convection, through ascent over high ground and within depressions.

Cloud types

Clouds are named in Latin in the same way as plants and animals:

> *Genus*: type of cloud.
> *Species*: its shape and structure.
> *Variety*: how the elements are arranged and the cloud's transparency.

Today, three basic cloud forms are named, being the same as those defined by Luke Howard in the early nineteenth century:

> *Cumulus*: Heaped with rounded tops, detached and dense with sharp outlines, growing vertically, familiar as fair weather clouds. Sunlit parts are brilliant white, the bases horizontal and darker (Fig. 14.2).
> *Stratus*: A grey layer of water-droplet cloud, identical to fog at ground level, with ragged base and top. If thin, the Sun can show through. They may hide the tops of buildings (Fig. 14.3).
> *Cirrus*: White, hair- or thread-like, detached clouds, composed of delicate filaments, patches, or narrow bands, forming high in the atmosphere and composed of ice crystals (Fig. 14.4). Halos and other optical effects can occur.

Howard's fourth type, *nimbus*, is not now used as a basic form, although it is retained as a composite name in two of the ten genera. It means *rain-producing*.

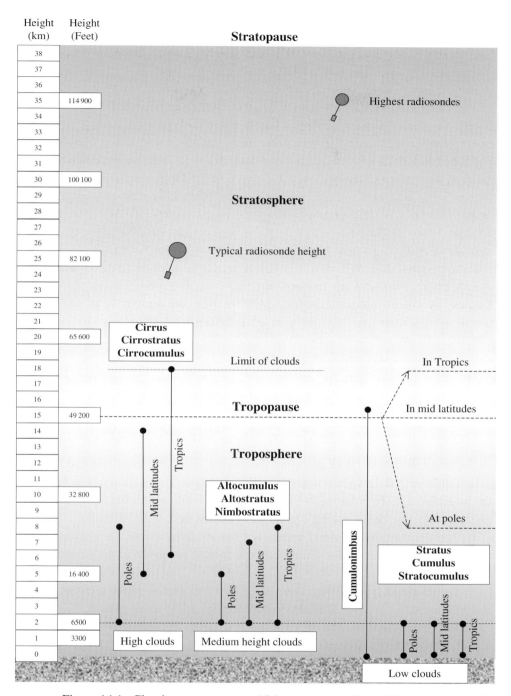

Figure 14.1. Clouds can occur up to 18 km, more usually to 15 km, in three main groups, as shown. Cumulonimbus clouds can extend from the ground up to and even slightly into the Stratosphere (Chapter 15). Radiosondes rise typically to 25 km with a maximum of around 35 km. The electrical activity known as *sprites* (Chapter 15) occurs at 40–80 km in the mesosphere, outside the range of the figure.

Figure 14.2. As the ground warms, cumulus clouds grow through convection of bubbles of warm air, as here by the Papua New Guinea coast. The near species are humilis, while further inland the larger species, congestus, are forming, which can subsequently grow into much larger cumulonimbus thunder clouds (Fig. 15.1). The effect of the warm ground in the formation of the clouds is well illustrated by the absence of any clouds over the cooler sea.

There are several ways of grouping the genera, one convenient method being to divide them into three height ranges (Table 14.1), the ranges shown here being for the mid latitudes. (In polar regions they are lower, while nearer the equator they are greater, reflecting the changing depth of the troposphere with latitude.)

As full descriptions of cloud genera are not straightforward, there being many variations (Table 14.2), only simplified descriptions are given in the several tables below.

To define cloud-type further, there are 14 species that describe a cloud's shape and internal structure (Table 14.3). Not all species apply to all genera.

A further sub-division into nine varieties describes the arrangement of elements and the cloud's transparency (Table 14.4).

Defining genus, species and variety is not always enough to fully describe a cloud for it may have *supplementary features,* such as trails of precipitation from its base. There are six such features (Table 14.5).

Sometimes clouds have *accessory clouds* in attendance, such as a cap on top. There are just three types (Table 14.6).

(b)

(a)

Figure 14.3. This stratus cloud (a) fills the valley in which Geneva is situated. It extends down to the ground, where it is seen as clearing fog at the airport (b). See also Chapter 13 on visibility.

Table 14.1 *Cloud genera, grouped by height*

Genera	Cloud base height, feet (m)	
Low level		
Stratus	0–2000	(0–600)
Stratocumulus	1000–4500	(300–1350)
Cumulus	1000–5000	(300–1500)
Cumulonimbus	2000–5000	(600–1500)
Medium level		
Nimbostratus	0–10 000	(0–3000)
Altostratus	6500–20 000	(2000–6000)
Altocumulus	6500–20 000	(2000–6000)
High level		
Cirrus	20 000–40 000	(6000–12 000)
Cirrostratus	20 000–40 000	(6000–12 000)
Cirrocumulus	20 000–40 000	(6000–12 000)

Figure 14.4. Cirrus cloud is made up of ice crystals falling from 'generating heads' at the tip of the clouds. As the crystals fall, they may encounter winds of lower speed, and so curve backwards producing a hook shaped tip, as here. Those with hooks are known as species uncinus. Sometimes, however, the cloud can have a fibrous appearance without any tufts or hooks when it is known as cirrus fibratus. Photographed in Wallingford on a summer evening.

Table 14.2 *Description of cloud genera (grouped by height)*

Genera	Brief description
Low level	
Stratus	Low, widespread, grey layer, ragged base. Fog is Stratus at ground level (Fig. 14.3)
Stratocumulus	Low, grey or whitish sheet of separate large rounded masses with dark shading
Cumulus	Brilliant white mounds. Sharp edged. Dark base. 'Fair weather clouds' (Fig. 14.2)
Cumulonimbus	High towers, top brilliant white, often anvil shaped. Base ragged. Precipitation
Medium level	
Nimbostratus	Dark grey, heavy, sheet with ragged base. Rain-bearing. Blots out Sun
Altostratus	Dull grey or bluish featureless layer. Sun can show through
Altocumulus	White and grey small rounded clumps with dark shading and clear sky between
High level	
Cirrus	Delicate white filaments. Straight or entangled. May end in a hook (Fig. 14.4)
Cirrostratus	High white transparent sheet, often thin, milky and invisible. Can form halos
Cirrocumulus	Very high thin white or bluish sheet of numerous small ripples or tufts

Note: 'Low, medium and high' refer to the height of the base, not the top, of the cloud.

Table 14.3 *Cloud species*

Species	Description
Calvus	Smooth top
Capillatus	Fibrus; striated
Castellatus	Turrets
Congestus	Growing vertically; sprouting
Fibratus	Nearly straight; no hooks
Floccus	Tufts
Fractus	Ragged shreds
Humilis	Flattened
Lenticularis	Lens shaped
Mediocris	Moderate depth
Nebulosus	Featureless thin layer
Spissatus	Denser cirrus
Stratiformis	Horizontal layer
Uncinus	Hook shaped

Table 14.4 *Cloud varieties*

Variety	Description
Duplicatus	Several layers
Intortus	Irregularly curved or tangled
Lacunosus	Thin cloud with regular holes
Opacus	Masks Sun
Perlucidus	Broad patches with small spaces
Radiatus	Bands, apparently converging
Translucidus	Sun shows through
Undulatus	Sheets with parallel undulations
Vertebratus	Like vertebrae

Table 14.5 *Supplementary features of clouds*

Name	Description
Arcus	Arched cloud
Incus	Anvil cloud
Mamma	Hanging pouches
Praecipitatio	Precipitation reaching ground
Tuba	Funnel cloud
Virga (Fallstreak)	Droplets falling under cloud

Table 14.6 *Accessory clouds*

Name	Description
Pannus	Shreds of cloud
Pileus	Cap cloud
Velum	Veil

A new cloud may develop as an extension of an existing cloud and may become a different genus in the process. The cloud that gives rise to the new cloud is called the *mother cloud*. The new cloud is classed as the new genus followed by the genus of the mother cloud with the suffix *genitus* (e.g. stratocumulus cumulogenitus). A cloud may also undergo transformation from one genus to another. The new cloud is then classified as the new genus followed by the genus of the mother cloud with the addition of the suffix *mutatus*.

Clearly, cloud identification is a far from simple matter, requiring training and experience. For more information, there are several books giving considerable detail, for example WMO (1975, 1987), Met. Office (1982a, b), Dunlop (1996). The

RMS magazine *Weather* also has regular examples of clouds on its covers with descriptive captions.

Units and terms

Cloud amount

Total cloud amount (or *total cloud cover*) quantifies the fraction of the celestial dome covered by all types and levels of cloud. Cover is estimated to the nearest eighth, expressed in *Oktas*, 0 representing no clouds at all, 8 indicating complete cover. Nine oktas is used to show that an observation was not possible due, for example, to fog or falling snow. *Partial cloud amount* is the amount of sky covered by each type or level of cloud alone. The sum of partials can, therefore, be greater than 8 oktas, since there is usually some overlap at different levels.

Cloud base

Cloud base or *cloud height* is the height of the base of the cloud above the ground at the observation site, expressed in metres, or for some aviation applications in feet. It is defined (by the WMO) as 'the lowest zone in which the type of obscuration perceptibly changes from that corresponding to clear air or haze to that corresponding to water droplets or ice crystals'. It is important to differentiate between cloud *height* (that is the height above the local ground) and *altitude* (height above mean sea level). This is usually done by adding the words height *above ground level*. Clearly this difference is important in aviation.

Estimating cloud height manually

Cloud cover and cloud type have to be estimated by eye, along the lines described above, and this takes considerable practice. Trials are, however, in progress to measure cover automatically using CCTV systems (see below). While techniques to measure cloud height automatically have been available for some time (below), at stations without measuring equipment height must also still be estimated by eye.

In mountainous areas it is possible to judge cloud height, if the base is lower than the tops of the hills, by reference to topographical features of known height. At all other locations, a table must be used that gives heights for the 10 genera of clouds, different tables being required for the three latitudinal regions. Table 14.1 is for temperate regions.

Measuring cloud height manually by balloon

The height of the cloud base can be estimated by measuring the time taken for a small rubber balloon, inflated with hydrogen or helium, to enter the misty layer before finally disappearing into the cloud. The rate of ascent is determined by the free lift of the balloon which can be adjusted by controlling the degree of inflation. Due to eddies near the ground, the balloon may not start to rise immediately and allowance must be made for this in the timing. In addition, the rate of ascent can be somewhat uncertain up to 2000 ft (600 m) and rain reduces the rate of climb. Vertical air currents and the shape of the balloon also affect ascent rate. Without binoculars, the method is limited to about 3000 ft (900 m) unless the wind is very light, because the balloon will disappear from view before entering the cloud. At night a light can be attached to the balloon. The method is clearly not particularly accurate and is not applicable to high clouds.

Measuring cloud height manually by searchlight

At night it is possible to make fairly precise manual measurements of cloud height using a searchlight, the angle of elevation of a patch of light produced by a vertically pointing searchlight being measured some distance away using an *alidad* (Fig. 14.5). The height of the cloud base is given by $h = L \tan E$, where L is the distance between the light and the alidad and E is the angle of elevation The best separation distance is from 650 ft to 2000 ft (180–600 m).

Accuracy and calibration

The largest error originates when measuring the angle of the light spot, $1°$ producing an error of 17 ft (6 m) when the cloud base is 1000 ft, and 450 ft (140 m) when 5000 ft, hence the importance of correct installation and of care in reading the angle.

Installation and maintenance

Ideally there should be an unobstructed line of site between the light and the alidad and both should be firmly mounted on stable permanent bases. If the two are at different heights, allowance must be made for this in calculating h. The beam must be adjusted to be vertical and this should be checked with a theodolite from two points at right angles, one being at the alidad site. Beam focus should be checked and adjusted regularly as well as each time the lamp is changed. (The 24-volt 500-watt lamp is overrun to boost its output, resulting in a life of only about 100 hours and hence frequent changes.)

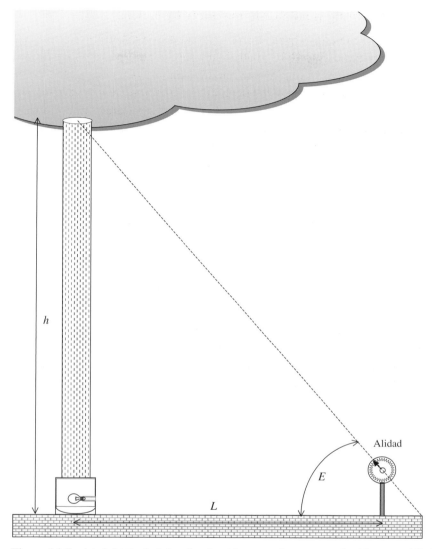

Figure 14.5. At night the height of a cloud base can be measured with a searchlight using the simple geometry shown, where L is best set at between 180 and 600 m. The alidad is a simple device for measuring the angle, E. This is the only measurement required.

Measuring cloud height automatically by searchlight

To obtain a continuous record of cloud height, to make measurements at an unattended site, or to get readings both day and night, a *rotating-beam ceilometer* was developed during the 1960s, although now largely replaced by laser methods (see below).

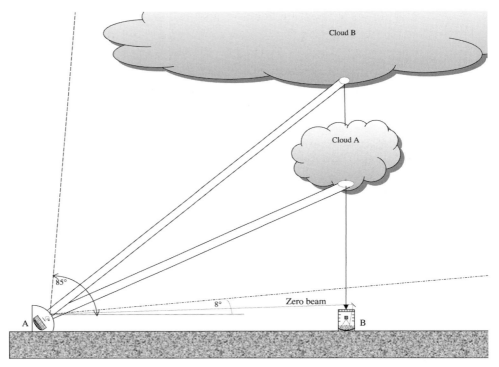

Figure 14.6. For unattended operation, for daytime observations, or if a continuous record is required, some form of automation is required. Before lasers were available, cloud height was measured with a searchlight that swept out an arc from the near horizontal (8°) to the near vertical (85°), taking about one minute to swing through the 77°. The beam, with a 2° divergence, is modulated at around 100 Hz. In the distant receiver, a parabolic mirror looks vertically, focusing the received light reflected from any clouds directly above on to a photocell. When received, a pen on a chart recorder is activated, its position up the chart being set by the angle of the searchlight beam (small for cloud A, larger for the higher cloud, B) and thus on the height of the cloud. At its lowest angle, the beam is detected directly at the receiver, producing a zero reference dot on the chart.

Whereas the manual searchlight beam is fixed vertically with the alidad movable, the automatic system is the reverse, the beam scanning in the vertical plane, from 8 to 85°, taking one minute to complete one full cycle. The detector, between 100 and 300 m distant, is fixed, receiving light from directly overhead (Fig. 14.6).

Typically the searchlight uses a 24-volt 200-watt quartz-iodine lamp at the focus of a 400-mm parabolic reflector, emitting mostly in the IR from 1 to 3 µm wavelength (small compared with cloud droplet size). The beam is chopped mechanically at around 1000 Hz by a rotating shutter between the light and the mirror, allowing phase-sensitive electronics to enhance the signal-to-noise-ratio.

A lead sulphide photocell at the focus of a second similar mirror, with baffles to exclude ambient light, detects light reflected from the cloud immediately above. As the searchlight beam scans, a servomechanism moves a pen across a paper chart in synchronism, the pen only marking the chart while light is being received back from the cloud. The lower limit of the trace indicates cloud base height, the upper limit either showing cloud-top height or simply the distance the beam penetrates into the cloud.

By an arrangement of prisms and mirrors, light is fed directly from the searchlight to the receiver when the beam reaches its lowest position, producing a *zero dot* on the chart, acting as a zero height reference. It also indicates that the system is operating correctly when there is no cloud, while in poor visibility the light fails to get to the receiver and no dot occurs.

Accuracy and calibration

Sources of error include optical misalignment and receiver electronics. An error in the order of 4 ft when the cloud base is at 100 ft and of 1300 ft when at 4000 ft has been reported by the UK Met. Office.

Installation and maintenance

The transmitter and receiver should be on level open ground fixed to firm bases. Optical alignment is critical to ensure that the searchlight projects light to the same area of sky at which the receiver is directed, and this needs to be checked regularly.

Measuring cloud height by laser

By timing the flight of a light pulse from a gallium arsenide semiconductor laser, transmitted to the cloud base and reflected back to a receiver, the height of the cloud can be determined. In a typical design, the laser is at the focus of a Newtonian reflecting telescope with a mirror of about 200 mm diameter, producing a beam of 8 min of arc. The laser emits around 40-watt, 100-ns pulses at a wavelength of 900 nm, with a repetition rate of 1000 Hz (Fig. 14.7).

The receiver has the same Newtonian construction, but in place of the laser a photodiode, with a narrow band optical filter (to remove natural radiation) and an angle of view of about 15 min of arc, senses the returned pulse. With these angles and when the transmitter and receiver are fixed side by side, their fields of view start to overlap at a height of 5 m with full convergence at 300 m.

The received pulse is passed to a series of sequential 100-ns time gates, each representing an increase of flight time equivalent to 15 m height. A pen starts to

(a)

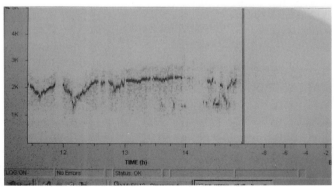

(b)

Figure 14.7. Modern automatic cloud height ceilometers use lasers that illustrate (a, upper) having a dual system for better accuracy and as back-up in the case of failure. This particular instrument is at the Rutherford Appleton Laboratory field site at Chilbolton in the UK as part of a joint ESA project. The trace (b, lower) on a CRT shows height variations over the last few hours interspersed with cloud-free periods and (right) two separate cloud layers.

move vertically up a paper chart as the pulse leaves the ground and when reflected light is received the relevant gate is activated, triggering the pen to record a trace on the paper. Because of the low speed of the pen compared with light, the laser is fired many times during one pen scan. The range covered is from 100 ft (30 m) to 5000 ft (1500 m).

Accuracy and calibration

Timing errors in the electronic gates will cause proportional errors in height indications. Offset from the vertical will also cause some error, although the WMO consider that up to 5° off vertical is acceptable. Because the base of a cloud is not clear-cut but diffuse, complex algorithms have been introduced to process the returned signal so as to obtain as good an estimate as possible of true cloud base, but some error will be introduced due to there being an ill-defined transition from no cloud to cloud. The transition also varies with cloud type, with time and from place to place, so some ambiguity is unavoidable. Precipitation can also be mistaken for cloud since it scatters the light pulses.

Calibration checks can be made by comparing the instrument's readings against measurements from balloon ascents, with aircraft measurements and from searchlights data. Laser instruments are now the best instrument available for cloud height measurement.

Installation and maintenance

Beyond ensuring a firm base and a clear overhead view and that the transmitter and receiver are adjusted so that the beam is vertical, this technique has no special installation needs. It is also one of the few meteorological instruments that can be installed on a roof if necessary. The optical windows must be kept clean and the stability of the oscillator which triggers the timing gates needs to be checked periodically against a frequency standard.

Cloud observation using CCTV

Developments are underway to evaluate the possibility of observing clouds remotely using CCTV (Hatton *et al.* 1998, Rowbottom *et al.* 1998). Thermal IR cameras sensing the 8–14 μm spectral range are the most useful since they are able to operate day and night using the natural radiation emitted by the clouds. Because clouds at different heights radiate according to their temperature, the cameras are able to locate the edges of the clouds more easily, unlike cameras working in the visible spectrum, and unlike observers eyes. IR cameras also penetrate haze.

At present, a major limitation to the general use, in meteorology, of CCTV is the high cost of the IR cameras (£20 000) since they are not widely used and so have a small market; however, this may well change. So far, cameras have been investigated for the remote *manual* observation of clouds, but the next step is to develop techniques for the automatic identification of cloud cover and type. This, however, is some time off because of the subtle variety of clouds, one type merging onto another, and the inevitable complexity of algorithms to model them; it is difficult to see how cloud species and varieties could be adequately distinguished automatically. But with the increasing move towards automating meteorological observations, the remote and automatic identification of clouds is becoming a pressing matter.

IR cameras also have other possible uses for remote observations, such as ground temperature measurement (Keogh 2000), although it is necessary to know the emissivity of the particular ground being observed. The protective windows of the camera housing also introduce some error which needs allowing for, and in consequence 1° is currently the limit of accuracy.

Cloud observation from satellites

Satellites, their orbits and the images collected by them are the topics of Chapters 12 and 19, where background information relevant to this section can be found. Here we will look specifically at cloud images, those generated by each geostationary satellite covering about one-quarter of the globe. These images are one of the satellite's greatest assets (EUMETSAT). They give an overall view of global cloud cover never before available, showing at a glance lines of convection clouds along the Intertropical Convergence Zone, depression over mid latitudes, thunder storms over equatorial Africa (Fig. 19.6), anticyclones and the paths of hurricanes. Being repeated every half hour, time-lapse animations can also be produced from these images, providing an indication of wind speed and direction (at cloud height). These observations are quite different from those made by an observer on the ground, and are a valuable new capability.

Geostationary satellite image pixels are coarse, at best 2.5 by 2.5 km. Because of their much lower orbital height, satellites in polar orbit produce somewhat finer resolutions (1 km), although they have a limited frequency of overpasses in equatorial regions. Despite these spatial and temporal limitations, it is possible to deduce, automatically, much about cloud cover and type from the images. For example, the UK Met. Office has developed techniques that involve applying threshold and comparative tests to the pixels of an image to determine if they contain cloud and what type of cloud it might be.

Using the IR images, any pixel colder than a certain threshold is assumed to represent cloud. By measuring the ratio between the near IR band (0.9 μm) and the visible band (0.6 μm) it is possible to establish whether the pixel represents cloud (ratio 1), water (0.5) or land (1.5). However, land surfaces have a wide range of emissivities so that these ratios are somewhat uncertain (this test is also used to detect the *absence* of cloud so that surface variables, such as sea surface temperature, can be measured). Tests of variance of temperature between adjacent pixels help further to establish whether the pixels contain cloud, high variance pointing to a mix of cloudy and clear conditions (or clouds at different levels), while small variance combined with low temperature suggests that the area is fully cloud covered.

The height of cloud tops can be estimated in several ways, the simplest being to use brightness temperature measurements. While this is reliable for extensive and heavy stratus and cumulus fields it is not effective for semi-transparent clouds such as cirrus. However, thin clouds have a brightness temperature which is higher in the 3.7 μm band than at 11 μm (AVHRR channels 3 and 4, see Chapter 19, Remote sensing), while the opposite is true for thick low clouds. Indeed this particular test is used to detect fog. By comparing the 11 and 12 μm bands (AVHRR channels 4 and 5) cloud thickness such as thin cirrus can be estimated (although this difference is also sensitive to the water vapour content of the atmosphere).

So while it is possible to go some way towards measuring cloud cover and classifying genera automatically from satellite images, and while it is possible for an experienced observer to deduce much about cloud cover and type from satellite images (Scorer 1986), it is not yet possible to replace the ground-based observer looking up at the sky and making judgements as to the subtleties and detail of the cloud cover overhead (Scorer 2001). But no doubt this will eventually be achieved as algorithms are developed and image resolution improved.

References

Dunlop, S. (1996) *Weather.* Harper Collins Publishers, London, ISBN 0 00 472272 8.

Hamblyn, R. (2002) *The Invention of Clouds.* Picador (Pan Macmillan), London.

Hatton, D., Jones, D. W. & Rowbottom, C. M. (1998) *Technical Experiences of Using a Video Camera to make Remote Weather Observations.* WMO Instruments and Observing Methods, TECO-98, May 1998.

Illingworth, A. & Hogan, R. (2002) Clouds: do they obscure the forecast? *Planet Earth.* Natural Environment Research Council, Summer 2002, pp. 12–13.

Keogh, S. J. (2000) *An Investigation of the Potential of Infra Red Cameras for the Determination of State of the Ground.* The Met. Office, OLA Technical Report, July 2000.

Met. Office (1982a) *Handbook of Meteorological Instruments.* HMSO, London.

Met. Office (1982b) *Observer's Handbook.* HMSO, London.

Rowbottom, C. M., Hatton, D. B. & Jones, D. W. (1998) *Operational Experiences in the Use of Video Camera Images to Augment Present Weather Observations.* WMO Instruments and Observing Methods, TECO-98, May 1998.

Scorer, R. C. (1986) *Cloud Investigations by Satellite*. Ellis Horwood Ltd., London, ISBN 0 85312 399 3.

Scorer, R. C. (2001) Private communication.

Weather (2003) Special issue on Luke Howard and clouds. *Weather*, **58**, 49–100.

WMO (1975) *International Cloud Atlas: Manual on the Observation of Clouds and Other Meteors*. Vol. I, WMO-No. 407.

WMO (1987) *International Cloud Atlas*. Vol. II, WMO-No. 407.

15

Lightning

On a second night we witnessed a splendid scene of natural fireworks; the mast-head and yard-arm ends shone with St. Elmo's light; and the form of the vane could almost be traced, as if it had been rubbed with phosphorus. The sea was so highly luminous, the tracks of the penguins were marked by a fiery wake, and lastly, the darkness of the sky was momentarily illuminated by the most vivid lightning.

Charles Darwin *Voyage of the Beagle* (on the sea journey from Rio to Montevideo).

The variable and its history

Lightning was probably the cause of the first fires seen by humans, but before Benjamin Franklin performed his experiments with a kite, it was just a mysterious and terrifying natural phenomenon, although others had been debating the matter before this. Born in 1706, Franklin was one of those rare creative people who encompassed great areas of knowledge and action – as printer, moralist, essayist, civic leader, statesman, diplomat, philosopher and, of course, as scientist and inventor. He lived in Great Britain from 1757 to 1762 and was long opposed to the American colonies separating from the British Empire. But after the Boston Tea Party, and Britain's harsh response, his efforts were doomed and in 1775 he became convinced that 'more mischief than benefit' would come from closer union. He helped draft the Declaration of Independence, and signed it at the age of 70.

In amongst all of this political activity he found time, aged 46, to perform his famous kite experiment, in which he established that lightning was due to static electricity, akin to laboratory-produced electrical discharges. This work was published by the Royal Society in London and four years later he was elected a Member. In 1772 he was also elected to the French Academy of Sciences. It is interesting

Measuring the Natural Environment, second edition, Ian Strangeways. Published by Cambridge University Press. © Ian Strangeways 2003.

to reflect that despite all of his other work, his world fame resulted from this single experiment and from his subsequent invention of the lightning rod. Some of his other scientific endeavours included measurements of the Gulf Stream and the tracking of storm paths. It is difficult to imagine a modern politician or diplomat having such broad capabilities.

The clouds that produce lightning

From Chapter 14, it will be seen that clouds can evolve from one genus and species into another. Cumulus congestus, for example, can grow into the more extensive cumulonimbus calvus, thence into C. capillatus and on to C. incus (anvil top), these being the most common lightning clouds (Fig. 15.1). This can grow even further into the multicell or still larger supercell form and reach a width of 80–100 km with cloud-top temperatures of -40 to $-50\,°C$. Strong updraughts, caused by solar heating of the ground, produce these clouds, with velocities that range from 10 to $50\ \mathrm{m\ s}^{-1}$ (the warm air often being collected from ahead of the cloud's motion, although the cloud as a whole moves along with the general air mass). Updraughts can be as a continuous column or as separate bubbles. They are accompanied by strong downdraughts of cold air that fan out in all directions when they reach the ground, sometimes undercutting the rising air producing wind shear (Dudhia 1996, 1997). Often due to meeting an inversion, typically at the height of the tropopause (10–20 km depending on latitude), the rising air flattens out to produce a fibrous shield of cirrus ice cloud, this being carried forward by the higher-speed upper

Figure 15.1. Over the Philippine Sea, this cumulonimbus cloud shows the classical flat top and anvil shape, rising to about the same height as the plane at 30 000 ft.

winds to produce an anvil-shaped top. If very active, the rising air can push through the inversion and form a dome above the anvil.

Lightning occurs mostly (99%) over the land, and rarely over the oceans, most probably because the sea does not generate such large updraughts. There are also dominant regions of lightning activity, in particular Africa (throughout the year). In other locations, activity is seasonal.

The charging process

During a thunderstorm, the upper regions of the cloud become positively charged with a negative charge near the base (Fig. 15.2). This in turn attracts a positive charge on the ground below, which travels along beneath the cloud as it moves. Pockets of different polarity can also be generated at different levels within the cloud and there is evidence that there can also be a small positive charge near the base of the cloud. There is, surprisingly, still no generally accepted theory as to the mechanism of charge separation, but in many storms worldwide, the negative region has been found to occur between temperatures of −10 and −25 °C, suggesting interactions involving the ice phase.

The view currently favoured by the RMS, in its Scientific Statement on Lightning (RMS 1994), is that during collisions between ice crystals and small hail pellets, in the presence of supercooled water droplets, the pellets become negatively charged while the ice crystals bounce off with a positive charge; this has been demonstrated to occur in laboratory tests. The updraught then carries the crystals aloft, where they produce the upper positive charge, while the pellets are heavy enough to fall against the updraught, forming a negative charge lower down. Lightning, however, rarely occurs if the updraught is less than 6 m s^{-1}. At warmer temperatures, the charging process reverses, giving the lower region a positive charge, although the mechanism involved is still unclear. There is also evidence, albeit limited, that clouds completely below the freezing level can become charged, and plumes from volcanoes and forest fires can also produce lightning, as can sandstorms. The charging processes are still being actively investigated (Dudhia 1996).

Initiation of a discharge

Lightning discharges are initiated when the difference in charge is sufficient to break down the resistance of the insulating air, but there is a problem. Although high, the charges are not high enough, under laboratory conditions, to cause electrical breakdown. So other processes must be involved. One current view is that transitory filaments of liquid, resulting from glancing collisions between supercooled raindrops, or protuberances from ice crystals or hail pellets (graupel), initiate the

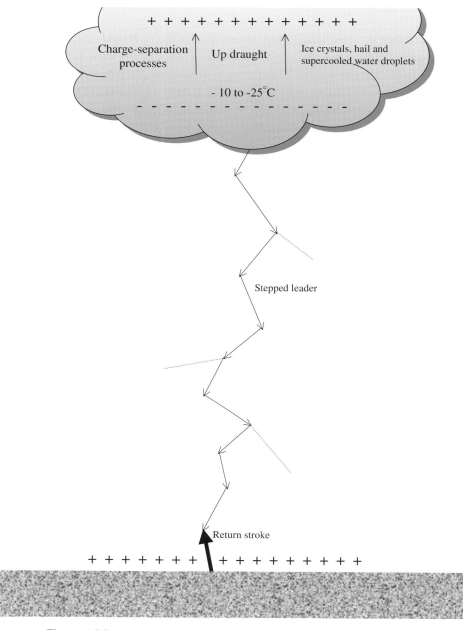

Figure 15.2. Although the charging process within thunder clouds is not fully understood yet, it is known that (usually) it becomes charged positively at its top and negative at the base, which attracts a positive charge on the ground beneath the cloud that travels along with it. The initial downward lightning strike is a faint trace travelling at about a third of the speed of light (100 000 km s^{-1}) in discrete steps of about 1 μs duration (100 m distance). Thus if the cloud base is at 1 km, there are about 10 steps and the time taken is 10 μs. When it has reached to within about 100 m of the ground, the return stroke occurs, the negative charge on the cloud discharging to ground, again at about one-third the speed of light, causing the familiar lightning flash and sound of thunder.

discharge. While this might explain the initiation process, it does not explain how the discharge spans a distance much greater to the ground than it does in the laboratory. However, it is known that initiation of a strike is also related to the nature of the ground surface and to the characteristics of the lower 1 km of the atmosphere. There is also a possibility that changing windspeed with altitude (wind shear) increases the possibility of flashes (Hardaker 1998). Further research is needed to understand both the charging process and the discharge itself.

Lightning strikes

Although the processes producing the charge and initiating the discharge are not yet fully understood, what actually happens in a thunderstorm is much better documented. Discharges can be between different parts of the cloud (intra-cloud), or from the cloud to the ground. There is no difference between the two, except that within the cloud the actual path of the discharge is hidden, illuminating the cloud internally, this being known as *sheet lightning*. Discharges to the ground allow the paths to be seen, as a network of tracks akin to the roots of a tree and this is known as *fork lightning*. There can also, more rarely, be flashes from cloud-to-cloud and into the clear air beside or above the cloud, the latter being known as *rocket lightning*. Again there is no clear explanation of these.

Over half the flashes (70% to 90%) are of the sheet lightning type, but it is the cloud-to-ground strikes that are of more practical concern since these are the ones that cause damage, although the former are of concern to aerospace activities. A ground strike starts near the base of the cloud, as a faintly luminous discharge, producing a stepped *leader*, which travels downwards in discrete intervals of about a microsecond, travelling at around 61 000 miles per second (98 000 km s^{-1}). When the leader approaches within about 100 m of the ground, a return stroke moves up from the ground (usually from protruding objects), the negative charge on the cloud flowing to ground along the ionised path thus established (Fig. 15.2). It is this return stroke that produces the visible lightning, the brightest part of the path moving upwards at about one-third the speed of light. There is often sufficient charge on the cloud to produce several strokes along the same ionised path in the space of less than a second, producing the familiar flickering effect. The combined series of strokes is referred to as the *flash*.

There are rarer occasions when positive discharges are initiated from the top of the cloud, often from the overhang of the cloud's anvil, positive discharges being more common in the Northern Hemisphere in winter. These tend to be one-stroke flashes. The increased use of lightning detectors (see below) is starting to show, however, that positive strokes are more common than were thought, perhaps as many as one in five. Leaders starting from the ground are rarer, but do sometimes occur from positively charged high buildings or mountains. Negative leaders from the ground are still less common.

Currents flowing along the ionised path range from 3 to 300 kA, typically 10 000–30 000 amps, releasing 100 megawatts per metre of channel. This causes violent and rapid heating of the air, to about 30 000 °C, and it is this which produces the sound of thunder. The sound is drawn out into a rumble in part because of there being several strokes in rapid succession, but also because the sound is not from a point source, but from the extended path from ground to cloud, sound arriving at different times from different points along the path. Echoing is also involved in drawing out the length of the sound yet further.

There are around 2000 storms in action at any one time worldwide, with about 100 flashes per second. In the UK there is an average ground strike of about one per square kilometre per year. Worldwide, thunderstorms cause a small negative charge to build up on the Earth's surface, with an equal number of positive charges in the atmosphere. This causes a voltage gradient of about 300 kV between the ground and the conducting layers at 60–70 km, decreasing with height due to the increasing conductivity of the air. Across the lowest metre, there is a potential difference of about 100 volts. Although each flash involves a very high current, each of the 2000 storms around the world produces an average current of only about 1 amp, as the negative discharges flow to ground. A slow, continuous, reverse, air-to-ground leakage current of 2000 amps brings positive charges to ground, balancing the overall current flow.

Ball lightning, St Elmo's fire and Sprites

Ball lightning is a rare phenomenon – a glowing sphere, ranging in size from a few centimetres to a metre. It is usually coloured red, orange or yellow with a lifetime of up to a minute. The ball can be stationary or move slowly, horizontally, at a few metres per second, ending either silently or with a small explosion. It can occur in- or out-of-doors. Many theories have been proposed to explain it, but none are yet widely accepted. The Royal Society (2002) recently brought together many unpublished sightings, of which there have been around 10 000 over the past few decades, demonstrating that it is real enough. The report contains seven papers in all and concentrates on the chemical effects of electrical discharges. One current hypothesis is that when lightning strikes the ground it may vaporise silica in the soil which condenses into a ball of fine dust held together by electric charge, although it is probable that no explanation is yet complete (Muir 2001).

St Elmo's fire is less problematic and consists of a continuous glowing electrical discharge from sharp objects such as masts, church spires, trees and even hair, caused by atmospheric electricity sometimes accompanied by a crackling sound. (I had personal experience of a similar phenomenon while working on meteorological instruments on the summit of Cairn Gorm in Scotland. Raising a screwdriver above my head to make an adjustment, a loud hissing sound came from it. Overhead was a large and dark cloud. I left the mountain without delay. It is probable that if it had been night time there would also have been a glow from the screwdriver.) The phenomenon is so named after the Patron saint of Mediterranean sailors, the discharge being said to be a sign of protection. (I felt it had the opposite significance on the mountain.)

Sprites are incompletely understood phenomena that occur high above thunder-clouds in the mesosphere between 40 and 80 km altitude, following a lightning strike from the positively charged top of the cloud. They take the form of towers of light lasting only milliseconds, but of large size, typically 10 km across to 30 km high. The most usual colours are red or blue. They are fairly common but were not photographed until 1989 and the mechanism of their production is still a matter of debate. There is some suggestion that they emit radio waves, as do the flashes, but this is not yet confirmed. Research continues, some of the latest with balloons (Williams 2001).

Units and terms

Some terms have already been defined above as appropriate, but there are others of a more general nature: The radio waves produced by a lightning flash are known as *atmospherics*, abbreviated to *sferics* (WMO 1996). The strength of a flash is defined as the peak value of current of the first return stroke, measured in amperes (A), typically tens of kiloamperes (kA). The intensity of a storm is defined in terms of: the flash rate in a given area; the average multiplicity (number of strokes in a flash); and the polarity of the flashes. Performance of sensing networks is expressed in terms of the accuracy of location of the flash in kilometres and detection efficiency as a percentage of flashes successfully detected.

As Heinrich Hertz discovered in the 1880s and Giuseppe Marconi harnessed for long distance communication when he made the first transatlantic radio transmission between Cornwall and Newfoundland in 1901, an electrical spark generates radio waves which can travel great distances. So, not surprisingly, lightning also emits radio waves, and it is the measurement of these emissions that form the basis of most detection systems, and of all long distance lightning detection. Lightning emits radio waves over a wide spectrum of frequencies from the extremely low frequency (ELF) to VHF (Chapter 12) bands, the longer waves being emitted most strongly by the return stroke of a ground strike.

Methods of detecting lightning at a distance are of two distinct types – those that measure the *direction-of-arrival* and those that measure *time-of-arrival* (TOR) of the sferic pulses. The former is often referred to as an LLP (the initials of the manufacturer) Direction Finding system. The latter is sometimes known as LPATS, this being the name given to the system (Lightning Position And Tracking System) by its manufacturer. A different design using the TOR principle is named ATD (*Arrival Time Difference*). This was developed by the UK Met. Office. A somewhat different system has also been developed that operates in the VHF band. All of these are examined below. The detection of local lightning is, however, done differently.

Local lightning detection

Local observations are made mostly for utilitarian purposes, at places vulnerable to lightning such as explosive stores, electricity installations, airports and rocket launching sites. Cape Kennedy, for example, is very liable to lightning, and shuttle flights can be cancelled at the last minute due to them (I personally witnessed such a cancellation, much to my disappointment). Since local measurements are not widely used for meteorology or scientifically, very little has been reported in the literature regarding the methods of measurement or their performance (Johnson & Janota 1982, WMO 1996). The techniques are thus only briefly mentioned here.

The visual observation of flashes and the timing of their distance by measuring the delay in the subsequent sound of thunder (every three seconds being equivalent to one kilometre, or five per mile) gives a rough measure without the need for instruments, but such observations are clearly very limited.

For automatic and more precise cover, the vertical static electric field in the atmosphere in the vicinity of a cumulonimbus cloud can be measured to give advanced warning of the likelihood of lightning. Since the electrical field decreases rapidly with distance from the cloud, the detection range is limited to about 20 km, although a network of such instruments can give cover for as wide an area as is instrumented.

There are several methods of measuring the vertical field, a common technique being the *electric field mill*, in which the potential gradient in the atmosphere is measured by an electrode fixed above a metal plate, about a metre below it, at earth potential. In one such design developed in France, a motor-driven rotating shutter alternately exposes and shields the sensing electrode from the field (in what is in effect a Faraday cage), producing a sinusoidal AC signal from the DC field. This minimises measurement drift and noise. (A similar mechanical modulation of a DC signal is used in the cloud height searchlight and the IR humidity sensor.) The 'source resistance' (the electrical impedance 'looking into' the sensor terminals) is very high (being made up of one metre of air). To prevent the sensor being 'loaded' by the measuring circuits, and thereby attenuating the voltage signal, a high input impedance circuit, such as an emitter follower, must be used; amplification may need to follow this. A switch on the motor shaft, synchronised with the rotating shutter, provides the means to sense the phase of the AC signal and thus allow the peak level of the sine wave signal to be sampled and stored. The field strength can be computed from this and related to the probability of a flash. In a similar instrument, the *Flash counter*, the rapid *change* of vertical electric field when lightning occurs is measured.

A high, sharp-pointed conductor, connected to the ground, emits a corona discharge when the local static field increases beyond a certain threshold. By measuring the discharge current, the strength of the field can be estimated and a warning given when a threshold is reached (Williams & Orville 1988). To be reliable, all of these field-measuring instruments must be well maintained, since leakage paths to ground due to dirt and insects act as a shunt and so weaken or destroy the signal.

Yet a further way of obtaining a local warning is to use a single, stand-alone, direction finding system of the type normally used in networks, as described next.

Direction-of-arrival measurement

For the detection of lightning at long distances, Automatic Direction Finders (DFs) are the most widely used method at present, measuring the apparent angle of arrival of the sferic radio waves. It is possible to use a single, stand-alone, DF station to measure the bearing of a flash, its distance being estimated roughly by sensing the attenuation of the signal, or the distortion of the pulse shape introduced during propagation. But for best results several DF stations, suitably spaced geographically, allow an intersection point to be computed by simple triangulation.

Radio waves, like all electromagnetic radiation, are made up of an electric field at right angles to a magnetic field, travelling at the speed of light in space (and almost as fast in air). Below about 10 MHz the ground acts as a good conductor (depending on its composition) and it is a law of electromagnetic radiation that the electric lines touching a conductor must do so at right angles. So in the case of a ground wave, the electric field is vertical and the signal is said to be vertically polarised. The field is radiated vertically in the first place since the flash is usually near-vertical.

The range of detection depends on the wavelengths measured and the bandwidth used. Very low frequencies (VLFs), in the range 10–16 kHz (30–18.75 km wavelength), travel distances up to megametres, propagating by way of the ground wave, which diffracts over the curved Earth, suffering attenuation at the higher frequencies. But in addition, a sky wave is propagated by successive reflections between the ground and the ionosphere (see Fig. 12.5). Due to the longer path that the sky wave has to travel, it arrives later than the ground wave, introducing distortion into the resulting composite wave. In addition the sky wave is not a single wave following one path, but many, due to reflections taking place over a range of angles and through the depth of the ionosphere, which is made up of several layers, which also vary during the day and night. Further, the polarisation of the sky wave is random making the final waveform unpredictable. Even lower frequencies, in the 1–10 kHz band (300–30 km), extend the range yet further, but a longer path causes progressively increased distortion of the received signal (Heydt & Takeuti 1977, Grandt & Volland 1988), in part due to the mix of ground- and sky-waves and in part to high frequency loss. By widening the bandwidth up to 1 MHz, more precise results are obtained, since the influence of the sky wave is lessened and the loss of higher frequencies in the ground wave is avoided, but at the price of reduced range, nominally 400–500 km, although many flashes beyond this can be successfully received. The shorter range can, however, be countered by installing larger networks of DF stations, as is now the case within the US, but at increased cost due to the larger number of stations needed and the communications cost from them to base. Worldwide about 500 stations are in operation at present (year 2003).

Lightning DF systems use the directional properties of a *loop antenna*, which sense the magnetic (horizontal) component of the sferic wave. A loop aerial is a closed-circuit antenna, in the form of a single turn or a coil of wire on a frame, often square in shape. Their value is that they are directional in sensitivity, with a maximum output when the incoming wave is in line with the frame, falling away to near zero at right angles to the coil. Two antennas are set at right angles to each other, oriented N/S and E/W, the ratio of their signals being proportional to the tangent of the angle of arrival of the sferic.

In addition, a flat plate antenna is used to detect the polarity of the discharge. The electric field component of a wave can point either upwards or downwards, depending on the direction of the current in the transmitting antenna – the lightning discharge in this case. This information automatically gives an indication of whether the magnetic component is left- or right-handed, and this removes a 180° ambiguity in direction as measured by the loops.

The UK Met. Office operated the first direction finding system from the 1940s to the late 1980s, known as the *Cathode Ray Direction Finder* (CRDF), working with a centre frequency of 9 kHz and a bandwidth of 250 Hz (approximately where the sferic spectrum is maximum) (Ockenden 1947, Maidens 1953, Horner 1954). The signals from the two antennas were applied to the x and y plates of a cathode ray tube, a line on the screen indicating the apparent direction of the flash. An operator at the station measured the direction and reported it to Beaufort Park in the UK where a fix was obtained by triangulation (Keen 1938). These measurements were reported on the WMO GTS (Chapter 12). Although of limited accuracy, CRDF was able to cover most of Europe, up to ranges of 3000 km.

Modern DF systems are improved and automated, by extending the bandwidth, through better communication from outstations, due to the arrival of computers and through an increasing understanding of exactly which point of the sferic signal to use, as will be described below.

A network of DFs needs a minimum of three receiving stations, the location of the flashes being calculated as the point at which the bearings of the stations intersect. The closer the intersection is to the perpendicular the more reliable the location.

A central Position Analyser (PA) receives the data by telemetry (Chapter 12) from the DF stations and computes the locations of the flashes. To do this it is necessary to be certain that the received signals are indeed occurring at the same instant, otherwise they may be signals from different, but similar and nearly coincident, flashes. If two or more stations receive pulses within a time window of 5–20 ms they are assumed to come from the same sferic. However, DFs may fail to detect the first return stroke, and may report the second flash instead, causing coincidence to be lost. If the window is widened to be 50 ms more flashes are successfully located, but by increasing the time beyond 100 ms false coincidences start to occur. If the telemetry link is fixed ('dedicated'), the coincidence of flashes is determined from the time of arrival of the data from the DFs, since the time delay in the link is constant and known. If the telemetry link introduces any delay, however, such as can occur in a packet-switched communications link, the PA must use the time of arrival of the sferic at the DFs as indicated by the DF clocks. When two or more stations indicate a coincident event, its position is calculated and its time recorded along with the amplitude and polarity of the peak of the waveform and the number of strikes per flash. The larger the distances covered, the more flashes received, and thus the greater the possibility of confusion over coincidence.

Installation and maintenance

A minimum of three stations is required, best placed in the corners of an equilateral triangle to reduce the possibility of near-parallel bearings. If four stations are installed, a square is best. For a larger network, stations lying on the same straight line should be avoided. The spacing of the stations should be even, and, to take best advantage of the nominal 400 km range of the stations, distances between adjacent sites should be around 150–250 km.

In practice, where the stations can be installed depends on factors additional to the above. Sites have to be found that are not screened by large buildings or geographical features such as hills, for these can distort the final angle of approach of the wavefront. The ground wave can also be moved off-vertical by local conductors such as cables and railway lines, causing further errors in the measured angle of arrival. Indeed these several site disturbances are the main source of error in DF systems. Nor should there be local sources of electrical discharge and the sites should be free of vandalism, the presence of local staff being useful in this respect. It should also be possible to establish telemetry links from the site to the base station.

The number of stations used, compared with the minimum essential (the redundancy of the network), can be usefully increased, to guard against the failure of a station or of its communications link. Increasing the number of stations to four, for example, is worthwhile. Additional stations also give a more accurate fix on the flash.

Once established, the field stations are very reliable and need little maintenance. At the base station, operation is also automatic and reliable. Of more importance, and of greater interest, is the establishment of what the site errors are; this is the subject of the next section.

Accuracy and calibration

Since the flashes are located by measuring the direction from which the sferic signals appear to arrive, any error in this angle leads to an error in flash location. Determining the site errors is, therefore, an important calibration procedure following installation. The distortions vary with direction and in a systematic way (MacGorman & Rust 1988). Once established, corrections can be introduced into the processing software as constants. Before correction, errors as high as ten degrees may occur in some directions.

One approach to estimating the errors is to compute the intersection point using one pair of station data and then to adjust the third to coincide, this process being repeated for all three pairs for the same event. The calculations are then re-run using the corrected readings, and the process repeated until the error is minimised. This is repeated for many events, and from all directions, to arrive at some kind of mean correction. However, there still remains a random scatter of errors, which can amount to several degrees in some directions. These are best dealt with separately, for each flash, using least-squares fitting (Orville 1987). The random errors are, of course, also present in the data that are used to determine the systematic errors and these produce a bias. A solution by Passi and Lopez (1989) uses complex statistical methods that decouple the two types of error.

Random errors can arise due to variations from the vertical of the flash and from pulse-distortion during propagation, for example due to mountains. Some of these difficulties were overcome (Krider *et al.* 1976) by gating the sferic analyser to measure the ratio of the two antenna signals during the first 1–5 μs of the return stroke, occurring when the strike is about 100 m from the ground and usually nearly vertical, minimising the polarisation error. The short 5-μs window also helps to exclude the delayed sky waves, which arrive slightly later, and which otherwise cause distortion. But the high frequencies of the ground wave are rapidly attenuated as they propagate over the curved Earth and this limits the range to about 400 km.

The accuracy of strike location varies depending on many factors and a single figure cannot be given. In general, however, it can be summarised by saying that close to the network strikes can be located to 0–10 km, while at ranges up to 250 km, accuracy reduces to 20–30 km due to the uncertainty of the exact angle of arrival. Bearing error is around 1–2° which gives an increasing error the greater the distance.

Another important accuracy characteristic of a lightning detection network is how effectively it detects the flashes in its quadrant. In tests carried out by MacGorman and Rust (1988), the detection efficiency of the DF systems (with a 100-ms window) was 60% to 70% at short range and also at longer range in one direction, but 40% to 45% at 250 km range in another direction. There was considerable variability from one time to another, depending, it seems, in part, on the type of lightning and the amount of intracloud flashes (Mach *et al.* 1986). Detection efficiency thus varies from 40% to 70%.

In these tests, ground truth measurements were obtained by operating video cameras at several locations, the downward-looking camera seeing an all-azimuth view reflected from a conical mirror beneath it. Time was synchronised to 1 ms from an accurate standard and encoded in the video images, giving the time of strikes to +1 to −32 ms.

Time-of-arrival measurement

An alternative to measuring the direction of arrival of VLF sferic signals from return stroke flashes is to measure the difference in their time-of-arrival (TOA) at several stations, thereafter solving for the intersection of the corresponding hyperbolas on a spherical surface. An advantage in locating flashes by this means is that whereas the polarisation, and so the apparent direction of arrival, of a sferic wave can be distorted by geographical and man-made features, the time of arrival is not influenced at all by them. Potentially, therefore, it is more accurate. First developed in the early 1980s, two such systems are in operation at present – a short-range (400 km) regional system, and a long range (12 000 + km) network developed by the UK Met. Office (Lee 1989).

Time-of-arrival network; a short-range system

Principle

A vertical, omni-directional whip antenna is all that is required for this type of receiver, sensing the (vertical) electric field of the sferic by capacitive coupling, in contrast to the twin loop aerials measuring the magnetic component by induction, as used in the DF systems. The stations include a time-signal receiver and time generator, which need to be accurate to a few tenths of a microsecond. One way of achieving this is to use the ground wave from a LORAN-C navigation transmitter (Bent & Lyons 1984) or GPS signals.

Operating in the 2–500 kHz range, the lightning stroke is detected at the field station by similar electronics to that at a DF system, the arrival time of the initial peak of the waveform being determined by the clock. The position of the flash is then identified by a technique similar to hyperbolic navigation (Lee 1986). As with the DF receiver, gating the system to look at just the start of the pulse (the first 5 μs) prevents sky wave distortion, since the sky wave arrives after this initial period. But over distances much greater than about 400 km, the waveform degenerates too much for accurate timing, through attenuation of the higher frequencies.

Installation and maintenance

The geometry of the networks is the same as for DF systems, with similar separations of 150–250 km between stations, although greater distances have been shown to be acceptable. However, because of the way in which the flash location is derived from the time signals, at least four stations are required (see under 'Arrival-time-difference; a long-range system' below). Because the TOA of the pulse is not affected by local structures, the local site is less critical.

Accuracy

Errors are introduced by anything that affects the timing of the peak of the signal. Random errors are caused by ambiguity in deciding on the exact position of the peak in the waveform or by drifts in the clock circuits. These can amount to 1 μs. Systematic errors are introduced by propagation effects, on both the incoming timing signals (GPS) and the flash waveform, the latter being affected by an increase in rise time caused by the loss of the higher frequencies and variations in the conductivity of the ground (farmland, desert or ocean) over which the wave passes. Intervening mountains can also affect path length through propagation delays, amounting to as much as 1.5 μs. If these various conditions are different between the flash and the several receivers, different delays will be introduced at each station, with a corresponding error in estimating the difference in arrival times (MacGorman & Rust 1988).

Overall, the detection efficiency of TOR systems varies between 40% and 60%, but there are many caveats to this and it can vary more widely still. There are too many variables to give a single precise figure; in one evaluation of a six-station network, reports suggest that the detection efficiency is from 80% to 85%. Location accuracy is, likewise, widely variable. However, tests seem to have shown that accuracy is the reverse of the DF system, in that location errors peak at 10–20 km when flashes are near to the network, while they fall to 0–10 km when the flash is 250 km from the nearest group of stations. Again much depends on the geometry of the network, the type of intervening land and the exact nature of the flashes. Quality control of the data (how the peak is determined), and allowances for known systematic errors need careful attention. But clearly DF and TOR systems still leave leeway for improvement. A comparison of the performance of the two systems is reported by Oskarsson (1989).

Arrival-time-difference; a long-range system

The arrival-time-difference (ATD) network was developed by the UK Met. Office as a replacement for the ageing CRDF system, to cover all of Europe and the Eastern Atlantic. It has now been in operation since 1988 (Lee 1986, Taylor 1996). TOR systems (as above) did not have a sufficiently long range to cover the area required without the use of a large number of stations and so high cost. Radar and satellite measurements were not suitable as an alternative (Lyons *et al.* 1985) (see also below). ATD hardware is similar to that of the TOA system and also uses the same VLF band, so as to obtain long range measurements. To avoid the problems of pulse distortion, inherent in long range propagation beyond 400 km, the whole waveform is used, instead of limiting attention to the first 5 μs of the pulse, thereby avoiding having to identify individual features. (The difference in names – TOA versus ATD – is not significant; they are used to allow the two methods to be differentiated from each other.)

The ATD receiver output is sampled at intervals of 10 μs and digitised to a resolution of 14 bits (see Chapter 11), the last 1024 samples being stored. When a signal judged to be sferic-like and of sufficient amplitude is received, sampling continues for a further 512 samples and then stops, freezing the waveform. The time difference between the same sferic signal received at different outstations is estimated by aligning the waveforms in a 'best fit' sense. For more details see Lee (1986).

When a sferic pulse is received at two different stations, the difference in arrival time does not indicate how far away the flash is; it could be 100 or 1000 km. For example, if both stations were the same distance from the flash, the ATD would be zero, however close or distant it was. But ATD does define a (hyperbolic) curve along which the flash could occur in order to give the observed ATD. A third station gives a second curve of equal ARD relative to one of the first two stations. The point of intersection indicates the flash location. However, the curves intersect at two points, and a fourth station and its hyperbolic curve are needed to select which of the two intersections is 'correct' (Fig. 15.3).

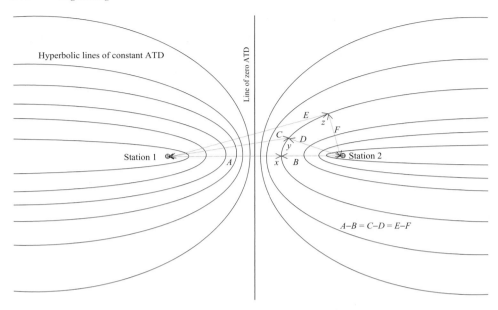

Figure 15.3. If a lightning flash is equidistant between two receiving stations (1 & 2), the ATD is zero and the sferic might lie anywhere along the straight line midway between the stations. If the flash is at a different distance from the two stations, there will be a difference in arrival time of the sferic. The same ATD can occur from a flash anywhere along a hyperbolic curve, for example at *x*, *y* or *z*. If a third station is operated, two more sets of curves are obtained and where they intersect is the location of the flash. However, there can be two possible intersection points and, to eliminate ambiguity, a fourth station can be introduced. By taking three stations at a time, three 'correct' fixes are obtained and the one common to all is selected. The best results are obtained by placing four widely spaced stations at the corners of a square, giving nearly ideal fixing of sferics within the square.

Installation and maintenance

The Met. Office network has five receivers in the UK, spaced between 300 and 900 km, with one in Gibraltar and one in Cyprus, 1700 and 3000 km distant respectively. Because local site characteristics do not affect the arrival time of the pulses, suitable sites are less difficult to find than in the case of DF systems. Because of the ability of the system to operate over long distances, fewer stations are also required, minimising costs.

Accuracy and calibration

Location accuracy is currently 1–2 km in the UK, 2–5 km in Europe and 5–10 km over the Atlantic. More distant flashes can be located to 1–2% of the range, as far distant as 12 000 km. The detection efficiency is around 70%. Evaluation of location accuracy and detection efficiency depends on having accurate ground truth data. This is not possible to achieve across the whole area covered but it is feasible to monitor selected areas carefully and to use this information to assess performance, such as the video system described earlier. It is not, therefore, possible to evaluate performance precisely. Limited case studies have to suffice, such as that carried out in France (Massif Centrale) for the early tests on the ATD system (Lee 1990).

Alternative networks

Other approaches have been followed to measure sferics, although they are not as widely used as the DF or TOA/ATD systems. Just two are noted below, but there is considerable activity at present and new commercial systems and networks are appearing regularly.

ELF system

A crossed loop magnetic DF system developed in the UK in the late 1980s operates in the Extremely Low Frequency (ELF) band working at 1–2 kHz (600–300 km wavelength) with a typical accuracy of 1–2 km. At this low frequency, only the ground wave is propagated without any interfering sky wave, eliminating one of the sources of error in magnetic DF higher frequency systems. This was developed by what was formerly the R&D centre at the Electricity Council Research Centre of the UK, but which is now an independent company. As throughout this book, names of companies are omitted where possible so that advertising is avoided, readers can generally find relevant companies by doing an internet search with the appropriate key words.

VHF system

A system operating in the VHF frequency band was developed by the French National Aerospace Research Agency in the 1980s. Known as SAFIR, the system uses an array of five vertically polarised dipole antennas which act as an interferometer to compute the direction and altitude of the flash by sensing the phase difference in the electric field (of the electromagnetic wave) at each dipole (Richard & Auffray 1985, Richard *et al.* 1988). It works in the 110–118 MHz band, and so is limited to line-of-site distances (Fig. 12.2). However, the measurement is very precise since the waveform is not modified by ground attenuation or by interfering sky waves, and also because it is the phase not the shape of the waves that is used. Manufacturers claim better than 1 km location accuracy and over 90% detection efficiency. As with

VLF DF systems, stations are deployed on a triangular grid, in this case between 20 and 100 km distant from a central processing station, giving it a range of location of 200 km, possibly more, depending on antenna height. At these frequencies it is possible to detect the initial cloud-to-ground stepped leader as well as the main return strikes and also lightning within clouds, the latter often being a precursor of ground strikes (Richard *et al.* 1988).

The VHF system also includes a conventional VLF direction finder working in the range 300 Hz to 3 MHz. Proctor (1971) and Krehbiel *et al.* (1984) describe a VHF system using the ATD method, although its range was limited because of the line-of-sight nature of VHF waves.

In selecting a system from amongst all of those described here, actual requirements generally point clearly to one or another. VHF systems are particularly useful for local and regional warnings for airports and launch sites where high local precision is vital. Electricity authorities and railway networks also find the precise information available from VHF systems useful, as do insurance companies. They would not, however, suit long distance applications.

Radar measurements

As already discussed in Chapter 8 ('Precipitation measurement using radar'), radar can be used to measure rainfall (Fig. 8.17). By applying the same techniques to examine thunderclouds in close detail, the evolution of the cloud can be followed closely. If two Doppler radars are used it is possible to measure the changing winds within the cloud by sensing the velocity of the precipitation particles towards or away from the radars. The returned signal can locate areas of high precipitation, while measurements of the angle of polarisation give information on the shape and fall-mode of the particles, and thereby information on the types of particle. From such measurements it has been found that hail is often present in those parts of the cloud where lightning is initiated.

If the radar pulse is circularly polarised, the returned signals indicate the handedness of the returned waves' polarisation and this tells us something about the alignment of the particles. In a thundercloud, the main determiner of particle alignment is the electric field and so it is possible to watch the build-up of electric charge and its collapse after the lightning strike.

Although radar can be used as a lightning warning system, it is most useful for investigating the processes of thundercloud electrification (Rutledge & Petersen 1994, Samsury & Orville 1994, Zipser & Lutz 1994, Mohr & Torancinta 1996). It also has advantages over the use of research aircraft for this purpose, being able to obtain a three-dimensional image with a resolution of about 100 m every few minutes. Upper air measurements are the subject of Chapter 16.

Satellite measurements

Although systems such as the ATD network can measure sferic activity at great range, virtually globally, there are large areas that cannot be covered by the shorter range methods. In remote areas such as this, satellite observations play an important role. Although Chapter 19 is concerned specifically with remote sensing, a brief word will be said here on how lightning is sensed remotely from satellites.

The sensing of lightning from space is new, a *Lightning Imaging Sensor* (LIS) having been launched in November 1997 aboard the Tropical Rainfall Measuring Mission (TRMM) satellite, which is in a nearly circular orbit inclined at 35° at an altitude of 350 km. A similar system is known as an *Optical Transient Detector*. The LIS comprises what is termed a *staring* imager, a camera with a wide field of view, a narrow-band filter centred at 777 nm (near infrared) with a high speed charge-coupled device (CCD) sensor. A processor determines in real time when a lightning flash has occurred. It can detect this easily at night, the flash spreading up through the cloud and laterally. In daylight pulses are more difficult to detect, but the processor goes some way to remove the background signal, allowing about 90% detection rate. Only rapid changes of illumination have to be detected, so the constant background can be removed. It is able to resolve storms 3–6 km in scale over an area of 550 × 550 km at the Earth's surface. The satellite is, of course, moving relative to the ground at a speed that allows any one point to be in the field of view of the camera for 80 s. Any flashes occurring during this time are detected, the radiant energy being measured and the location estimated (Christian *et al.* 1992). The records from the LIS are compared with ground truth measurements, made with ground-based lightning detectors. Comparison can also be made with cloud images and rain measurements made both on the ground and from satellites. In this way the correlation of lightning with other phenomena can be studied while also checking the efficiency and accuracy of the LIS.

As the satellite orbits, a map is built up of lightning flashes during the day. Of course a snapshot impression of just over a minute is not sufficient to give very full coverage. It will miss many storms that come and go while the camera is not over-head, but it is a start, and gives a summary of activity if not full cover. Indeed, bearing in mind the number of strikes occurring worldwide at any one time, full cover might prove difficult. A system for use on geosynchronous satellites is being developed (it is understood) which would give continual cover (Christian *et al.* 1989). At this distance the signals will be that much smaller and more difficult to detect.

A number of false strikes can be indicated, since the LIS also responds to non-lightning and even non-optical signals. Software filtering has been developed to remove many of these that can be caused by solar glint off the sea or lakes, electronic noise in a system that has to detect very small light pulses, or energetic particles in the Van Allen radiation belts striking the CCD sensing surface. As with so much work now in satellite sensor development, the pace of change is high and the instruments very specialised. For those interested, the best up to date information is to be found on NASA web pages.

References

Bent, R. B. & Lyons, W. A. (1984) Theoretical evaluation and initial operational experiences of LPATS (Lightning Position And Tracking System) to monitor lightning ground strikes using a time-of-arrival (TAO) technique. *Reprints of the Seventh International Conference on Atmospheric Electricity*. Albany, New York, AMS, pp. 317–24.

Christian, H. J., Blakeslee, S. J. & Goodman, S. J. (1989) The detection of lightning from Geostationary Orbit. *J. Geophys. Res.*, **94**, 13329–37.

Christian, H. J., Blakeslee, S. J. & Goodman, S. J. (1992) Lightning Imaging Sensor (LIS) for the Earth Observing System. NASA Tech. Mem. 4350, MSFC, Huntsville, Ala.

Dudhia, J. (1996) Back to basics: thunderstorms: part 1. *Weather*, **51**, 371–6.

Dudhia, J. (1997) Back to basics: thunderstorms: part 2 – storm types and associated weather. *Weather*, **52**, 2–7.

Grandt, C. & Volland, H. (1988) Locating thunderstorms in South Africa with VLF sferics: comparison with METEOSAT infrared data. *Proceedings of the Eighth International Conference on Atmospheric Electricity*. Upsala, Sweden, Institute of High Voltage Research, pp. 660–6.

Hardaker, P. J. (1998) Lightning never strikes in the same place twice – or does it? *Proceedings of Lightning Protection 98. Buildings, Structures and Electronic Equipment*. Birmingham.

Heydt, G. & Takeuti, T. (1977) Results of the global VLF-atmospherics analyser network. In: Dolezalek, H. & Reiter, R. (eds.) *Electrical Processes in Atmospheres*. Steinkopf, Darmstadt, pp. 687–92.

Horner, F. (1954) New design of radio direction finder for locating thunderstorms. *Meteor. Mag.*, **83**, 137–8.

Johnson, R. L. & Janota, D. E. (1982) An operational comparison of lightning warning systems. *J. Appl. Meteorol.*, **21**, 703–7.

Keen, R. (1938) *Wireless Direction Finding*. Iliffe and Sons Ltd, London (Chapel River Press, Andover).

Krehbiel, P. R., Brook, M., Khanna-Gupta, S., Lennon, C. L. & Lhermitte, R. (1984) Some results concerning VHF lightning radiation from the real-time LDAR system at KSC, Florida. *Seventh International Conference on Atmospheric Electricity. American Meteorological Society*, pp. 388–93.

Krider, E. P., Noggle, R. C. & Uman, M. A. (1976) A gated wideband magnetic direction finder for lightning return strokes. *J. Appl. Meteor.*, **15**, 301–6.

Lee, A. C. L. (1986) An experimental study of the remote location of lightning flashes using a (VLF) arrival time difference technique. *Q. J. R. Meteorol. Soc.*, **112**, 203–29.

Lee, A. C. L. (1989) Ground truth confirmation and theoretical limits of an experimental VLF arrival time difference lightning flash location system. *Q. J. R. Meteorol. Soc.*, **115**, 1147–66.

Lee, A. C. L. (1990) Bias elimination and scatter in lightning location by the (VLF) arrival time difference technique. *J. Atmos. Oceanic Technol.*, **7**, 719–33.

Lyons, W. A., Bent, R. B. & Highlands, W. H. (1985) Operational uses of data from several lightning position and tracking systems (LPATS). *Tenth International Aerospace and Ground Conference on Lightning and Static Electricity.* Paris, June 10–13, AAAF, pp. 347–56.

MacGorman, D. R. & Rust, W. D. (1988) An evaluation of the LLP and LPATS lightning ground strike mapping systems. *Proceedings of the Eighth International Conference on Atmospheric Electricity.* Upsala, Sweden, Institute of High Voltage Research, pp. 668–73.

Mach, D. M., MacGorman, D. R. & Rust, W. D. (1986) Site errors and detection efficiency in a magnetic direction-finder network for locating lightning strokes to ground. *J. Ocean Atmos. Ocean Tech.,* **3**, 67–74.

Maidens, A. L. (1953) Methods of synchronising the observations of a sferics network. *Meter. Mag.,* **82**, 267–70.

Mohr, K. I. & Torancinta, E. R. (1996) A comparison of WSR-88D reflectivity, SSM/I brightness temperature and lightning for Mesoscale Convection System in Texas. Part II: SSM/I brightness temperature and lightning. *J. Appl. Met.,* **35**, 919–31.

Muir, H. (2001) Puzzle balls. *New Scientist,* **2322/23**, 12.

Ockenden, C. V. (1947) Sferics. *Meteor. Mag.,* **76**, 78–84.

Orville, R. E. Jr. (1987) An analytical solution to obtain the optimum source location using multiple direction finders on a spherical surface. *J. Geoph. Res.,* **92**, (D9) 10877–86.

Oskarsson, K. (1989) *En jmfrande studie mellan blixtpejlsystemen LLP och LPATS.* Meteorological Institute, Upsala University, Sweden.

Passi, R. M. & Lopez, E. L. (1989) A paremetric estimation of systematic errors in networks of magnetic direction finders. *J. Geophys. Res.,* **94** (D11), 13319–28.

Proctor, D. E. (1971) A hyperbolic system for obtaining VHF radio pictures of lightning. *J. Geophys. Res.,* **76**, 1478–89.

Richard, P. & Auffray, G. (1985) VHF-UHF interferometric measurements, applications to lightning discharge mapping. *Radio Sci.,* **20**, 171–92.

Richard, P., Soulage, P., Laroche, P. & Appel, J. (1988) The SAFIR lightning monitoring and warning system, application to aerospace activities. *International Aerospace and Ground Conference on Lightning and Static Electricity.* Oklahoma City, NOAA, pp. 383–90.

Royal Meteorological Society (1994) Scientific statement on lightning. *Weather,* **49**, 27–33.

Royal Society (2002) Ball lightning. *Philos. Trans. A.,* **360**, No 1790. [Compiled by Abrahamson, J.].

Rutledge, S. A. & Petersen, W. A. (1994) Vertical radar reflectivity structure and cloud-to-ground lightning in the stratiform region of Mesoscale Convective Systems: further evidence for in situ charging in the stratiform region. *Monthly Weather Rev.,* **122**, 1760–6.

Samsury, C. E. & Orville, R. E. (1994) Cloud-to-ground lightning in Tropical Cyclones: a study of Hurricanes Hugo (1989) and Jerry (1989). *Monthly Weather Rev.,* **122**, 1887–96.

Taylor, P. (1996) When lightning strikes: split-second timing and thunderstorm forecasting. *GPS World,* November 1996, 24–32.

Williams, E. R. & Orville, R. E. (1988) Intracloud lightning as a precursor to thunderstorm microbursts. *International Aerospace and Ground Conference on Lightning and Static Electricity.* Oklahoma City, NOAA, pp. 454–9.

Williams, H. (2001) Rider on the storm. *New Scientist*, **2321**, 36–40.
WMO (1996) *Guide to Meteorological Instruments and Methods of Observation*. Sixth edition, WMO-No. 8. WMO, Geneva.
Zipser, E. J. & Lutz, K. R. (1994) The vertical profile of radar reflectivity of convective cells: a strong indicator of storm intensity and lightning probability? *Monthly Weather Rev.*, **122**, 1751–9.

16

The upper atmosphere

At the place where we slept water necessarily boiled, from the diminished pressure of the atmosphere, at a lower temperature than it does in a less lofty country ... Hence the potatoes, after remaining for some hours in the boiling water, were nearly as hard as ever. The pot was left on the fire all night, and next morning it was boiled again, but yet the potatoes were not cooked.

Charles Darwin *Voyage of the Beagle* (high in the Andes).

The variables and their history

For use in numerical weather prediction models, for climate change and pollution studies, to predict radio propagation behaviour, for aviation and for the launch of space vehicles, it is necessary to know how conditions change with altitude, in particular temperature, humidity, pressure and wind.

Upper air measurements started around 1650 when Pascal and Perrier carried barometers up a mountain (see Chapter 6). But because of the lack of suitable platforms to carry instruments aloft, there was no further progress until about 1749 when Alexander Wilson made one of the first attempts using a kite, this being followed in 1784 by a balloon flight over Paris, giving the first observation of temperature lapse rate. Balloons and kites developed alongside into the nineteenth century when Gay-Lussac made a balloon ascent to 23 000 feet while in 1865 Glaisher flew to 11 km in a coal gas balloon, almost dying in the process. Kites continued to be used into the early twentieth century, tethers of piano wire allowing heights of 10 000 ft to be reached, but this resulted in some disasters and a move was made to free-flying, un-manned balloons, much aided by the introduction of extensible rubber balloons in 1901 (Pettifer 2002). The first radiosonde was designed by Molchenov in 1928 and independently in France and Germany at about the same time, while from around

Measuring the Natural Environment, second edition, Ian Strangeways. Published by Cambridge University Press. © Ian Strangeways 2003.

1930 the UK operated radiosondes using sensors that varied in inductance. In 1931 Duckert and Vaisala introduced capacitive sensors, which proved, eventually, to be the best method.

On a routine basis today, upper air measurements are made mostly by radiosondes carrying sensors aloft by balloon, although lately measurements made aboard commercial aircraft have been added to the sonde data. In addition, upward-looking, ground-based 'sounders' have also recently been developed measuring profiles using radar, lidar and sonar. This chapter examines each of these techniques, with the emphasis on radiosondes.

Radiosondes

Radiosondes (abbreviated hereafter to 'sondes') are disposable instrument packages, lifted through the atmosphere by balloons, measuring profiles of temperature, relative humidity and barometric pressure. They rise to an altitude of between 25 km and 35 km (16–22 miles or 82000–115000 ft), which includes all of the troposphere and much of the stratosphere (Fig. 14.1). As the balloon rises, readings are taken and transmitted back to a ground station. Since the instruments are rarely recovered and never reused, they are designed to operate for a few hours and then to be lost. Cost must, therefore, be low. Wind is also measured, either from the ground by radar or theodolite or aboard the sonde using navaid systems.

The same sensor types as are described in Chapters 3, 4 and 6 are used in sondes, although the actual form of the sensors may be different so as to miniaturise them, while the electronics that interface them to the sonde's telemetry system, and the telemetry itself, are covered by Chapters 11 and 12. Beyond a few additional comments, therefore, to illustrate the differences, these topics need not be repeated here. The main concerns of this chapter are the construction of the platform and the problems encountered by sensors exposed to the rapidly changing and extreme conditions as they rise quickly up through the atmosphere. In their short journey they encounter rain, cloud, snow, ice, sub-zero temperatures and very low atmospheric pressure, all in the space of an hour, presenting a much more extreme environment than occurs at ground level, even in the Antarctic.

Since radiosondes are not 'installed' in the usual meaning of the word, that is permanently (because they are mobile and expendable), or maintained (due to their transitory life), the question of exposure concerns how the sensors are deployed during their brief flight. Calibration also differs in that it is a once-and-only matter in two stages. The manufacture will usually test every sensor at production to ensure that they meet their specifications, probably at only one or two points on the curves, and the user must also do tests just prior to launch to check them and to calibrate them against local instruments.

Today there are many hundreds of radiosonde launching stations around the world including mobile stations on ships, and there are many different makes and designs, some old, many not meeting the required modern standards; as so often occurs, there is a variation in the quality of data from different regions of the world. It is not proposed to describe all models old and new, good and bad, but rather to concentrate on typical good modern designs that ideally should be aspired to worldwide.

Units and terms

The same units are used as at ground stations, that is hectopascals for pressure, degrees Celsius for temperature and percent for relative humidity. The ranges covered, however, are much wider than on the surface – 1050 hPa to 5 hPa for pressure, +50 to −90 °C for temperature and 1 to 100% for relative humidity.

In addition there is the unit of *geopotential height*. The geopotential height $H = R z/(R = z)$, where z is the geometric height and R is the radius of the Earth (both in the same units). This compensates for the decrease in gravity with height. The difference is very small and in the troposphere the *geopotential height* is approximately the same as the *geometric height* in metres.

Temperature

Temperature sensors used in sondes need to respond rapidly and should have a time constant of around one second to ensure that errors from thermal lag are less than 0.1 °C while rising through any 1 km layer of atmosphere. This calls for a small device with maximum surface-area-to-volume ratio. While radiation screens perform adequately at the high pressures of lower altitudes, screens present considerable problems as the pressure falls, their surfaces becoming much warmer or cooler than the atmospheric temperature, depending on the terrestrial and solar radiation conditions. Although a screen made of two concentric polished aluminium cylinders with a spacing of 1 or 2 cm might be acceptable, measurements have been shown to be more accurate when the sensor is exposed directly without any protection, although this only applies to the smaller modern sensors.

To minimise the radiation errors that result from having no screen, the sensor must have a surface that absorbs as little solar and IR radiation as possible and this can best be achieved by depositing a very thin metallic film on it. Many of the white paints used in the past have high absorption in the IR. An additional consequence of having no protection is that the sensor will be exposed to rain and cloud droplets and will be subject to ice formation. The sensors must shed these if large errors are to be avoided. Sensors are best exposed above the sonde, to avoid heating, or cooling, by the package. This is achieved by mounting the sensor on a short arm

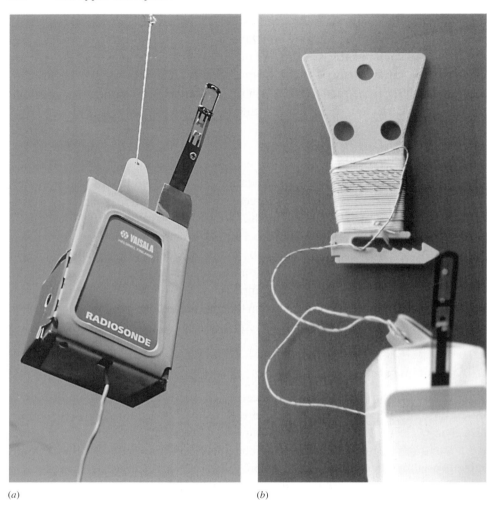

(*a*)   (*b*)

Figure 16.1. Radiosondes (a) are throw-away packages of cardboard and poly-styrene with a cavity for the battery, a pressure sensor and an arm on which the temperature and humidity sensors are held away from the body of the sonde to avoid its heating or cooling effects, a radio transmitter (antenna wire below) and a small electronics board, the whole costing only £100. To avoid entanglement with obstructions at launch, the cord is released slowly from an unwinder (b).

just higher and to the side of the package (Fig. 16.1(*a*), (*b*)). The sensor also needs to be robust enough to survive rough handling during launch, often done under difficult field conditions.

In the past, nearly every type of temperature sensor has been used for sondes, even bimetallic strips (despite their slow response and the problems of shielding them). But today the preferred sensor is a thermistor, thin wire or thermocapacitor, each of which can be made very small and suitably protected from radiation by a thin reflective film. Thermocapacitors were developed especially for radiosondes and so were not included in the chapter on temperature. They can take various forms but that illustrated in Fig. 16.2 is of the more recent thin-wire type made of two thin platinum wires separated by a glass-ceramic dielectric, the value of which changes with temperature causing a change of capacitance between the wires. The whole is sealed in a glass coating which is covered in a reflective aluminium film.

Figure 16.2. Recently temperature sensors have become very miniaturised, as in the case of this thermocapacitor, made of two very fine wires of platinum separated by a glass-ceramic dielectric, the whole covered in a reflective aluminium film. Its very small size (compare with needle) and reflective coat minimise heating by solar or IR radiation, reduce the retention of water droplets from cloud and allow a rapid response to changes of temperature.

Accuracy and calibration

Each make of radiosonde has its own range of errors, although it would be inappropriate to list them here for all makes. For those who need such detail, WMO (1996a) includes tables comparing the various makes of sonde, old and new, currently in use. A summary of errors that all sondes are subject to, and of the performance of just the latest sondes, will suffice here.

General sources of error that affect all sonde sensors are errors in factory calibration/production, errors in pre-flight check readings (see below), and problems arising from interpolating between readings when transmission is intermittent due to poor reception. Specifically to *temperature*, sources of error (additional to any error inherent in the sensor itself) are several. The electrical contact resistance of the removable arm can introduce electrical resistance additional to that of the sensor. IR heating occurs day and night, from the atmosphere above and from the atmosphere and the ground below, the latter being the greater. As the sonde rises, the IR flux changes, and it also varies from flight to flight depending on the cloud present and the temperature of the atmosphere and of the ground. But the sensor also emits IR itself, and to obtain the correct reading absorption and radiation must be in balance; under some conditions the sensor can cool by IR radiation well below the air temperature. The balance varies with pressure and so with height, being most skewed at the lower pressures, but the balance is also dependent on the actual air temperature, so the situation is quite complex and variable. To complicate matters further, latitude is also a factor and because conditions vary so much from flight to flight the errors of individual flight are difficult to correct for. The older sondes, with white rod thermistor sensors, are more prone to these errors. Screens do not improve matters since they, too, are liable to similar radiation-exchange effects and these are passed on to the sensor inside. In the older designs, thermal lag is a further source of error, but again the newer models are free of this.

All sensors will also have heating errors in daytime flights due to solar radiation, which introduces a systematic bias. (Completely effective shields or reflective films are not practicable.) These errors can be partly corrected by software during processing, corrections being based on tests comparing differences between day/night flights, adjusted for solar angle. Rate of ascent is also taken into account, since this affects ventilation. At sunrise and sunset, the correction must take into account sonde height, as errors are very sensitive to solar elevation. The presence of cloud must also be allowed for in correcting for solar heating. The albedo of the surface (see Chapter 2, 'Reflected solar radiation') is an additional and important factor, low cloud and snow reflecting a large proportion of the incident solar radiation, while in contrast the sea has a low albedo. Sondes also swing and rotate beneath the balloon and this can allow air warmed by the sonde body to contact the temperature sensor, this effect increasing rapidly in the low pressure of the stratosphere. The motion of the sonde beneath the balloon also varies the orientation of the sensors to the Sun and thus the amount of radiation incident on it.

If water or ice accumulate on the temperature sensor, further errors are caused due to cooling (wet bulb effect) after the sonde has left the rain or cloud belt and the RH has fallen below 100%. Ice forming over the metalised reflective skin acts as a black body and leads to radiative cooling. If water, formed on the sensor at low altitude, freezes, the released latent heat will warm the sensor towards $0\,°C$. Conversely, if an ice-coated sensor moves into warmer air, the reading will stay at freezing point until all the ice has melted.

To minimise errors, sonde readings must be checked manually *just* prior to launch. This applies to all sensors (temperature, humidity and pressure). Any error in these readings is included in the sonde reading, and errors in manual observations may not be small, as Chapters 3, 4 and 6 have shown. The sonde's accuracy thus depends very much on the quality of the instruments used in pre-flight tests and on the observer's skill in reading them. Although pre-launch checks give confirmation at one point on the sensor's response curve, the performance at other points cannot be checked, and the manufacturer's calibration has to be relied on. Reliability varies from make to make.

It will be very apparent, therefore, that no simple universal statement about temperature accuracy is possible, given the wide range of factors that affect it and also the wide range of instruments manufactured. However, as a rough guide, the better, modern, sondes measure temperature to a resolution of $0.1\,°C$, to an accuracy of $\pm0.2\,°C$, with a response time of 0.2 s at ground level slowing to 1.0 s at 10 hPa. This includes the variability introduced by virtue of the fact that every flight is different.

Relative humidity

It is difficult enough to measure RH at ground level, but as temperatures and pressures fall, it becomes increasingly problematic. To provide acceptable measurements, the sensor must be able to measure to 1% of saturated water vapour pressure from 46 hPa at $30\,°C$ to 0.06 hPa at $-50\,°C$. While most modern RH sensors agree well in dry conditions down to $-10\,°C$, at lower temperatures and as the pressure falls, performance is degraded and agreement diverges. If the exchange of water molecules is slowed by low temperature, the molecules acquired in the lower atmosphere are not shed, giving incorrect, higher, readings. The time constant of RH sensors increases much more than that of a temperature sensor during ascent. Of the several RH sensors discussed in Chapter 4, the fastest reacting is the thin-film capacitor sensor, being 20 s at $-30\,°C$, whereas at $+20\,°C$ it will be less than a second. Down to temperatures of $-30\,^{\circ}\mathrm{C}$ the carbon hygristor's speed of response is comparable to that of the thin-film sensor, but at lower temperatures its time constant (in response to falling humidity) decreases to as low as 10 min. At similarly low temperatures, the thin-film sensor time constant is about 2 min. Most of the sensor readings are also temperature dependent and need correction for this (during processing by the ground system). It is WMO's view (1996a) that no radiosondes give reliable RH measurements at low temperatures and pressures.

The smaller sensors, such as the thin-film capacitor, can be deployed from an outrigger arm with a protective cap. Larger sensors, however, have to be housed internally or on the top or the side of the radiosonde body, in a duct that allows air to flow freely. The duct or cap must shed water and ice quickly.

Recently a new design of thin-film capacitive sensor became available with twin sensors, both with miniature, built-in, heating elements (Fig. 16.3). In operation, one sensor is used to make the measurements while the other is heated to remove water or ice, followed by a recovery period, whereupon it replaces the second sensor, which then undergoes the heating/recovery cycle.

Accuracy and calibration

Not every sensor can be completely checked fully during manufacture but they are assumed to fit a standard curve. There is bound, however, to be some departure from this (Wade 1995) and sensors can vary by several percent from batch to batch. Calibration may also change during storage. WMO performed an extensive international comparison of sondes and the results are summarised in several papers (1987, 1991, 1996b) with details in Nash *et al.* (1995). The best humidity sensor was found to be the thin-film type, the worst goldbeater's skin and lithium chloride. As with temperature sensors, it is necessary to check their performance just prior to launch and in doing this it is useful to include a check at very low humidity; this produces more consistent results between flights. A record of readings as the sonde passes through low cloud also gives a useful reference for 100% RH.

The speed of response of most humidity sensors is less than ideal, particularly at low temperature and pressure. This is mostly so for the older sensors while thin-film and carbon sensors fare better to about $-20\,^{\circ}\mathrm{C}$, but the carbon types then become less reliable as $-40\,^{\circ}\mathrm{C}$ is reached. Thin-film sensors operate reliably below $-40\,^{\circ}\mathrm{C}$, although the calibration deteriorates a little. The two-sensor, warmed type obviously performs better than the standard model.

Because the RH can rise to 100% and then fall rapidly, perhaps a number of times as a sonde rises thorough cloud layers, hysteresis is more of a problem than with temperature. However, the error only amounts to a few percent RH, although they are much greater for older sensors such as goldbeater's skin. The effect will

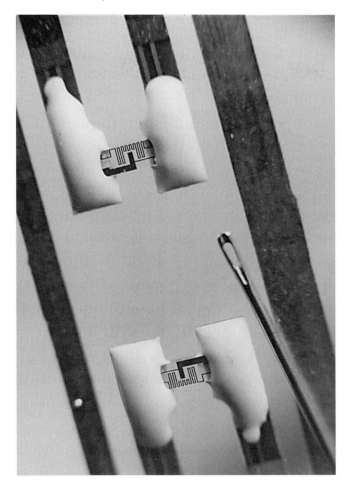

Figure 16.3. This latest form of thin-film humidity sensor has two elements with integrated heaters. While one is making observations, the other is heated to drive-off any moisture which may have collected during passage through cloud or retained from a previously higher RH condition due to slow response at low temperatures. The sensors alternate between heating and reading every minute or so.

be lessened in the case of the heated, thin-film sensors, for it is the loss of water molecules rather than their gain that is slowed up at low temperatures. Since RH is only meaningful if the temperature is also known, errors will be introduced if the RH and temperature sensors are not at the same temperature. This will occur if the thermal lag of the two are different or if they are differently exposed or have different radiation properties.

As in the case of the temperature sensors, but to a greater extent, RH sensors are affected by water or ice forming on their active surfaces. Only those with heating elements will be less prone to these problems. With sondes without such remedial action, RH measurements in the upper troposphere after the sonde has passed through layers of cloud need to be treated with caution, for they can be considerably in error.

The better, modern, sondes measure RH to a resolution of 1%, to an accuracy of $\pm 2.5\%$, with a response time of 0.5 s at ground level slowing to 20 s at $-40\,^{\circ}$C.

Barometric pressure

Unlike ground-based pressure sensors, those in sondes must measure over the much wider pressure range of 3 to 1000 hPa. A resolution of 0.1 hPa is needed to obtain the required height discrimination, sensitivity sometimes being arranged to increase with decreasing pressure to obtain better performance at high altitude. As with ground-based barometers, there is no need to expose the sensors directly to the environment, and so the sensors are enclosed in the body of the sonde, in some cases with additional thermal insulation, and even heating.

Until recently only aneroid sensors could be used in sondes, due to a lack of any viable alternative, and indeed they are still widely used (Chapter 6). Although designs are varied, a typical design will consist of two sensor plates inside the aneroid capsule fixed to the membranes, the capacitance varying between the plates as they move slowly apart as the sonde rises and the capsule expands. Recently, however, sensors using silicon diaphragms (Chapter 6) have been greatly miniaturised making them very light and with a negligible response time, removing any time lag problem. An alternative to measuring barometric pressure is to use the altitude as measured by radar (for measuring upper winds – see below) as a proxy. Conversion from height to pressure follows the curve of Fig. 16.4.

Accuracy and calibration

Unless there is a radar check of altitude for comparison, errors are not easily determined, although the accuracy of the sensors themselves, in laboratory conditions, can be specified (Chapter 6). Pre-flight checks are essential to give a ground level reference, although this will not tell us what happens over the rest of the range. Hysteresis error will be small because the pressure changes in just one direction during ascent – downwards. But if readings continue to be taken as the sonde eventually falls, the pressure will rise and hysteresis will be involved (in older sensors). If temperature corrections must be applied to the pressure sensor, errors will arise if the pressure sensor is not at the same temperature as the temperature sensor. Since the former is protected within the sonde and the latter is outside of it, such differences will certainly exist and could be large. If aneroid capsules are the primary

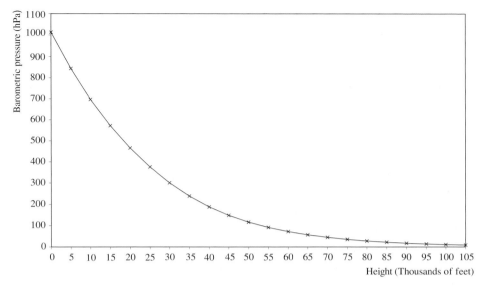

Figure 16.4. The *International Standard Atmosphere* (*ISA*) as defined by the International Civil Aviation Authority shows the variation of pressure with altitude, altitude being in geometric not geophysical units. At mean sea level the temperature is taken to be 15 °C, the pressure 1013.25 hPa and the density 1225 g m^{-3}. The rate of fall of temperature with height is practically equivalent to 6.5 °C km^{-1} (or 1.98 °C per 1000 ft) up to 11 km (36 090 ft). From there to 20 km (65 617 ft) the temperature is constant at −56.5 °C. From there to 32 km (104 987 ft) the temperature rises at 1 °C km^{-1} (0.3 °C per 1000 ft).

sensing element, the errors of those that measure displacement capacitively are much smaller (±1.5 hPa) than those involving mechanical linkages. Errors can also be large if the sonde suffers thermal shock at launch (coming from a warm room to cold condition without gradual conditioning).

As a general guide, the better, modern, sondes measure pressure to a resolution of 0.1 hPa with an accuracy of about ±1 hPa, although this varies with altitude.

Conversion to geopotential height from the pressure measurements is a complex function involving the surface pressure, the true pressure at any particular time into the flight, the actual pressure as indicated by the sonde and the errors in the sonde's pressure and temperature measurement.

Telemetering the data

Equipment onboard the sonde

Techniques for telemetering data collected by ground-based instruments are covered by Chapter 12 and need not be repeated here. Because sondes usually remain within line-of-site of the ground receiving station throughout their flight, UHF frequencies are well suited to the transmission of their data. (In high upper wind conditions and when the sonde does not reach full height, the angle of the sonde above the horizon can be low and transmission may be impaired.) The frequency used is similar to that over which data collection platforms transmit their data to satellites such as METEOSAT, that is around 400 MHz, with a power output of 250 mW; 1680 MHz is also used, at a power of 330 mW.

In earlier designs, the sensor signals were transmitted by modulating a fixed frequency carrier with audio tones, which vary with sensor readings, typically between 7 and 10 kHz. Reference signals may also be transmitted to calibrate the circuits. There is some advantage in this simple method, in that by listening to the pattern of the tones on a speaker, the quality of the signal can be monitored as well as the performance of the sonde overall. An alternative, in older models, was to convert the sensor readings into a Morse-type code using a mechanical converter on each sensor, while yet another older type mechanically generated a train of pulses proportional to the amplitude of the signal, which were counted at the ground station. In these early models, multiplexing and conversion to coded signal were mechanically driven by (sideways-mounted) anemometer cups or a (vertical) fan, turned by the wind (which resulted from the sonde rising through the atmosphere) (Fig. 16.5). These caused switches to rotate.

In modern sondes (as in loggers), the analogue sensor readings are handled first by a multiplexer that switches each sensor sequentially to an analogue-to-digital converter and thence, via a modem, to be transmitted in digital code. Scanning is rapid and repeated about every second (giving a height resolution of about 5–10 m). Because in digital format the signal-to-noise ratio can be smaller, this method allows the bandwidth and radiated power to be reduced, reducing battery size. It is possible to convert the sensor readings directly into meteorological units in the sonde's microprocessor and to transmit them in this form, with the calibrations already applied. Alternatively the sonde may transmit its data in raw, engineering units, along with the calibration figures for the particular sonde, the processing being done at the ground station. The transmission can also include an identification code, thereby preventing confusion if several sondes are within range.

The power supply for a sonde needs to maintain its voltage for up to three hours, should be light weight, environmentally non-polluting and be able to operate at low temperatures. Two types are used today: dry cell batteries, which have the advantage of low cost, but are heavy, and water-activated batteries (cuprous chloride and sulphur) which are lighter and can be stored for a long time. The action of the water battery also generates heat which helps stabilise the internal temperature of the sonde and its electronics. Water batteries are the preferred type.

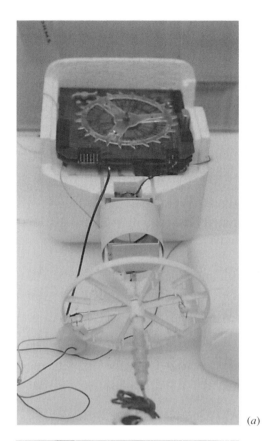

(a)

(b)

Figure 16.5. These two early radiosondes are at the Science Museum, London. The upper model shows how the rise of the sonde through the atmosphere was used to turn a fan that activated the cam switch above it, sending the sensor readings in turn for transmission. The lower model used cups to produce the rotation. Both sondes were kindly made available by Jane Insley, Curator, Environmental Science and Technology.

Receiving equipment on the ground

The ground station comprises an antenna, radio receiver, logic to decode the received signals and software to convert them into meteorological units (unless this was done by the on-board electronics in the sonde). The received data then need to be disseminated over the GTS (Chapter 12). At some stations (see below) there may also be additional equipment for wind measurement (for example by radar or theodolite). In modern sondes, the decoded messages are input to a PC for processing, but older designs may either record the readings on a strip chart or print them out, although these are now becoming obsolete.

Each manufacturer has their own method of coding the transmissions and they sell integrated ground systems to handle their unique design. These are ideal when the system is maintained by the manufacturer, but if there is in-house technical and software capability, a modular system might be preferable, since it allows several different decoders to be operated without duplicating the rest of the system, thereby enabling different types of sonde to be operated.

The software at the ground station, if produced by the manufacturer, will be free of major sources of error, for example errors that can be introduced by interpolation between points when some data are lost through transmission problems; too much interpolation is dangerous. If users decide to write their own software, it needs to be based on a sound knowledge of radiosondes; there are many pitfalls for the unwary.

The cost of a typical ground station is about £50 000 and that of a sonde is £100.

Comparison of different makes of sonde

The WMO has been particularly concerned with radiosondes and has organised international comparisons of various makes (WMO 1987, 1991, 1996b). The first series was held in Switzerland in 1950. The ongoing tests helped identify major errors, which initially were large, and thus their correction. While laboratory tests allow comparisons to be made of sensor performance under conditions similar to those used by manufacturers, it is not possible to simulate many of the conditions experienced during actual flight, for example wetting, icing and radiation. These can only be evaluated during actual ascents. If it is wished to compare two makes of sonde, it is necessary to ensure that they rise at a similar rate, close together. Because modern sondes give fairly reproducible results, systematic biases can generally be detected and quantified after 10–15 flights, at one solar elevation; it being necessary to repeat these for different elevations as well as at night. For more information on this complex task, the reader is referred to WMO (1987, 1991, 1996b). The good performance of modern radiosondes is undoubtedly in debt to this extensive work.

Balloons and their launching

Today's balloons are not just used for lifting sondes but also for the visual measurement of upper air winds (below) and for visually determining cloud base height (Chapter 14). These are known as *pilot balloons* and since they carry no load they are smaller than sonde balloons. Like sonde balloons they are usually spherical and of the extensible type. Balloons for lifting an instrument payload at a reasonable rate of ascent are larger. There are other types of balloon, such as those made of inextensible material (polyethylene) which rise to a fixed altitude and float there, while tethered balloons are used for obtaining profiles (Kilburn *et al.* 1999, 2000). These specialised tasks will not be discussed here. For measuring upper winds by radar (below), larger or sonde balloons are used. Pilot and sonde balloons are made from natural rubber or synthetic latex, the former being stronger and less affected by temperature, but with a shorter shelf-life and vulnerable to UV radiation and ozone.

The lift of a balloon is given by $T = V(\rho - \rho_g)$, where T is the total lift, V is the volume of the balloon, ρ is the density of the air and ρ_g the density of the gas in the balloon (assumed to be spherical). The *buoyancy* $(\rho - \rho_g)$ at ground level for hydrogen is about 1.2 kg m^{-3}. All the figures, of course, change with height. The free lift of the balloon is the difference between the total lift less the weight of the balloon and of the payload. The rate of ascent, however, is related to many factors beyond the value of lift alone, including drag coefficients (and thus the Reynolds number). Gas leakage is another factor that affects performance while rainfall adds a further complication. Because of these many and varying factors, it is not possible to derive a simple universal formula to calculate speed of ascent; this must be determined by experiment. A precise knowledge of speed is, however, not essential (except in determining cloud base height visually – Chapter 14), but a rate of around 300 m min^{-1} is best, to minimise the time required for an observation and also to provide adequate ventilation. In practice it is necessary to reach a compromise between ascent rate and burst height by adjusting the amount of gas at inflation. This in part depends on the temperature at launch, so clearly it is not at all a simple matter.

Balloons range in weight from 10 g with a diameter of 30 cm, to the largest at 3000 g and 210 cm diameter, with equivalent rates of ascent from 60 to 300 m min^{-1}, rising to altitudes of 12 to 38 km respectively, lifting payloads of from zero to 1 kg. There is a range between these two extremes of about eight other sizes and performance. Thickness of the skin is 1–2 mm at release allowing for an extension of up to four times its original size at ground level before bursting. Usually a balloon of about 350 g will be used, costing around £10. A larger balloon, say 2000 g, will cost more than the sonde.

Hydrogen or helium is used to fill the balloons, the former being the commoner because it is the cheaper, although the cost is not very high and the situation is changing. Hydrogen also gives a slightly greater lift (about 8%). With hydrogen, however, there are many safety precautions to be followed due to the danger of explosion, from which helium does not suffer. The gases are usually held in cylinders, although chemical and electrical generators are also available. Launching can now be done entirely automatically and eight of the ten UK sonde stations are now automated, with the remaining two due to be converted shortly. However, worldwide the great majority of launches are still manual. Care is necessary in windy conditions and even automatic launchers are limited to winds below 20 m s^{-1}. *Unwinders* (Fig. 16.1b) allow the sonde to be lowered slowly after launch to minimise the risk of it catching on obstructions during the early part of the flight, which can be affected by turbulence. To achieve the required lift, weighted valves can be used to stop filling automatically at a pre-set value. For descent, a parachute is used to slow fall to a maximum of 5 m s^{-1} to avoid accidents or damage.

Upper air wind measurement

Wind measurements through the depth of the atmosphere are essential for all scales of weather forecasting, for all aircraft operations, for accurate military operations and for the launching of any form of rocket. As for surface measurements, the units used for upper wind are now mostly metres per second and degrees clockwise from north. The height is expressed as the geopotential height (see above).

Rawinsondes (radiosondes equipped to measure windspeed) are the main vehicle for measuring upper winds, although pilot balloons tracked visually or by radio (*radiowind* methods) can also be used for cheapness. To these can be added observations from aircraft and measurements made by *sounders* or *profilers* (below). Over the sea, wind data derive mainly from civil aircraft at cruise height (below) and from rawinsondes launched from both ships and from remote islands. The movement of clouds and water vapour patterns can be tracked using geostationary satellite images (Chapter 19). In the future, space-borne active remote-sensing *lidar* (light detection and ranging) and radar measurements will become available. Lidar and radar methods are already operated from ground stations along with *sodar* (sound detection and ranging). Anemometers on kites give high-resolution measurements for specialised applications and small pilotless aircraft are being developed (Holland *et al.* 1992).

Pilot balloons and optical theodolites

The simplest and cheapest method of tracking a balloon is with a single optical theodolite, the compass direction and angle of elevation of the balloon being measured minute-by-minute. Provided the rate of ascent is known, the height and so the plan position of the balloon can be calculated by trigonometry. For more precise results two theodolites are used spaced at least 2 km apart on a line as near as possible at right angles to the line of flight (Fig. 16.6). The method is of course unworkable in cloud and under such circumstances radio direction finding must be used. This takes three basic forms.

Radar

Primary radar systems can be used to track balloons, which need not have radiosondes attached, a radar corner reflector being all that is required. Since elevation, bearing and slant distance are measured by the radar, the height can be calculated and thus the target's plan position, and so its speed (Fig. 16.6). If the balloon is fitted with a transponder that transmits a pulse when it receives one from the ground, simpler, secondary radars are sufficient since the returned signal is much stronger than a reflected pulse; this also gives a greater distance cover. But radar is expensive to install and complex to maintain and other methods of direction finding are more commonly used.

Radiotheodolites

A radio direction finder can be used to track the movement of a radiosonde, by following the sonde through its radio transmissions. Since sondes transmit measurements of pressure, their height is known, so unlike optical methods no assumption need be made about ascent speed. There are two types of radiotheodolite, the first automatically adjusting the angle of its single antenna by a servomechanism to obtain the maximum signal in the vertical and horizontal directions. A faster response can be obtained, however, by detecting the phase difference of the incoming wave at four separate, but physically linked, antennas. By using this interferometer approach, the direction of the sonde can be measured quickly and accurately without the need for the precise, and slow, physical alignment of the antennas, required by the single antenna. Only an approximate pointing is needed to achieve a reasonable signal level and this uses simpler servomechanisms. (Compare this with the interference method of detecting the position of lightning, Chapter 15.) A problem with radiotheodolites arises from the interference of the signal reflected from the ground with that received directly, making measurements below about 10° difficult. This is minimised by making the antenna polar diagrams as narrowly focused as possible, with small side lobes.

Radio and optical theodolites, as well as radar, measure the *azimuth* of the target (position degrees clockwise, or eastward, from north – a term derived from the astronomical co-ordinate 'Horizontal System' for defining the position of a star).

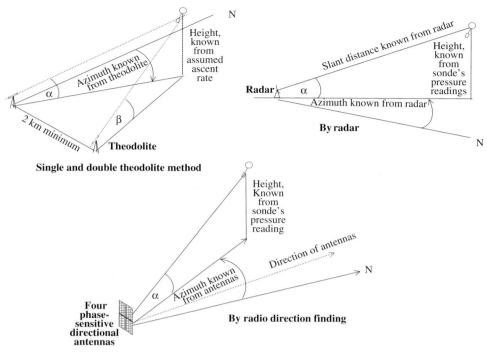

Figure 16.6. If the sonde's rate of ascent is known or assumed, its height can be estimated minute-by-minute. Using one theodolite (upper left), the angle α and the sonde's bearing can be measured, allowing the sonde's plan position to be estimated. Repeated observations allow the wind's direction and speed to be estimated. For greater accuracy, two theodolites are used. With radar, the slant distance, angle of elevation α and compass bearing (azimuth) are measured (top right). The sonde's pressure reading gives its height (Fig. 16.4). From these observations, the plan position can again be estimated and thus the windspeed and direction. With a radio direction finder, such as the phase-sensitive antenna system as shown (bottom), the plan position can again be estimated, and thus windspeed and direction.

All three methods also measure the angle of elevation above the horizon, while in addition radar measures the slant distance to the target, which gives height and plan position by trigonometry (Fig. 16.6). To derive the plan position of the balloon (and thus its speed) using theodolites, height is also required and this can be derived either from a sonde pressure measurement or an assumed ascent speed.

Radio navigation systems

If the sonde can measure its own position directly, the need for optical or radio direction finding and range from the ground is eliminated. This can be achieved by using either ground- or space-based navigation systems, such as Omega, Loran-C or GPS (Global Positioning System). It was recently announced that Europe will install *Galileo*, a new satellite navigation system that can pinpoint location to a fraction of a metre, and no doubt this can also be used. The difference in time of arrival of the signals from the radio navigation beacons can be used to establish the sonde's position, in a way similar to the location of lightning flashes by arrival-time-difference (Chapter 15). Location information is then transmitted from the sonde along with temperature, humidity and pressure readings. Details of how these position-finding techniques work can be found in WMO (1985) for ground-based systems and WMO (1994) and Kaisti (1995) for satellite systems. The ground-based Omega and Loran systems were not designed for meteorological use but for aircraft and shipping, Loran-C being essentially for shipping around coasts, especially in the northern hemisphere. Consequently they may not operate in some locations. They are also being rapidly replaced by satellite systems, and may well be phased-out completely. While satellite systems are more complex to use, because the satellites are moving (unlike the ground-based beacons), they are worldwide in cover and have the potential to give position measurements better than primary radar. The added complexity is thus worthwhile and is easily handled.

Radar height as a substitute for pressure measurement

If radar is used to measure upper winds, the height of the sonde is directly available and the pressure sensor can be omitted (to reduce cost), pressure being calculated using the surface pressure as the starting point. However, unless the radar is precisely manufactured, carefully installed and properly maintained, errors in height can be too great to meet the needs of pressure measurement accuracy. Radar measurements can, however, be used as a useful cross-check of pressure measurements. But if the position of a sonde is measured in future by GPSs, the radar measurement will become unnecessary, and so not available for height estimation.

Accuracy and calibration

Errors are introduced through imperfect tracking of the balloon and due to the balloon 'slipping' with respect to the actual windspeed. Tracking errors vary depending on the make of equipment and on the position-location method. Because of the wide variety of combinations of these, accuracy varies considerably. Using GPS location, WMO (1996b) give a systematic error of ± 0.1 m s^{-1} with random errors of ± 0.4–2 m s^{-1}. Using Omega and Loran-C these increase to ± 1 m s^{-1} for systematic errors and ± 0.6–5 m s^{-1} for random errors. But it must be appreciated that these are only very approximate guides. The error in height estimate is ± 1 hPa.

Installation and maintenance

Where GPS is being used, the only requirement is a clear site for the reception of the sonde signals, as already discussed earlier. If radiotheodolites are used, there is advantage in placing them on a symmetrical hill top since this gives a clear view of the horizon while also reducing the reflected signal from the ground. Siting in a valley with hills rising to one or two degrees is also acceptable, and also reduces the ground reflections. Maintenance of radiotheodolites and radars is complex but can be achieved by interchanging modules. The use of GPS for location, however, eliminates these needs, making the system much simpler to install and maintain.

Dropwindsondes and rocketsondes

Sondes dropped from high-flying aircraft on parachutes are becoming more common. They carry the same sensors as normal radiosondes and it is mainly the method of deployment that is different. Lithium batteries are more conveniently used than water batteries when working in aircraft. The sondes are launched from a chute from the pressurised cabin, the parachutes opening in about 5 s. The telemetered data are received on the aircraft, one system being able to handle several sondes operating at the same time. Uses include research into hurricanes, sondes being dropped into the required areas. Sondes can also be flown in small rockets attaining a height of up to 1 km before the sensors are deployed. The same types of sensor are used as in a standard, balloon-launched radiosonde, with the same accuracy.

Aircraft measurements

Aircraft make useful platforms from which to deploy instruments and they are used in two distinct ways. Research planes are used for specific investigations with special instruments installed for the purpose, while civil airliners are now used as platforms from which to make routine measurements using their *existing* instruments and measurements. The research aircraft are similar to the research ships (Chapter 17) in that they carry out detailed and specialised research concerning basic processes for short periods. But they do not make routine measurements. As on the ships, the type of instruments used on research aircraft are the same as those already described for more conventional use on the ground.

Research flights

Almost as soon as planes began to fly, measurements started to be made from them. A wet and dry bulb psychometer was used in 1921, fixed to the wing in the airstream and visible from the cockpit. Later, dew and frost point measurements were made (Chapter 4) and infrared absorption techniques were adapted for use on aircraft (Chapter 7). But it is specialised measurements, such as cloud particle observations, that are the speciality of research flights. These will not be discussed here.

In the UK, the Met. Office formed its Met. Research Flight (MRF) in 1942 at Boscombe Down using two Spitfires and four Bostons, the purpose being to research physical, chemical and dynamic atmospheric processes. A Mosquito was used in 1944 to measure profiles of temperature and humidity. The MRF moved to Farnborough in 1946, where it has remained ever since, when a Halifax bomber was used for boundary layer research, up to 30 000 ft. In the 1960s Canberras were used for turbulence measurements, while a C130 Hercules (known as Snoopy), which can stay aloft for 12 hours, served the MRF for 28 years, its final flight being in March 2001, making ice observations in Norway. It is to be replaced by a BAE 146 aircraft which can operate up to 35 000 feet (Foot 2002). The MRF has also changed its name to the Facility for Airborne Atmospheric Measurement.

Civil airline commercial flights

Measurements made by civil commercial aircraft have recently (mid to late 1990s) been made available in real time for meteorological use by telemetering their data to the ground, the system being known collectively as AMDAR (Aircraft Meteorological Data Relay). Modern aircraft have sophisticated navigation and sensor systems on board that give information on air speed, ground speed, air temperature and air pressure, along with data on bearing, position, acceleration and orientation. The aircraft also carry airborne computing equipment that manages the flight and navigation systems. These systems are already there, purely for the purpose of the aircraft itself. An AMDAR system simply processes them further to give appropriate meteorological data, which are then fed automatically to the existing Aircraft Communication and Reporting System for transmission to ground (Painting 2001, Truscott 2001). Nothing new has to be added. (In the early 1990s, steps were taken to develop a system to transmit the meteorological data via a geostationary satellite, but this required additional, special equipment, and although some systems are still in use, it is probable that the standard aircraft VHF communications system, rapidly gaining worldwide acceptance, will be the only system in operation after about 2003. It will not, therefore, be considered further here.)

Wind speed and direction

Imagine an aircraft flying on a heading of 55° east of north with a speed relative to the air of, say, 500 mph. If the air mass in which the aircraft is moving is stationary

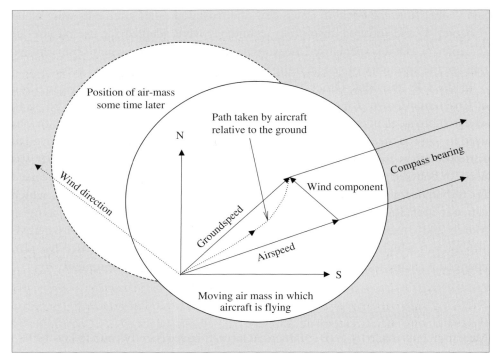

Figure 16.7. Routine on-board navigation measurements of windspeed and groundspeed, bearing and plan position, made on commercial passenger flights, allow the speed and direction of the wind in which the aircraft is flying to be calculated.

relative to the ground (that is, there is no wind at the plane's height), the aircraft's groundspeed and heading will be the same as that relative to the air. If the air mass is moving, however, relative to the ground (that is, there is a wind at the altitude of the aircraft), the windspeed and direction will alter the plane's direction and speed relative to the ground. Thus from the aircraft's routine measurements of groundspeed, airspeed and heading, it is possible to derive windspeed and wind direction at the height of the aircraft (Fig. 16.7). Airspeed is measured by a Pitot tube (Chapter 5). While the aircraft is manoeuvring or when climbing to cruise height or during landing, the calculations of wind are more complex and may be omitted when angles are above a certain threshold. At low windspeeds, direction measurements can be in large error. However, if expressed in terms of the wind vector (that is just its magnitude) accuracy is in the order of 1–3 m s^{-1} (Nash 1994). (WMO use the term *uncertainty* rather then accuracy in the context of AMDAR. See Chapter 1, 'Definitions of terms'.)

Air temperature

Temperature readings are required to correct indicated airspeed and so are already measured by aircraft. The sensor is usually a platinum resistance thermometer (Chapter 3) housed in a special screen on the outside of the fuselage in the airstream (Lawson & Cooper 1990). The temperature that the sensor actually measures is the *total air temperature* (TAT). The *static air temperature* (SAT) is the temperature of the free undisturbed air stream, but it differs from the TAT because the temperature of the thermometer is raised by compression and viscous heating of the air when it is slowed down by the screen and sensor. The two temperatures are related by a complex equation involving the specific heat of dry air, the Mach number (true airspeed divided by the speed of sound in the free air) and a *probe recovery factor* which includes the effects of air viscosity and slowing-up of the air in the screen (Dommasch *et al.* 1958). For a typical sensor and screen at a Mach speed of 0.8 (a typical cruising speed of a modern civilian jet) the SAT is lower than the TAT by a factor of 1.124 (temperatures in K). For example, if the SAT is 223 K ($-50\,°C$) then the TAT is 251 K ($-22\,°C$) which is too high by 28 °C. If the sensor is wet it will cool from evaporation by about 3 °C. Despite all of the complex processing required of temperature measurements to allow for the effects of speed, an uncertainty of around 1 °C is achieved.

Pressure

Static pressure is measured directly by an electronic pressure sensor of the kind described in Chapter 6, such as a silicon diaphragm type. This is connected directly to the static pressure port of the windspeed Pitot tube (Fig. 5.8), the pressure being converted to the equivalent altitude. The uncertainty is about 2 hPa.

Humidity

Although sensors of the thin-film type have been used experimentally on aircraft, the WMO state that there are not, as yet, any sensors that meet all the requirements, in particular the frequency of maintenance required. They would also suffer from the same problems that afflict temperature sensors due to the effects of the high speed of the aircraft. However, Fleming and Hills (1993) describe some developments in progress and no doubt RH will be measured eventually on aircraft.

Sounders

Upper air variables can also be measured by remote-sensing instruments operated from the ground. This approach has already been touched on in the use of radar for measuring rainfall, while remote sensing from satellites is the subject of Chapter 19. As in these two related topics, remote sensing by *sounders* can either use passive methods, which make use of naturally occurring radiation, or active methods that radiate artificially generated radiation into the atmosphere. The latter are the most usual in sounders, the radiation being sound, radio waves or light, or a combination of them.

Acoustic sounders

Sodar (sound detection and ranging) operates by emitting sound pulses into the atmosphere and measuring the scattered echoes. There are two forms: Sodar and Doppler Sodar.

Sodar

Eddies in the atmosphere scatter sound waves because slight changes in temperature, humidity and airspeed within them cause minute differences in air density and thus in refractive index. (The direct, in situ, measurement of these is discussed in 'The eddy correlation method' in Chapter 7, for evaporation measurement.) The process is particularly apparent at inversions where there are large gradients in these variables. Backscatter at $180°$ is caused only by temperature eddies. At other angles both temperature and wind fluctuations produce a detectable scatter, while at $90°$ there is no scatter (Brown & Hall 1978, Gaynor 1990). Sodars are also selective regarding the turbulent scales that produce echoes, peaking at a size of around half a wavelength of the radiated sound wave. This is the Bragg condition or law, discovered by Sir Lawrence Bragg in 1912 at Trinity College, Cambridge, initially concerning the spacing between atoms in a crystal and the angle at which X-rays are scattered when they strike the crystal, for which he received a Nobel prize (jointly with his father, Sir William Henry Bragg). Amongst many things, this made it possible to discover the helical structure of DNA. What is of interest here is the size of eddies that cause scatter. The same principles also apply to radar and lidar.

In a *monostatic* sodar (transmitter and receiver at the same place), sound pulses are radiated vertically with a beam width of about $15°$, a very small fraction being scattered back to the adjacent receiver (Fig. 16.8). The time delay of the round trip gives an indication of the height of the 'target'. An analysis of the sound spectrum of the emitted pulse and the received backscatter gives information about the detailed structure of the boundary layer, of the height of inversions and of changes in the stability of the atmosphere (see later under 'Radar'). However, interpretation of the time-height graphs plotted by the sounder needs skill and experience.

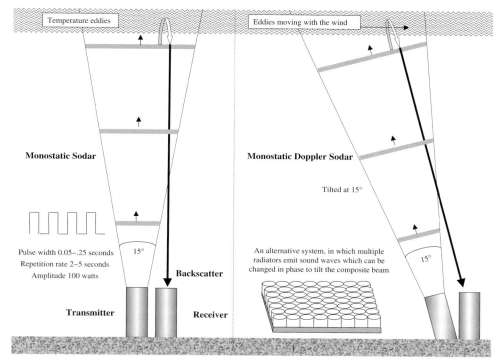

Figure 16.8. A vertically pointing sound emitter (left) radiates a pulse and detects the echo returned from eddies caused by temperature variations, giving information on the turbulent structure of the lower troposphere and the height of any inversions – where turbulence mixes the airs at different temperatures giving stronger echoes. By tilting the emitter (right), the returned echo includes Doppler shift information on the speed of the wind at various heights because the eddies travel with the wind. An array of many smaller piezo-electric transducers allows the beam to be tilted by changing the phase of each emitter. The same transducers also double as detectors of the returned echoes. The power and frequency emitted depend on the height to be reached varying from 2 kHz at 5 watts for 1 km altitude to 480 Hz at 3.5 kW for 5 km.

In monostatic *Doppler* sodar, wind profiles can also be detected by measuring the Doppler shift of the reflected backscatter caused by the temperature eddies, which move with the wind. To achieve this, three beams are transmitted, one vertical, the other two offset at a north and east angle of about 15°; wind components in the vertical and horizontal can then be determined. Alternatively, an array of around 200 small sound emitters can be used, suitably phased to tilt the beam electronically (Fig. 16.8, right).

Most sodars are operated in the monostatic mode, but by moving the receiver some distance away from the transmitter, the Doppler frequency shift due to the horizontal movement of the eddies with the wind can be measured without the need to tilt the beams. This is not widely used.

Because sodar systems are selective in the size of turbulent eddies that they detect (following the Bragg law as above) multi-frequency systems are being explored that will sense a range of eddy dimensions (Bradley & Nerbein 2001). Sodars have a range from about 500 m to 5 km with a windspeed accuracy of ± 0.3 m s^{-1} of the speed and $\pm 3°$ for direction. Apart from the limitation of height, sodar systems are affected by loud noises, such as from traffic, strong winds and precipitation. The noise they themselves produce can also cause annoyance to people.

Wind profiler radars

Just as sound waves are reflected from discontinuities in the atmosphere due to turbulence and the resultant variations in refractive index, so too are radio waves. For the measurement of wind, the frequencies most commonly used are 50 MHz (6 m wavelength), 400 MHz (0.75 m) and 1000 MHz (0.3 m). As with sound, these radar frequencies are backscattered from eddies with scales around half the radio wavelength, obeying the Bragg law. As eddies move with the average windspeed, their velocity gives a direct measure of the mean wind vector. As with sonar, the radar profilers measure the velocity vertically and off-vertical by about 15° north and east, allowing the horizontal and vertical wind components to be derived (Gossard & Strauch 1983, Hogg 1983, Weber & Wuertz 1990).

The frequency used depends on the altitude and resolution required, the lower frequency of 50 MHz, with a transmitting power of 500 kW, reaching up to 30 km with a resolution of 150 m (Vaughn 2002). At the higher frequency of 1000 MHz, with an output of 1 kW, the altitude is limited to 2 km with a resolution of 50–100 m. The size of the antenna increases as the frequency falls so that at 50 MHz the antennas cover an area of 100 × 100 m, while at the shorter wavelengths a dish may suffice, or a 10 × 10 m array of Yagi elements (see Chapter 12). With increasing height, the returned signal becomes progressively smaller, ultimately setting the limit to height. To reach as high as possible, transmissions are made over several minutes and the small return signals integrated; typically data may be collected over 6–12 min to get the three sets of readings required for windspeed and direction. Because the signals returned from eddies are very small, requiring the gain of the antenna and receiver to be high, aircraft, birds and insects can produce false indications of wind. Rainfall is also a source of interference at high frequencies although it is much smaller at 50 MHz, while at intermediate wavelengths light rain is tolerated reasonably well.

The larger profilers are expensive and require large areas of land to deploy the antenna (figure in Vaughn 2002). They are, however, able to make measurement up into the stratosphere and even beyond into the mesosphere, also they are little affected by precipitation and are automatic. For everyday meteorology, however, the medium-frequency and medium-power systems are most appropriate and are of modest size. The high-frequency profilers monitor just the boundary layer, are less costly and use small antennas. But all radar profilers emit radio waves and it is necessary to get approval from the telecommunications authority in the country concerned and this can often be difficult. Profilers do, however, have the advantage over balloons that they operate unattended and measure conditions directly overhead in real time, all the time.

Radio-acoustic sounding systems (RASS)

Whereas in the foregoing systems, natural variations in refractive index in the atmosphere, due to turbulence, are used as targets for measurement by scattering, from which wind data can be extracted, in RASS systems the profiler generates its own markers by transmitting a short sound pulse, just as in the sodar system. Sound waves, being small variations in air pressure and thus air density, provide refractive index anomalies that scatter radio waves, just as turbulence does. By measuring the variations in the speed of propagation of the pulse as it travels upwards, the temperature of the air at each level can be calculated, as this is proportional to the square of the speed, with allowance for any vertical air speed (May 1990, Lataitis 1992, 1993, Angevine 1994, Oakley *et al.* 2001).

Sodar and radar systems have been developed for independent use on their own, as above, but it is possible to combine them to produce a RASS system, the sodar providing the marker, the radar measuring the speed. To meet the optimum conditions, the sound wavelength is arranged to be one-half of the radar wavelength to meet the Bragg condition. This results in a sound frequency of 110, 900 and 2000 Hz being used with the radar frequencies of 50, 400 and 1000 MHz. Acoustic attenuation limits height coverage at 2000 Hz to about 2 km, while at 110 Hz heights between 4 km and 8 km are possible under good conditions. Tests have shown that sondes and RASS agree to about 0.3 °C with a resolution of 100–300 m, but strong winds and precipitation can affect readings. A RASS system can thus measure both wind and temperature profiles, but they are expensive, are not all-weather systems and can be a noise hazard.

Radar

Weather radar (Chapter 8) is now fully operational, but research continues, for example at the Chilbolton Advanced Meteorological Radar system in Hampshire. It is more relevant to include comments on the developments here than in Chapter 8, because it is part of a wider, experimental, upper-atmosphere, remote-sensing group of instruments at Chilbolton, an old airfield from the second world war. The laboratory was specifically set up to investigate the effects of weather on radio communication, both terrestrial and to satellites. The 3-GHz, 25-m antenna (Fig. 16.9) (Kilburn *et al.* 2000) is the largest steerable meteorological radar in the world and has been operating for 30 years. It produces a very narrow beam (0.25° – much narrower than operational weather radars), transmitting half-microsecond, 700-kW pulses at a repetition rate of 610 per second, the returned signals being averaged every quarter of a second to produce a picture of the meteorological conditions up to 50 km distant. As the dish rotates a snapshot of the atmosphere is obtained at the level of interest.

Large rain drops become much flattened as they fall. By arranging for the polarisation of the beam to be switched rapidly from the vertical to the horizontal, the ratio of the returned signal from the two polarizations, the *differential reflectivity*, allow drop sizes to be determined, since there is greater reflection when the polarisation is in line with the longer axis of the drops. Small drops are nearly spherical, producing little difference in reflectivity (Hall *et al.* 1980, Hall 1984). Because falling rain follows the horizontal wind, the Doppler signal allows windspeed to be measured to 0.1 m s^{-1} (when it is raining – although other scatterers can also be used). In-line speed is measured, but the other components can be inferred from it.

Reflections can also occur from clear air turbulence, especially in the summer, although the returned signals are much smaller than from rain. Reflected signals can also be produced by insects and from particles generated by fires. Reflections from the top of the convective boundary layer at the temperature inversion are very common. This happens because turbulence disturbs the interface, thereby mixing air of different temperatures, and so of different density, creating anomalies in the refractive index which reflect the radar pulses; similar turbulence above and below the interface does not produce echoes because the air is at the same temperature.

Rain radar operates by sensing the reflections from raindrops, the signal strength depending on drop size and number. However, a few large hailstones can give as large a signal as heavy rainfall, although the hail represents much less equivalent water and this can be a problem if the radar data are used for flood forecasting. An experimental *differential phase* method operates by measuring the difference in speed between the vertical and horizontal polarised pulses as they pass through the precipitation. If it is heavy rain, the horizontally polarised pulses are slowed more than the vertical, because the longer axis of the large drops is horizontal, causing a shift of phase which increases with range. There is a unique relationship between differential reflectivity and differential phase because the drops always have a clearly defined shape, set by size, while the phase-shift for hail is much less because the hail is random in shape and alignment (Smyth *et al.* 1999).

Figure 16.9. The 25-m, 3-GHz, 700-kW radar dish at Chilbolton in the UK has been in operation for 30 years and is the largest steerable meteorological radar in the world. To its right are the much smaller cloud radar antennas operating at 94 and 35 GHz. The large dish detects raindrops and clear air turbulence, the smaller dishes detect the drops that make up clouds. (Photograph courtesy of the Council for the Central Laboratory of the Research Councils (CCLRC)).

Figure 16.10. The 94-GHz cloud radars (Fig 16.9) can also be operated independently of the large dish, looking vertically, seen here being serviced in the workshops.

Cloud radar is also operated at Chilbolton, in co-operation with Reading University, the much higher frequencies of 94 GHz measuring particles smaller than the 3 GHz radar, but larger than the lidar systems, thereby giving an overall picture of the atmosphere at many particle-scale sizes (Hogan & Illingworth 1999). These small antennas can be attached to the large dish, operating in-line with it at the same time, or they can be operated separately, looking vertically for long periods (Fig. 16.10). In the latter form, they have been used to evaluate how successfully forecast models predict clouds. Comparisons of observations with model predictions suggest that the models are not yet able to simulate clouds in detail, although large-scale features are more precisely modelled. A radar with a frequency intermediate between that of the large dish and the small cloud radars has also been operated.

Lidar sounders

Using Rayleigh (or elastic) scattering

Light detection and ranging systems (lidar) also work on the principle of scattering/reflection, in this case not from refractive index variations but by scattering from the atmospheric gases themselves at molecular scales and also from small suspended particles, allowing radar principles to be used. But in addition they can also use absorption effects. (Scattering of light is also used to measure the visibility of the atmosphere, Chapter 13, and the turbidity of water, Chapter 10.) The source of light is a laser, operating in the visible or near-visible (UV and IR) wavelengths. The question of scattering at these atomic and small particle scales has already been touched on in Chapter 13, on visibility, where the work of Rayleigh and Mie in particular was noted. The type of scattering they investigated is what is now termed *elastic* – scattering without change of wavelength. However, there is another type of scattering termed *inelastic* or Raman scattering, in which the incident radiation is absorbed by the molecules and re-radiated at a different frequency. Both types can occur together and all are used in lidar sounders (Hinkley 1976, Thomas 1991).

Typically a lidar sounder is monostatic (the laser and detector being adjacent to each other), the laser emitting pulses of coherent light. The average power of the emissions is from a few milliwatts to several tens of watts. An optical telescope receives the backscattered energy which is detected by a photomultiplier or photo-conductive diode (Fig. 16.11a, b), the output being displayed visually and input to a

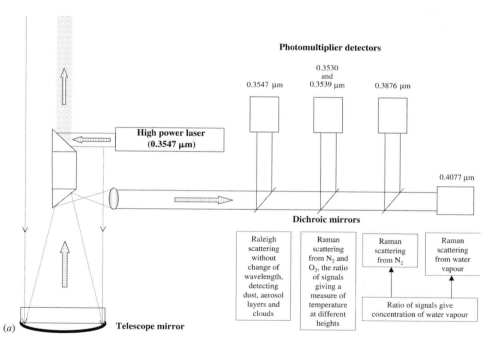

Figure 16.11. The laser (a) (top right, b) emits UV light, reflected by the 45° mirror (top left, b), through an opening in the roof of the laboratory. The optical telescope mirror (bottom left, b) receives the backscattered energy which is reflected, by the lower 45° mirror, to focus on a lens producing a collimated beam that is directed by three mirrors onto four photodetectors (centre right, b). Figure a indicates the wavelengths measured and the uses made of the information.

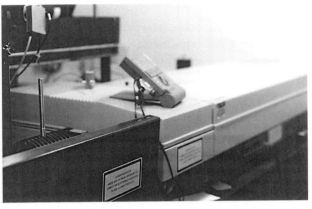

(b)

Figure 16.11. (*Cont.*)

PC for detailed analysis. The magnitude of the received signal depends not only on the degree of scattering from the target itself but also on the degree of attenuation of the beam, due to scatter and absorption by the atmosphere, and by the aerosols, in the round trip from the laser back to the telescope. These various characteristics can all be used to extract different types of information.

Lasers using Rayleigh scattering are useful mostly for cloud and particulate content. The measurement of cloud height by laser has been described in Chapter 14, the sudden increase of backscatter from the cloud base being easy to detect. This is a widely used and fully operational application (Fig. 14.7). But in many conditions there is no clear-cut sudden transition, as at a cloud base, the particles or droplets often being suspended evenly and sparsely in relatively clear air. The large changes that occur at inversions can, however, be detected and natural particulate content in the lower atmosphere can be high enough to measure the velocity of the air, while man-made, industrial particle emissions can also be sensed and mapped. In the stratosphere, aerosols due to volcanic activity, which affect global radiation balance, can be measured with laser techniques based on Rayleigh scattering.

The absorption coefficients of different gases vary considerably with wavelength. In a differential absorption lidar system, a laser is tuned alternately to two closely spaced wavelengths, one being strongly absorbed, the other not; the difference giving a measure of the concentration of the particular gas. This has been used to measure the concentration of water vapour, sulphur dioxide, nitrogen dioxide and ozone.

Using Raman (or inelastic) scattering

The Indian Nobel prizewinner, Chandrasekhara Venkata Raman, born in 1888, discovered what is now termed *Raman Scattering*, a quantum mechanical process in which light emitted at one frequency excites the atoms that it strikes, which then re-radiate or scatter the incident photons at a different energy/wavelength (Raman & Krishnan 1928, Plaezek & Teller 1933). In this type of scattering, the returned frequency depends on the species of molecule that has been activated by the light, the monochromatic source today being a laser (in Raman's experiments it was focused sunlight used over a short distance). The laser need not be tuned to a specific wavelength to excite the re-radiation, and so a wavelength can be selected that is not absorbed by the atmosphere. By analysing the spectrum of the returned signal, measurements can be made of specific atmospheric components.

Such a Raman scattering instrument has been developed at the Rutherford Appleton Laboratory at Chilbolton (Fig. 16.11(*a*), (*b*)). A powerful laser radiates a collimated, vertical beam, at a frequency of 0.3547 µm (in the UV band) and receives back radiation scattered at the same frequency (Raleigh scattering) from air molecules and aerosols (as above), but also Raman-scattered wavelengths at

0.3530 and 0.3539 μm from nitrogen and oxygen. The temperature at different heights can be calculated from the ratio of these two signals. Water vapour returns a wavelength of 0.4077 μm, while 0.3876 μm originates from nitrogen, the latter being a constant constituent of the atmosphere. A humidity profile can be estimated from the ratio of the two returned wavelengths. The work is directed at studying the effects of water vapour, air temperature variations, clouds and raindrops on radio propagation, but the instruments and methods have far wider applications for all environmental work.

At the moment, such instruments are intended only for research and, apart from cloud height measurement, lidar systems have not so far been widely used operationally, due to their fairly high cost, the need for specialised operators and because some modes of operation require special conditions such as no precipitation. However, lidar systems have a powerful role to play in boundary layer meteorological research; for example, at Salford University Pearson and Collier (1999) and Sandiford and Collier (2001) are developing techniques using CO_2 Doppler lidar to estimate temperature gradients in the boundary layer. This is just a glimpse of an expanding field.

Microwave radiometers

Natural thermal radiation from the atmosphere in the microwave band (Fig. 19.1) originates mostly from oxygen, water vapour and liquid water (cloud droplets) and carbon dioxide, the intensity of the radiation depending on the temperature and the quantity of the species. The distribution of oxygen with altitude is well known, and it is constant over time, so its radiation contains information on temperature, allowing vertical temperature profiles of the lower atmosphere to be made by ground-based radiometers working at around 60 GHz (Fig. 16.12). By making measurement at different frequencies, temperatures at different heights can be obtained, although this simple statement belies a complex process that cannot be explained here (see Westwater 1990). Height discrimination is, however, coarse (500 m) and the method only operates up to about 3 km. (See also Sounders, Chapter 19.)

Unlike oxygen, the amount of water vapour and liquid water (cloud and rain droplets) varies widely with time. These two atmospheric components radiate at around 22 and 29 GHz, respectively. Although it is not possible to distinguish amounts at different height in the same way as for oxygen (the reasons are again complex) radiometer measurements at these frequencies do give information on the total quantity of water integrated over the vertical path (Hogg 1983).

Radiometers have the advantage that they can operate continuously, but their costs and the complexity of interpreting their measurements make them unsuitable for replacing radiosondes. However, attempts to improve the vertical resolution of

Figure 16.12. Typical of radiometers making measurements of natural microwave radiation are these two at Chilbolton. Outside the laboratory, a metal plate at 45° reflects incoming radiation into the detectors. Radiation from oxygen at frequencies around 60 GHz allow temperature profiles to be obtained up through the atmosphere. To measure liquid water or water vapour content of the atmosphere, integrated over the full vertical path (rather than profiles), frequencies of 29 and 22 GHz, respectively, are measured.

the temperature profile that they can achieve are being made (Gaffard *et al.* 2001) and their development is ongoing.

Combining several techniques

All of the above methods have the same purpose – to measure the characteristics of the upper atmosphere, for whatever reason – meteorological, radio propagation, pollution or climatology. Combining a number of these techniques has advantages. For example, when radiosonde stations are all automated, about 5 % will fail. Gaffard (2000) at the UK Met. Office is working towards supplementing some of the sonde stations with sounders to obtain independent profiles of temperature, humidity and wind in the lower troposphere, to allow gaps to be filled in the event of failure, despite the limited vertical resolution of sounders compared with sondes. Oakley *et al.* (2001), also from the Met. Office, describes how the UK upper air network is currently undergoing considerable change, using automatic sondes combined with sounders.

It will be clear from the above that the methods used to make measurements up through the atmosphere by sounders are many and are developing rapidly. However, sondes are likely to continue to be used for a long time yet. In situ measurements, be they on the ground or through the depth of the atmosphere, will always be required since they are precise and specific. All forms of remote sensing complement rather than replace in situ techniques.

The above is just a glimpse of the new types of ground-based, remote-sensing instruments that are now being operated experimentally. Some of these are shortly to be replicated on board satellites (such as the cloud radars and lidars). They will develop over the coming decades into powerful tools both for research as well as for routine operational use.

References

Angevine, W. M. (1994) Improved radio acoustic sounding techniques. *J. Atmos. Ocean. Technology.*, **11**, No. 1 42–9.

Bradley, S. & Nerbein, S. von H. (2001) Multi-frequency acoustic sounding of the turbulent boundary-layer properties. *Book of Abstracts*. Royal Meteorological Society, 2nd National Conference, Manchester, September 2001. p. 111.

Brown, E. H. & Hall, F. F. (1978) Advances in atmospheric acoustics. *Rev. Geophys. Space Phys.*, **16**, 47–109.

Dommasch, D. O., Sherby, S. S. & Connolly, T. F. (1958) *Aircraft Aerodynamics*. Pitman, New York, pp. 560.

Fleming, R. J. & Hills, A. J. (1993) Humidity profiles via commercial aircraft. *Proceedings of the Eighth Symposium on Meteorological Observations and Instrumentation*. Anaheim, California, September 1993, J125–J129.

Foot, J. (2002) The history of Snoopy. *Proceedings of the RMS Specialist Group Meeting*, 13 March 2002, Science Museum, London.

Gaffard, C. (2000) Initial evaluation of signal power characteristics of a boundary layer profiler in the UK. *Proceedings of the 9th International Workshop on Technical and Scientific Aspects of MST Radar* (Combined with COST76 Final Profiler Workshop). Toulouse, March 2000, pp. 431–3.

Gaffard, C., Caddedu, M. & Nash, J. (2001) Development of operational multi sensor ground-based remote station within UK. *Book of Abstracts*. Royal Meteorological Society, 2nd National Conference, Manchester, September 2001. p. 111.

Gaynor, J. E. (1990) *The International Sodar Intercomparison Experiment. Acoustic Remote Sensing*. McGraw Hill, New York.

Gossard, E. E. & Strauch, R. G. (1983) *Radar Observations of Clear Air and Clouds*. Elsevier, Amsterdam.

Hall, M. P. M. (1984) A review of the application of multi-parameter radar measurements of precipitation. *Radio Sci.*, **19**, 37–43.

Hall, M. P. M., Cherry, S. M., Goddard, J. W. F. & Kennedy, G. R. (1980) Rain drop size and rainfall rate measured by dual-polarization radar. *Nature*, **285**, 195–8.

Hinkley, E. D. (1976) *Laser Monitoring of the Atmosphere*. Topics in Applied Physics. Springer Verlag, New York.

Hogan, R. F. & Illingworth, A. J. (1999) Analysis of radar and lidar returns from clouds: implications for the proposed Earth Radiation Mission. *CLARE'98 Final Report, ESTEC International Workshop Proceedings WPP-170, ESA/ESTEC*. Noordwijk, The Netherlands, Sept. 1999 75–82

Hogg, D. C. (1983) An automatic profiler of temperature, wind and humidity in the troposphere. *J. Climate Appl. Meteorol.*, **22**, 807–31.

Holland, G. J., Mc Geer, T. & Youngren, H. (1992) The autonomous aerosonde for economical atmospheric soundings anywhere on the globe. *Bull. Am. Met. Soc.*, **73**, 1987–8.

Kaisti, K. (1995) New low cost GPS solution for upper air windfinding. *Ninth AMS Symposium on Meteorological Observations and Instrumentation.* Charlotte, N. Carolina, pp. 16–20.

Kilburn, C. D. D., Price, J. D., Hardaker, P. J. & Pilditch, A. (1999) Turbulent clear air boundary observations with Chilbolton radar and a tethered instrumented balloon. *Proceedings of the 29th International Conference on Radar Meteorology.* Montreal, Canada, pp. 476–9.

Kilburn, C. D. D., Chapman, D., Illingworth, A. J. & Hogan, R. J. (2000) Weather observations from the Chilbolton Advanced Meteorological Radar. *Weather*, **55**, 352–6.

Lataitis, R. J. (1992) Signal power from radio acoustic sounding of temperature: the effects of horizontal winds, turbulence and vertical temperature gradients. *Radio Sci.*, **27**, 369–85.

Lataitis, R. J. (1993) Theory and application of a radio acoustic sounding system (RASS). *NOAA Technical Memo ERL. WPL-230.*

Lawson, R. P. & Cooper, W. A. (1990) Performance of some airborne thermometers in clouds. *J. Atmos. Ocean Tech.*, **7**, 480–94.

May, P. T. (1990) Temperature sounding by RASS with wind profiler radars: a preliminary study. *IEEE Transactions on Geoscience and Remote Sensing*, **28**, 19–28.

Nash, J. (1994) Upper wind observing systems used for met. operations. *Ann. Geophys.*, **12**, 691–710.

Nash, J., Elms, J. B. & Oakley, T. J. (1995) Relative humidity sensor performance observed in recent international radiosonde comparisons. *Ninth AMS Symposium on Meteorological Observations and Instrumentation.* Charlotte, N. Carolina, pp. 43–8.

Oakley, T., Smout, R. & Nash, J. (2001) Automation of the upper-air observing network in the UK. *Book of Abstracts.* Royal Meteorological Society, 2nd National Conference, Manchester, September 2001, p. 108.

Painting, D. (2001) Accuracy of AMDAR measurements. *Book of Abstracts.* Royal Meteorological Society, 2nd National Conference, Manchester, September 2001, p. 111.

Pearson, G. N. & Collier, C. G. (1999) A compact pulsed coherent CO_2 laser radar for boundary layer meteorology. *Q. J. R. Meteorol. Soc.*, **125**, 2703–21.

Pettifer, R. (2002) The history of radiosondes. *Proceedings of the RMS Specialist Group Meeting, 13 March 2002.* Science Museum, London.

Plaezek, F. & Teller, E. (1933) The rotational structure of the Raman bands of a polyatomic molecule [Translated by A. J. Gibson from Die Rotationstruktur der Ramanbanden mehratomiger molekule]. *F. Phys.*, **81**, 209–58.

Raman, C. V. & Krishnan, K. S. (1928) A new type of Secondary Radiation. *Nature*, **501**, 121.

Sandiford, K. J. & Collier, C. G. (2001) A proposal for the measurement of boundary layer temperature gradients using Doppler lidar. *Atmos. Sci. Lett.*, **1**, 1–6.

Smyth, T. J., Blackman, T. M. & Illingworth, A. J. (1999) Observations of oblate hail using dual polarization radar and implications for hail-detection schemes. *Q. J. R. Meteorol. Soc.*, **126** (N0 555, Part a), 993.

Thomas, L. (1991) Lidar probing of the atmosphere. *Ind. J. Radio Space Phy.*, **20**, 368–80.

Truscott, B. (2001) EUMETNET AMDAR – The European aircraft data project. *Book of Abstracts.* Royal Meteorological Society, 2nd National Conference, Manchester, September 2001. p. 111.

Vaughn, G. (2002) The UK MST radar. *Weather*, **57**, 69–73.

Wade, C. G. (1995) Calibration and data reduction problems affecting National Weather Service radiosonde humidity measurements. *Ninth AMS Symposium on Meteorological Observation and Instrumentation.* Charlotte, N. Carolina, pp. 60–4.

Weber, B. L. & Wuertz, D. B. (1990) Comparison of rawinsonde and wind profiler radar measurements. *J. Atmos. Ocean. Tech.*, **7**, 157–74.

Westwater, E. R. (1990) Ground-based radiometric observations of atmospheric emission and attenuation at 20.6, 31.65 and 90.0 GHz: a comparison of measurements and theory. *IEEE Transactions on Antennas and Propagation*, **38**, 1569–80.

WMO (1985) Meteorological observations using navaid methods [A. A. Lange]. *Technical Note No. 185, WMO-No 641.* Geneva.

WMO (1987) WMO International radiosonde comparison (UK 1984, USA 1985): final report [J. Nash & F. J. Schmidlin]. *Instruments and Observing Methods Report No. 30, WMO/TD-No 195.*, Geneva.

WMO (1991) WMO International radiosonde comparison – Phase III – Dzhambul, USSR, 1989: Final report [A. Ivanov *et al.*]. *Instruments and Observing Methods Report No 40, WMO/TD-No 451.* Geneva.

WMO (1994) A new GPS rawinsonde system [D. B. Call]. Papers presented at the WMO Technical Conference on Instrumentation and Methods of Observation (TECO-94), 28 February–2 March 1994, *Instruments and Observing Methods Report No. 57, WMO/TD 588.* Geneva, pp. 159–63.

WMO (1996a) *Guide to Meteorological Instrumentation and Methods of Observation*, sixth edition. WMO No. 8.

WMO (1996b) WMO International radiosonde comparison – Phase IV – Tsukuba, Japan, 1993: Final report [S. Yagi, A. Mita and N. Inoue]. *Instruments and Observing Methods Report No. 59, WMO/TD-No 742.* Geneva.

17

The oceans

Nobody – not even Captain MacWhirr, who alone on deck had caught sight of a white line of foam coming on at such a height that he couldn't believe his eyes – nobody was to know the steepness of that sea and the awful depth of the hollow the hurricane had scooped out behind the running wall of water.

Joseph Conrad Typhoon.

The oceans: a history of their measurement

The entire heat capacity of the Earth's atmosphere is equivalent to about the top 3 m of the ocean's water (Houghton 1997). And below these 3 m lies another 2000–5000 m, with trenches down to 11 000 m, with an average depth of 2000 m. In addition to their great depth and thermal mass, the oceans also cover 71% of the Earth's surface. A glance at a globe demonstrates that the Pacific Ocean alone covers almost half of the planet, and, viewed from the south, the 81% ocean-cover of the southern hemisphere is clearly evident. Given this dominance by the seas, it seems a reasonable expectation that the weather, and changes in climate, may well be driven as much, if not more, by the oceans than by the atmosphere and the continents. At the very least we should pay as much attention to measuring the weather above the oceans and the conditions down into their depths as we have done to measuring that over the land and up into the atmosphere. Being land animals, it is perhaps natural that we have been somewhat biased towards the continents and the atmosphere. As a result, the ocean depths are less well known than the surface of many planets.

Comment on the significance of the link between land and sea and atmosphere as regards climate is far from new. Charles Darwin in *The Origin of Species* (Chapter 12, 6th edition, 1872) comments on the importance of a suggestion made by a fellow Englishman, James Croll, that glacial conditions result from an increased

Measuring the Natural Environment, second edition, Ian Strangeways. Published by Cambridge University Press. © Ian Strangeways 2003.

eccentricity of the Earth's orbit, in particular the indirect influence this has on oceanic currents. Darwin also comments on Sir Charles Lyell's conclusion that the relative position of land and water were important in this respect. (The idea of the influence of orbital changes on climate was later taken up by Milankovitch in 1920, who elaborated on it and brought it into prominence – in particular regarding changes in seasonal solar radiation.) Croll also concluded that whenever the northern hemisphere passes through a cold period, the southern hemisphere temperature is raised (and conversely), chiefly through changes in the direction of ocean currents. Charles Darwin comments that 'This conclusion throws so much light on geographical distribution (of species) that I am inclined to trust it'. And just recently, Charles Keeling of the Scripps Institute of Oceanography, San Diego, proposed that long-term changes in the Moon's orbit influence the relative strength of tides in such a way as to increase, periodically, the vertical mixing of water in the oceans, bringing cold water up from the depths to the surface when tides are stronger (Pearce 2000). This may, Keeling says, have caused the ice ages. This is entirely speculative and so is written off by many as unfounded. Voicing a similar view that the oceans are all powerful, Dean Roemmich, also of the Scripps Institute, has said that 'If there is global warming going on, it has to be because the oceans are warming. They have the most heat storage capacity – the atmosphere can't do it' (Feder 2000). These views, and the above ideas expressed by Darwin and his colleagues, highlight just how little is understood about the oceans. As ever, it is only measurements that will instruct us as to what is happening. In this, the development of the Argo floats (see later) promises to be very useful; indeed it is a vital new development.

From the voyages of the seventeenth century onwards, meteorological and oceanographic observations (as well as astronomical, cartographic and geomagnetic) have been made at sea (Kenworthy & Walker 1997). But it was not until the first International Meteorological Conference was held in Brussels in 1853, at which ten maritime nations met, that the scientific and commercial value of weather observations from ships was fully recognised. The conference was largely at the instigation of Lieutenant Matthew Fontaine Maury of the US Navy.

As a direct result of this conference, the British Met. Office was established as a department of the Board of Trade in the UK in 1854. It was first headed by Captain (later Vice-Admiral) Robert FitzRoy (Barlow 1994, 1997), earlier Captain of the Beagle which took Charles Darwin on his voyage of discovery. By 1855 FitzRoy had obtained the co-operation of 105 merchant ships and 32 ships of the Royal Navy in making weather observations at sea, equipping them with the necessary instruments. Measurements were initially recorded in a weather register agreed at the Brussels conference. This was improved in 1874 by the Marine Superintendent of the Meteorological Office, Captain Henry Toynbee, who drew up the meteorological log, which was then recognised internationally. These logs remained unchanged for about 45 years, but after about 1921 they were gradually discontinued in favour of observations made at synoptic hours (00.00, 03.00, 06.00, etc.) and transmitted by radio. Until this time observations had been made six times a day, at the end of each watch. In 1953 and again in 1982 the layout was changed, resulting in today's meteorological logbook, which combines observations made and radio messages sent. Small changes were made in 1988, 1990 and 1994. Because of the importance of maritime meteorology, the International Meteorological Organisation was set up in 1872; this later became the World Meteorological Organisation (WMO) in 1951.

Sensor compatibility

In view of the powerful influence and physical dominance of the oceans, and our relative ignorance about its role in weather and climate, one chapter would seem rather insufficient to do the subject justice. However, all the sensors covered under the topic of fresh water measurement (Chapter 10) can also be used in the sea. These include water level staff gauges, float and pressure water level sensors for tidal records (Fig. 17.1), all the water quality sensors (Warner 1972) and current meters. The same sensors that measure atmospheric variables over the land are equally suitable for their measurement over the sea, although there may be problems with how to expose them. Also applicable to use at sea without any modification are data loggers and telemetry systems. So rather than there being a completely different

Figure 17.1. In most cases, the same instruments used for measuring fresh water can be used to measure the sea, the example shown here being one of the earliest sea level monitoring stations in Antarctica, at the BAS base at Faraday; a conventional float-operated chart recorder is operational in the hut over the stilling well.

range of instruments for the ocean, it is more a matter of how the instruments are deployed that differs between land and sea applications. There are two main ocean surface platforms on which instruments are operated – ships and buoys (to which might be added small islands). In addition, new instruments (Argo floats) are being developed for subsurface measurements that give regular profiles down to 2000 m.

Ships as instrument platforms

Ships are used both for specialised research, when the ship is dedicated entirely to that purpose (see later), and also for routine, long-term measurements for meteorology and weather forecasting.

All routine observations from ships today are made on a voluntary basis, so as to get the required dedication and high quality required – it is better to have no data than poor data. The ships are known as Voluntary Observing Ships, jointly as the Voluntary Observing Fleet (VOF) and the observers as the Corps of Voluntary Marine Observers (VMO). Each NWS recruits its own ships, and there are regular visits to the ships by foreign NWS to supply materials and attend to instruments. There are over 6000 merchant ships in the VOF. For a time there were also dedicated ocean weather ships (OWSs), the sole purpose of which was to observe and report the weather from fixed locations, such as in the mid-Atlantic. But cost has meant that, in the North Atlantic, numbers have fallen from eight in the 1970s to just one today (a Norwegian ship).

There are three classes of ship in the VOF. Selected ships observe wind, 'weather' (a summary of present and past weather in a specially coded form), barometric pressure (and its tendency), air and sea temperature, humidity, clouds and waves. Supplementary ships make the same observations less pressure-tendency, sea temperature and waves. Auxiliary ships are similar to supplementary ships but do not report cloud (under 'weather', see above). The Marine Observer's Handbook (Met. Office 1995) describes all these in detail, as well as the observation of sea-ice and other marine and atmospheric phenomena.

Of these observations only barometric pressure, air temperature, humidity and sea surface temperature are actually measured by instruments. The rest are non-instrumental observations and are, therefore, to some extent beyond the terms of reference of this book, but, as they overlap instrumental observations, something will be said about them.

Barometric pressure

There is evidence that Robert Hooke, in 1667, was the first to see the advantage of a ship having a barometer, and he presented a paper to the Royal Society on his experiments to overcome the problem of oscillations of the mercury due to the ship's motion (Banfield 1976). Edmund Halley took such a barometer on one of his voyages of exploration in the Atlantic in 1698. One of Captain FitzRoy's initial tasks after taking up his position as head of the Met. Office in 1854 was to supply ships with mercury barometers. He also designed a barometer that would resist the harsh conditions on-board a sailing ship, including the damage that the firing of a ship's guns could do (Barlow 1994). The present-day Kew-pattern marine barometer (Chapter 6), a derivation of the FitzRoy design, reduces the effects of 'pumping' of the mercury due to wind and movement of the ship by making part of the tube in the form of a fine capillary, opening out to full bore only towards the top, to allow the meniscus to be read easily. This design remained virtually unchanged since it was first made in about 1900, but it has been replaced by aneroid barometers, which can now be made having a similar accuracy to the Kew marine instrument while being more compact and easier to read. As on land, on-board a ship a barometer is installed in an office, since it requires the same precautions against wind effects.

Air temperature and humidity

Temperature and humidity are measured in the same way on a ship as on land, but using a smaller Stevenson screen, just large enough to hold the wet-and-dry thermometers and nothing more. No chart recorders or maximum and minimum thermometers are used, just spot observations at the specific locations at the time. The screen must be exposed, so as to be well away from any heat sources such as air from the boilers or the living quarters, the rails on the bridge being a typical choice. However, there cannot be the same consistency of exposure as is possible on land, since ships differ in size and structure. Also, while the wind will sometimes be directly off the sea, at others it will blow across the warmed deck. And, on larger vessels, the temperature tens of metres above the sea surface will be measured and this is bound to introduce some uncertainty. The same type of mercury thermometers used at land stations are suitable at sea, as are electrical resistance thermometers if a remote display is needed. In the nineteenth century, most thermometers on-board ships were not screened and so there is doubt as to the reliability of readings from the days of sail (Chenoweth 1996).

Measuring sea surface temperature manually

Because of the influence that the sea has on the world's weather and climate, sea surface temperature is of great interest to both meteorologists and climatologists. This has been measured since marine observations began in the nineteenth century, and there are several ways of making the measurement.

Figure 17.2. James, one of my grandsons, being tutored in the art of collecting sea water from a moving boat for the measurement of SST. It is not a simple matter to avoid collecting just the spray or the top few centimetres of water, if the bucket is allowed to skip along the surface. As soon as it is submerged, the pressure on it can be considerable and it is difficult to restrain and withdraw the bucket, even at a few knots speed. The art is to throw the bucket forward, let it sink, and pull it out as it passes vertically below, while its speed, relative to the water, is still zero. How the bucket and water are then treated and how the thermometer is held and stirred are crucial to good accuracy and this is very dependent on each individual observer's skill and commitment.

Manual measurements are made by collecting a sample of surface water in a bucket on a rope (Fig. 17.2). Upon withdrawal the thermometer is inserted immediately into the water sample, away from sunlight, and submerged up to the top of the stem, stirring all the time, being withdrawn after about a minute just far enough to take the reading (Fig. 17.2). The observer's hand must not warm the water. A thermometer of the type used to measure air temperature is suitable. Buckets vary in design, but a typical one is made of double-skinned canvas or rubber. Single-skinned buckets are not suitable since evaporation on the outside surface can cause cooling. As the speed of the ship or its height increase, the use of a canvas bucket becomes increasingly difficult and the UK Met. Office now supplies a heavier rubber model. As the ship's speed increases, care is also needed to prevent even the heavier buckets from skimming along the surface, collecting a sample just from the very top layer or even of sea spray, neither of which will be at a representative temperature. To avoid this, by ensuring adequate mixing, the bucket must submerge 'cleanly', ideally to a depth of about one metre. This is not easy to ensure every time.

Hull temperatures

There is an advantage in reading the thermometer while it is immersed in the sea, and although it is possible to lower an electrical resistance thermometer on a cable from the bridge into the sea, it is difficult to control the depth it takes up. Instead, it is more usual to fix a platinum resistance thermometer to the inside of the hull, a metre or so below the normal water line, the conduction of the steel being good enough for it to take up the temperature of the outside water. As the hull has a large mass, the mean temperature is sensed and this compensates for any rolling or pitching of the ship.

Engine-room intake temperatures

If hull temperatures cannot be read, an alternative is to measure the water as it is pumped into the ship via the intake to the engine-room. However, as the ship rolls and pitches water will be drawn in from different depths and this can cause some error. Because the water passes along pipes within the warm ship, there can be heating if the temperature is not measured very close to the intake. Much work has been done recently to assess the accuracy of sea surface temperature data, with regard to climatic change (Folland & Parker 1995, Parker *et al.* 1995, Chenoweth 1996), and to assess the quality of data collected by the various methods over the last 150 years.

Windspeed and direction

Although anemometers and wind vanes of the type used on land operate perfectly well on ships, there are two problems. The ship's forward motion, and its rolling and pitching, affect the readings and correction must be made for this, but more problematic still is finding a suitable place to fix the sensors, since the ship disturbs the airflow. On research vessels (Birch & Pascal 1987) duplicated sets of sensors can be sited at several places around the ship to ensure a fair exposure of at least one set at any one time. But this is not practical for the ships of the VOF, and for these the preferred method is to estimate wind force and direction by observing the sea state. Even though this is not a measurement but an observation, it is worth saying something about it.

Wind force expressed on a scale of 0 to 12 was originally devised by Admiral Sir Francis Beaufort in 1806, based on the sail canvas carried by a frigate of the Royal Navy (Chapter 5). As ships changed the scale was adapted, but, with the passing of sail, the practice arose of judging wind force by the appearance of the sea corresponding to each Beaufort number; this new scale came into use in 1941 (Met. Office 1995). This is the method currently used by the VOF. However, judgement and experience are necessary since the state of the sea is also dependent on such additional factors as the closeness of the shore, the state of the swell (which results from winds a long way from the point of observation or from more local winds now ceased) as well as tides and currents. Also, the difference in temperature between the air and the water affects the waveform, while rainfall can have a smoothing effect on the sea surface. How long the local wind has been blowing also affects the sea state; it takes time for waves to build up. The quality of the observations is thus very observer dependent and is nothing more than an estimate.

Rainfall

Many factors reduce the effectiveness of a raingauge when operated on a ship rather than on land, and even on land its exposure is difficult enough. Rolling and pitching will require that the gauge is kept horizontal by some form of gimbal – not just to ensure reliable tipping bucket operation, but also to collect the correct amount of rain initially. But gimbals are expensive and not completely effective. The higher winds experienced at sea, together with the forward motion of the ship and the very open exposure, make the wind-induced errors even greater than those experienced on land, further worsened since the gauge has to be exposed amongst the ship's superstructure that causes turbulence. To these problems must be added the possibility that sea spray will be collected along with the rain. There are no simple answers to these problems. One possibility is the use of aerodynamic collectors (Folland 1988, Strangeways 1996, Hasse *et al.* 1997), but this can only be a partial solution, since it does not address many of the problems. Yet a further problem is that the rainfall total will refer not to one geographical location but to the whole area across which the ship has travelled during the collection period, compounded by the 'fair weather bias' – the tendency of ships to select routes that avoid storms, thereby not obtaining a representative sample.

Large ships have radar systems for navigation, but even land-based radars used specifically for rainfall measurement need regular calibration against ground truth raingauges and ships' radars are susceptible to drift which cannot be corrected in this way. Even if the calibration was good, there are many sources of error in deriving rainfall from the reflected signal, as is discussed in Chapter 8, such that measurements can be in error by 50% (Lebedev & Tomczak 1999). A ship's radar is thus not a realistic option.

Observing ocean waves

The study of ocean waves was only recently put on a scientific footing, with the development of automatic wave recorders producing quantitative measurements. These have shown up the limitations of the previous observational methods. Ocean waves are complex and impossible to analyse simply by inspection, but when electronically recorded they can be analysed into the simple component waves that combine to form them. But wave recorders can only be effectively used on stationary installations such as oil rigs, weather-research or oceanographic-research ships, or buoys. For the VOF, wave recorders are not, therefore, an option and the old observational procedures still have to be used.

Apart from the few waves caused by tidal effects, all waves are due to wind – although how the wind produces them is still not precisely understood (see later section 'Measuring ocean-atmosphere fluxes'). Waves raised by the local winds blowing at the time of the observation are referred to as 'sea'. Those due to local winds now ceased or to winds a large distance away are known as 'swell' and are long in wavelength (hundreds of metres) compared with locally generated waves. Although there can be swell from two directions at once it is usually just from one, but this will be combined with the 'sea', which as a rule will be from a different direction.

It is difficult to observe this complex motion of the surface, but certain rules have been developed. Three characteristics have to be noted in a ship's report. The direction from which the waves come is relatively easy to estimate by sighting either across the wavefront or along the crests. The period is measured with a watch by selecting a patch of foam or small floating object and measuring the time between the larger waves (of each successive wave group), the process being repeated 20 or more times to get an average. The height is not so easily estimated and there is no suitable method for general use on merchant ships. However, if the wavelength is less than the ship's length, the height can be estimated roughly by the appearance of the wave on the side of the ship, while if it is longer than the ship the observer can take up a position such that the top of the wave is level with the horizon. The height of the wave is then equal to the height of the observer's eyes above the ship's water line.

Estimating ocean currents

Knowledge of ocean (surface) currents has been acquired over 100 years or so, mostly through observations from ships on-passage. Much remains to be discovered, however, especially in areas remote from the main shipping lanes and about the variations in currents over time.

Currents are measured from a moving ship by estimation of its position using dead-reckoning (that is, calculating its position by measuring its speed, direction and transit time between two points), with correction for leeway-drift due to winds, and comparing this with the actual position, today more easily known from Global Positioning System (GPS) instruments. The vector difference between the two positions is that caused by currents. The method is assumed to give the current at a depth of about half the draft of the ship, but is not used over distances of more than 400 miles, or travelling times of more than a day, as several different currents may then be involved. But too short a time is also undesirable because of the small differences involved. This method is not very precise, and it is also liable to error particularly because it is difficult to estimate the drift due to wind precisely enough. We are still very short of good ocean-current measurements.

Moored buoys as instrument platforms

The UK situation will again be used as a basis for a more general consideration of the worldwide position. The UK Met. Office operates three types of buoy – moored buoys in deep water, moored buoys in shallow coastal waters and drifting buoys, these taking two forms.

Currently there are seven deep-water moored buoys on the edge of the European continental shelf west of the UK and Ireland, in water depths from 2000 to 3500 metres, including one off the coast of Brittany, with French co-operation. There are also two in the North Sea. The US and Canada operate similar buoys off their Atlantic and Pacific coasts and across the Pacific, America currently having about 65 such buoys. At present there are four shallow-water moored buoys in the Irish sea and the English Channel. In addition, lightships are used as moored platforms, of which currently four are equipped with sensors; all these are in the English Channel. Use is also made of small islands and oil rigs to deploy similar instruments, and around the UK coasts there are at present seven, to the east and north.

Figure 17.3 illustrates a deep-water moored buoy of the UK design being visited by the Salmaid, a ship used by the UK Met. Office to install and service this type of station. The hull of the buoy is 3 m in diameter with a sensor ring 4.5 m above sea level (Met. Office 1997). The variables measured are barometric pressure, air temperature and humidity, sea surface temperature, windspeed and wind direction as well as wave height and period. With the exception of the wave sensor, all sensors are duplicated in case there are failures, as service visits are expensive and so limited to a set schedule.

Figure 17.3. The UK Met. Office operates nine buoys of the type illustrated, moored along the edge of the European continental shelf; this one is receiving a six-monthly service visit. The wind sensors and temperature screens can be seen on the upper ring, with solar panels on three sides. Sea-surface-temperature sensors are fixed to the hull 1 m below average sea level, and the barometric pressure sensors are protected within the superstructure. An inertial wave sensor is housed in the main central cavity. Measurements are telemetered via Meteosat. (Figure reproduced by permission of the UK Met. Office.)

Barometric pressure

A ceramic diaphragm sensor, vibrating cylinder or aneroid capsule (all described in Chapter 6) is used to produce a 10-s average reading of pressure at the synoptic hours. The sensor is housed on the outside of the buoy's superstructure, protected from the effects of wind by a static pressure head (Chapter 6). While some countries deploy their sensors inside the buoy's superstructure whenever possible, the UK policy is to deploy them externally to simplify interchange.

Air temperature and humidity

A platinum resistance thermometer (PRT), housed in a miniature screen of the AWS type (Fig. 3.8), measures air temperature to $\pm0.2\,°C$ with a resolution of $\pm0.1\,°C$. The humidity sensor is housed in a separate but identical screen, to allow its exchange independently of the temperature sensors. Unlike most land-based AWSs, the buoys use an ion-exchange humidity sensor (Fig. 4.6). Along with all sensors except the sea-surface-temperature and wave sensors, the humidity sensor is changed every six months (weather permitting). Experience has shown that after 9–12 months the calibration of the humidity sensor starts to drift, probably due to the salty environment; they are not reused. The sensor is protected from direct contact with the atmosphere and sea spray in a cavity covered by a PTFE membrane (Clarke & Painting 1983), which is permeable to water vapour but not to liquid water. This type of sensor is preferred to the capacitive type because the humidity is usually high over the sea (75% to 100%) and the resistive sensor works well in this range (having an accuracy of $\pm 5\%$ below 85% RH and of $\pm 3\%$ above 85% RH). When the RH measurement is combined with the temperature measurement, dew point can be calculated to $\pm 0.4\,^\circ$C with a resolution of $0.1\,^\circ$C. The value measured is an instantaneous reading at the observing time.

Sea surface temperature

As on ships, a PRT sensor, fixed to the hull of the buoy, measures the water temperature at 1 m depth to $\pm 0.2\,^\circ$C with a resolution of $0.1\,^\circ$C. Because it is possible to locate the sensor more precisely at the required depth than on a ship, the data are more truly representative of temperatures at 1 m.

Windspeed and direction

Conventional cup anemometers and vanes are used (Chapter 5), although the vane is of the self-referencing type, with a built-in magnetic compass providing an automatic reference to magnetic north, making it unnecessary to determine the orientation of the buoy. The anemometers are oil-filled to protect the bearings from contamination by salt. In accordance with standard synoptic procedure, wind speed and direction readings are averages for the 10 min preceding the observation time, made with an accuracy of ± 2 kn below 40 kn and ± 5 kn above 40 kn, and with a resolution of 1 kn. Gusts are also measured.

However, the exact height is debatable, since although the anemometer is deployed at 4 m above the buoy, the buoy may be changing height by more than this amount because of the wave movement. In a trough the wind speed may be lessened, while the swaying of the tower introduces some further uncertainty. However, perforce, we have no alternative.

Wave measurements

Because it is not possible to use a pressure sensor to measure waves from a floating platform, it is necessary to use an inertial sensor, consisting of a passive accelerometer which measures the vertical movement and accelerations of the buoy. Since the buoy has a considerable inertia, the smaller waves will not be detected as they will be averaged-out. An algorithm converts the detected direction and amplitude of movement into wave height and period. The significant wave height is defined as the rms (root-mean-square) value of the water level above the average level, measured over a period starting 17.5 min before the observing time; accuracy is ±10%, or ±20 cm if greater, over a range of ±10 m relative to the mean level. The wave period is defined as the average time interval, during the 17.5 min, between successive passages through the mean water level in an upward direction; accuracy is to ±0.5 s. Waves can now be measured from satellites using altimeter readings (Chapter 19).

Rainfall

Because rainfall is not required for forecasting purposes, it is not at present measured on buoys, the main purpose of which is to provide data for weather forecasts. It is, however, a difficult variable to measure on a buoy, for the same reasons that it is on a ship (see earlier). In Chapter 8, two types of gauge that might be more suited to buoy operation in place of the tipping bucket were mentioned. In one, the collected water is measured capacitively and not with a tipping mechanism, the other is the optical raingauge, which is not particularly sensitive to levelling; however, it is very expensive and consumes rather a lot of power. In addition, for any type of raingauge there is the possibility that sea spray will be collected or sensed and read as rainfall; spray can also contaminate the windows of optical gauges. Because of the problems of exposing a raingauge on a ship and because rain is not measured on buoys, our knowledge of rainfall over a very large part of the Earth's surface is limited in the extreme.

Experiments are, however, being undertaken to measure rainfall at sea by the sound of the raindrops, not the sound as they impact the surface, but the sound of the bubbles produced as the rain impacts the water. These sounds are analysed by an algorithm, the bubble sound-signals being distinguishable from amongst the myriad other noises of a stormy sea and even from the sounds of sea-spray bubbles, owing to their different drop-size spectra (Nystuen & Selsor 1997). But such systems are not currently beyond the experimental and evaluation stage. Because of the difficulties of measuring rainfall on ships and buoys, remote sensing from satellites is an attractive possible alternative. To what extent this is possible and the means used are discussed in Chapter 19. Quartly *et al.* (2002a, b) give a useful summary of the question of rainfall measurements at sea.

Evaporation

Nor is evaporation from the oceans measured on a regular basis, since there are no calls for this to be done by organisations willing to pay for it. However, research is in progress (see 'The eddy correlation method', p. 119 in Chapter 7; and Yelland *et al.* 1994, 1998) as how best to measure the fluxes of momentum, heat, water vapour and carbon dioxide between the ocean and the atmosphere, and these may ultimately lead to routine measurements from ships.

But three of the four variables for estimating evaporation (Chapter 7) by Penman-type methods are already measured on moored buoys; the fourth, radiation, is not. While it would not be a simple matter to measure net radiation (Chapter 2) on an unstable platform liable to sea spray, nevertheless solar radiation, or sunshine duration, probably could be measured with fair accuracy. Further, since the environment is open water, uncomplicated by the concerns on land of water availability and vegetation, the estimate should be more accurate. This lack of one sensor means that a second very important hydrological and climatic variable goes unmeasured over two-thirds of the Earth's surface. This highlights the problem that because NWSs are not asked to measure rainfall or evaporation over the oceans, there is little funding either to undertake the measurements or to develop the means to do so. It might be argued that since NWSs are not asked to make such measurements, there is apparently no need for them; but such data would be of undoubted value to climatology if available. This is an obvious gap in our knowledge.

Logging and telemetry

Just as with any AWS in a remote location, sending data in real time, a logger and the means to telemeter the measurements are required (Chapters 11 and 12); the moored buoys on the edge of the UK continental shelf use Meteosat (EUMETSAT 1995) to relay data to shore. The logger and DCP, like the sensors, are duplicated, each DCP transmitting the measurements from both sets of sensors in engineering units (not SYNOP). Upon reception, the two sets of data are combined to give optimum results, which are then coded in SYNOP for transmission on the GTS. As with land-based AWSs, lead–acid gel batteries charged by solar panels power the buoys.

Location systems

Although moored in deep water, buoys on the surface do not move by more than a mile from their point of fixing and so need no method of location. However, if one should break free from its mooring and drift, a means to locate and retrieve it is necessary, owing to the high cost of replacement. Two methods are used, the first being the inclusion of a satellite GPS receiver as one of the channels transmitted by the DCP, the second being an Argos transmitter (Chapter 12). Two radar reflectors are fitted, allowing the buoy to be detected by local shipping for safety reasons.

Inshore moored buoys

The moored buoys installed in shallower waters (around the UK) are simpler than the deep-water type, although similar (Fig. 17.4). The sensors are the same, but, since the buoys are more easily visited, there is no need to duplicate them. Similar stations are also operated on four lightships in the English Channel and also on some islands and oil rigs. Telemetry is by whichever means is most suitable, via either ground-based UHF or DCPs (for example on oil rigs).

Figure 17.4. The UK Meteorological Office also operates moored buoys in the shallower coastal waters around Britain, these being a scaled-down version of the larger, deep-water buoys, having just one set of sensors and telemetering by ground-based UHF radio links or by satellite, whichever is the most suitable. (Figure reproduced by permission of the UK Met. Office.)

Drifting buoys as instrument platforms

By far the greatest number of buoys in use around the world are not moored, but drifting, the greatest concentration of moored buoys being in the Atlantic and Pacific oceans. There are two types of drifting buoy used in the Atlantic – those that drift under the combined effects of wind and current (Fig. 17.5) and those mostly driven by currents only, these being in the form of small floating spheres (not illustrated) attached to a large drogue (parachute) which causes them to act as Lagrangian tracers embedded in the fluid, moving with the currents rather than with the winds.

Figure 17.5. Drifting buoys are by far the most common type, worldwide, there being many hundreds at any one time. They either drift under the combined influence of wind and currents or are largely submerged and fitted with drogues to move with the current only. The one illustrated is of the former type, on test at the UK Met. Office. The screen at the top of the buoy contains an air temperature sensor; a barometric pressure sensor is housed in the central column; and a sea-surface-temperature thermometer is fixed to the hull at 1 m depth. (Figure reproduced by permission of the UK Met. Office.)

Launch programme and area of operation

Both types of buoy are deployed from ships of convenience. In the case of the North Atlantic, ships sailing from Iceland to the eastern coast of the US, ships en route from Denmark to Cape Farewell and ships from the UK to the Caribbean cover an area from the Canary Islands up to Iceland and from the European and African coasts to 45° west. The launch programme is co-ordinated by the European Group on Ocean Stations (EGOS 1996), set up in 1988 with participants from eight countries. EGOS maintains a continuous operational network of drifting buoys, numbering from 25 to 40, with numbers set to rise as the cost of the technology falls. They are not recoverable for reuse, but have a long lifespan, averaging 200 days. Their cost varies from about £3000 to £5000 for the basic model to £15 000 for the full complement of sensors. The paths taken by the buoys vary considerably, some moving a long distance and others circling in a small area, the paths being quite unpredictable.

While there are few buoys of the moored type (beyond those operated by the UK and France in the North Atlantic, the North Sea and the Baltic, and those operated by Canada and the US, many of the latter being in the Pacific), drifting buoys are quite commonly used worldwide. South Africa has 60 at present in the South Atlantic, while several countries, including the UK and France, operate drifting buoys in the Indian Ocean. Australia and New Zealand deploy their own buoys in the South Pacific and Indian Oceans (EUMETSAT 1995). Because there is less land in the southern hemisphere to interfere with their movement, the life of these buoys is longer, often allowing them to circumnavigate the globe.

Variables measured

Both types of drifting buoy measure barometric pressure (and its tendency) and sea surface temperature. The larger type, as illustrated in Fig. 17.4, also measures air temperature (the screen is seen on top of the column). A few measure wind speed and direction. The temperature and pressure sensors are the same as on the moored buoys. Although this is not included on the buoy illustrated, windspeed may be measured by a Slevonius rotor (aerofoils that rotate on a vertical axis in a way similar to cup anemometers) mounted beneath the screen. A vane, fixed to the central column of the buoy (not on the example illustrated), causes the complete structure to align itself with the wind, a compass-sensor measuring the bearing of the buoy and thus of the wind. The holes in the vertical mast are ports to the barometric sensor. Methods of measuring windspeed by detecting the sound signature of sea waves in the band 8–16 kHz are being investigated.

Logging and telemetry

Because the buoys are moving, their measurements are telemetered via the Argos satellites instead of Meteosat, since Argos can also fix their position without the need to include a GPS receiver. However, while the cost of using Meteosat is covered by the contributions made by the individual participating countries of the member states of EUMETSAT (Chapter 12), Argos has to be paid for, currently at €3640 per year per buoy.

The telemetered data can be received directly from the satellite at three local user stations located in Oslo, Söndre Stömfjord and Toulouse. By using these three receivers, instead of the usual one, the area over which buoy transmissions can be received directly from the satellite in real time is extended (both transmitter and receiver must be in view of the satellite at the time of transmission if the data are to be received directly). The primary method of receiving the data, however, is by the store-and-forward option, via Argos ground stations in the US and France (Chapter 12). Ice buoys are also operated in a similar way, allowing the movement of pack-ice to be monitored, although of necessity they are deployed from the air by parachute. This technique is also starting to be used for drifting ocean buoys, since it is then possible to place them in the remoter areas not covered by shipping.

Sea level monitoring stations

As part of the international Tropical Ocean Global Atmosphere (TOGA) project and the Antarctic Circumpolar Current Levels from Altimetry and Island Measurements (ACCLAIM) project, measurements of sea level are made at a number of coastal sites worldwide; this is in order to study the interaction between air and sea, to provide ground-truth for comparison with satellite altimetry data and to obtain long-term measurements of sea level for global-climatic-change studies. At these sites, sensors fixed to the sea bed or to a harbour wall measure the pressure of the sea (Chapter 10, 'Measuring water depth by pressure') and its temperature, while on land barometric pressure is measured (Chapter 6). Readings are taken every 15 min, the logger being programmed to give hourly averages. Sea level is obtained by subtracting the air pressure reading from the sea pressure, whereas river level pressure sensors, which are vented to the atmosphere, require just one reading to be made. The processed data are stored in the logger as back-up and are telemetered every 12 h via Meteosat, GOES or GMS, whichever is geographically appropriate. Each 12-h transmission is of the last 24 hours' data, so that if one transmission is lost the data can be retrieved at the next transmission. The data are sent to the TOGA Sea Level Centre at the University of Hawaii and to the British Oceanographic Data Centre at the Proudman Laboratory, where they are made available to the international scientific community. Meteosat receives data telemetered from 12 sea level stations located in the Indian and Atlantic oceans (EUMETSAT 1995).

Argo – a global network of free-drifting, profiler floats

Information about the oceans into their depths is as important as meteorological measurements into the upper atmosphere. Until recently, however, most observations were surface measurements from ships, buoys and more recently satellites, but very few subsurface data have been collected. Historically, an instrument known as a bathythermograph has been used to measure the change of temperature with depth. In the past this took the mechanical form of a bimetallic strip (of the type used for measuring air temperature) that recorded a trace on a smoked-glass slide, the whole being contained in a bronze, torpedo-shaped housing and lowered and recovered by winch. After recovery, the slide was read against a graduated scale.

With the development of the expendable bathythermograph (XBT) in the 1960s, it became possible to collect many more data, and during the 1990s around 40 000 profiles per year were collected to depths of 450 and 750 m under the TOGA and WOCE projects (the World Ocean Circulation Experiment), with up to an additional 20 000 profiles from other sources (particularly from TAO, the Tropical Atmosphere Ocean array). But XBT datasets have many shortcomings, only temperature being measured and with observation limited to the shipping lanes, with few measurements south of 30 °S.

Origins of Argo

The idea of a neutrally buoyant float for measuring sub-surface ocean currents was first put forward in the 1950s, both in the US (Stommel 1955) and in the UK (Swallow 1955) independently. Their designs remained unchanged until the early 1970s. The floats were tracked by monitoring the continuous 10-kHz sound they emitted.

The TOGA and WOCE experiments started in the 1980s have shown how important the oceans are in the coupled climate system, in particular the powerful influence of the El Niño Southern Oscillation (ENSO) and the several other 'oscillations' (North Atlantic, Arctic, Pacific Decadal, Indian Ocean Dipole and Antarctic Circumpolar Wave). Each oscillation is associated with different weather and climate changes in different parts of the globe and each has its own periodicity and predictable behaviour. The experiments also showed that ocean currents carry vast amounts of heat from the tropics to the mid-latitudes, amounting to 2×10^{15} watts in the northern hemisphere alone (Bryden *et al.* 1991), with large interannual variations of up to 30% (Roemmich & Owens 2000), the reasons for which are unknown. It is clear that global measurements of heat storage and transport are essential in any climate-observing system. In addition to this compelling scientific reason for better measurements, innovative developments were emerging that made an improved instrument network practical.

Profiling floats had been developed during the 1990s and precision satellite altimeters measuring sea level globally every 10 days were about to be put into orbit, the two being essential partners. The Argo project brought these independent developments together through two separate proposals – for an '*A*rray for *R*eal-Time *G*eostrophic *O*ceanography', named Argo (in part as an acronym, but also after the ship of the Greek hero Jason – emphasising the close relationship of the new float network with the new satellite-based *Jason-1* altimeter mission) coupled with proposals for a programme for *G*lobal *O*cean *Sa*linity *M*onit*or*ing (GOSAMOR). In 1998 the International Steering Team for GODAE (Global Ocean Data Assimilation Experiment) undertook to develop the idea of an integrated system combining these two sources of data along with the satellite altitude measurements (Stammer & Chassignet 2000). The Upper Ocean Panel of CLIVAR (Climate Variability and Predictability Experiment) of the World Climate Research Programme (WCRP) agreed that the proposals should be given high priority. From out of this jungle of acronyms (there are many more), and pursued by diverse influential working groups, came the Argo project.

Such is the complexity of modern science that no longer can the lone mind explore new ideas with simple observations or by thought experiments, written down with quill or nib; it is corporate and huge.

Some countries fear the floats may be used to spy with, but their suspicions should be dispelled by the fact that the Argo project has been endorsed by the WMO and the Intergovernmental Oceanographic Commission. I see it as the most important recent advance in instrumentation for the natural environment. It promises to clarify a lot of things.

The hardware

The Argo project aims to have around 3000 free-drifting floats in operation by 2003/5, measuring temperature and salinity profiles of the *upper ocean* (the top 2000 m). The floats will be deployed across all of the world's oceans, separated by distances of around 3° latitude/longitude (about 350 km at the equator).

The floats will normally be 'parked' at a depth of about 2000 m, in sleep-mode but moving with the currents at that depth, which are believed to be slow (Fig. 17.6). Because of the slow movement, gaps between, and clusters of, floats will hopefully not form (unlike on the surface when currents and wind can move buoys large

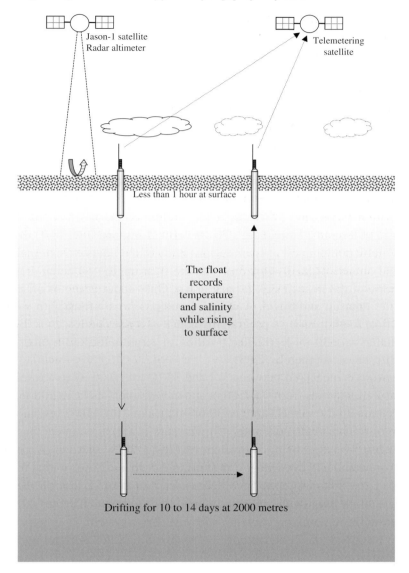

Figure 17.6. The 30-kg, 2-m-long Argo float automatically descends to its pre-programmed depth, usually 2000 m, where it remains in sleep-mode for 1–2 weeks, whereupon it becomes active and rises to the surface, taking readings at about 5-m vertical intervals. At the surface, it transmits the data collected during its ascent via satellite to base, thereafter sinking back down to its parking depth where it repeats the cycle over and over for several years until its power is exhausted or it fails. The height of the sea surface is also measured, by the radar altimeter aboard the new Jason-1 satellite.

distances). After 10 days, they will awake and ascend quickly to the surface, measuring temperature and salinity as they rise. On arrival at the surface, where they will remain for half an hour or less, their measurements will be transmitted either via the ARGOS or the Orbcomm satellite network, whereafter they will sink down to their 2000 m parking depth, repeating the cycle until their power is exhausted, which is estimated to be after about four to five years (or around 150 round trips to the surface and back). The cost, including data transmission and processing, will be about $US 20 000 per float.

Floats are currently made by two companies, one in the US, which worked with the Wood Hole and Scripps oceanographic institutes in their development, and one in France (although this situation will probably change; it is a statement to give a feel for the position at present). Floats are made of aluminium, weigh about 30 kg and are around 2 m long in all including a 70 cm vertical antenna (Fig. 17.7). Buoyancy is controlled by pumping oil from within the body of the float into external flexible bags on the outside of the cylinder to send the float up, the oil being drawn back into the body of the float to sink it back down. Only about 250 g of oil is needed to move the float up and down. A recent development is a float with wings that allow the float to glide along a programmed track as it rises or descends. Buoyancy is controllable to ± 30 m depth. The position of Argo is located by the ARGOS satellite, if used, if not then by GPS.

Temperature is measured between -2 and $35\,^{\circ}\mathrm{C}$ to $\pm 0.05\,^{\circ}\mathrm{C}$ to a resolution of $0.01\,^{\circ}\mathrm{C}$, at 2-m intervals from 0 to 500 m and at 5-m intervals from 500 to 2000 m. Salinity (conductivity) is measured to 30 or 70 ms cm^{-1} to ± 0.05 ms cm^{-1} with a resolution of 0.01 ms cm^{-1} (Chapter 10). Pressure is measured in decibars (dbar), that is 0.1 bar, 100 dbar being equivalent to 100 m depth of water. Argo sensors measure between 0 and 2000 dbar (2000 m) to ± 5 dbar (± 5 m) with a resolution of 1 dbar (1 m). The aim is to have floats spend an hour or less on the surface to minimise biofouling of the salinity (conductivity) sensor (Chapter 10). This is one of the main remaining problems with the Argo system. It is very difficult to operate a conventional conductivity sensor for four years without the surfaces becoming covered in algae, thereby affecting the readings. Other methods are being investigated. The magnetic method, described in Chapter 10, might be suitable. Currently, in mid 2003, floats have to be on the surface for around nine hours to ensure that their data are collected reliably, but the long transmissions draw on a lot of battery power. A new Argos satellite will offer two-way communication and so allow the float to be instructed to dive when the data have been successfully received, reducing the time on the surface. With newer satellites such as Iridium, the time to transmit a full profile will be less than one minute and two-way communication will also be available. The situation is, therefore, developing. Nonetheless, the basic concepts of Argo remain.

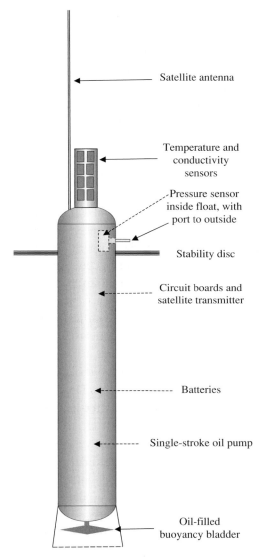

Satellite antenna

Temperature and
conductivity
sensors

Pressure sensor
inside float, with
port to outside

Stability disc

Circuit boards and
satellite transmitter

Batteries

Single-stroke oil pump

Oil-filled
buoyancy bladder

Figure 17.7. The Argo float has the appearance of a large aqualung, with a rod antenna on top. The temperature and conductivity sensors are deployed within a U-shaped tube, protected within the cage on top of the main cylinder, the water being pumped through the tube while readings are being taken, while for the rest of the time, the tube is filled with an anti-fouling solution to protect and clean the conductivity sensor. The pressure (depth) sensor is inside the float with a port to the outside. At the base, a flexible buoyancy bladder can be filled or emptied from an oil pump inside the main cylinder, causing the float to rise or sink on command. Also contained internally are all the other requisites, including the satellite transmitter, all of the necessary electronics and the batteries.

One reason for the parking depth of 2000 m is to minimise biofouling, but also because salinity and temperature are fairly stable at this depth over most of the globe and so can be used as a calibration check. Anti-fouling chemicals are one possibility for preventing biological growth, another is to isolate the sensor during the long period between profile-taking. A further possibility is to have two sensors to help identify problems – unless both suffer equally to the same degree! One problem in all of this is the time required to demonstrate any improvement, since tests inevitably take years.

The parking depth can be chosen independently of profile depth, deeper parking producing less dispersion due to slower currents, as well as less biological fouling. A shallow parking level can be selected if required, for example by placing the float at or above the thermocline (the thin dividing layer between the warm, shallow, top layers of 100 m or so depth and the cold water below, extending down to 4000 m, across which there is a relative large change in temperature). Placed here, the float can be dispersed more rapidly while also measuring conditions at the level of interest. With two-way communication, the float can be instructed to go to any specified depth.

Apart from a low quiescent current, power is drawn from the batteries mainly while taking readings on the way to the surface, less than 10% being required for transmitting the readings to the satellite. Improvements have also been made to the efficiency of the single-stroke pump used to inject the oil into the external bag to initiate ascent, thereby reducing power consumption. Data are transmitted in compressed form taking up about two kilobytes per profile. There are to be processing centres in the US, France and Japan.

Floats are launched from merchant and research ships and by airdrops. About half of the floats will be deployed and paid for by the US, other countries contributing to varying degrees; the UK for example, via the Met. Office and the Southampton Oceanographic Centre, will install about 50 a year and maintain 150–200 floats at any one time (Turton & Cattle 2000). Floats are already deployed in the Imanger Basin and Rockall Trough near Iceland. At the time of writing (mid 2003), the UK has deployed 62 floats, with a worldwide count of around 450. Turton (2002) also gives a good overall review of Argos.

Jason – satellite altimetry

Measurements of global sea level were initiated by NASA/CNES, through the TOPEX/Poseiden satellites, in 1992, measuring topography to about ±4 cm. This was supplemented on 7 December 2001 by Jason-1, a satellite-based radar altimeter that measures sea level globally every 10 days to a similar accuracy to TOPEX, with a goal of 2.5 cm (1 inch). Other NASA satellites, and those of the JMA and

NOAA, also measure sea surface temperature and the speed and direction of winds over the oceans. In addition NASA's Quiksat and the Japanese Advanced Earth Observing Satellite (ADEO-2) will carry a sea wind scatterometer to measure wind at the surface of the sea.

But it is the Argo/Jason combination that is most important, since the altimeter and the float readings are closely related. Having both sets of data helps explain the causes of sea level anomalies through sub-surface processes. Peaks and valleys in sea surface height are caused by wind-driven currents and by heating and cooling or changes in salinity. Variations in sea level, as read by the altimeter, are due to these two factors. Firstly to changes of pressure at a reference depth (known as *reference pressure height*), derived indirectly from the float's speed of drift at its parking level (Davis 1998), taking into account the Coriolis acceleration, in much the same way that winds in the atmosphere are due to pressure differences. This gives the horizontal pressure gradients at the reference depth, which measure changes of the weight of water above the parking level, such as that caused by wind-driven components of ocean circulation.

Secondly, changes in level due to changes in the temperature and salinity profile of the column of water above the float (known as the *steric* or *dynamic height*) are derived from the measure profile during ascent (Patullo *et al.* 1955). Thus as the column warms it expands, as with air in the atmosphere, for example a rise of 30 cm in level occurred in the Eastern Pacific during the 1997 El Niño due to sub-surface warming and freshening. Altimeter height and steric height have a high correlation over a wide range of distances (Gilson *et al.* 1998), the reference pressure height making only a small contribution, especially in the tropics, although becoming more important at higher latitudes.

The uses of Argo data

It is not too much of an exaggeration to say that this new instrument initiative will revolutionise our knowledge of world weather and climate. The oceans move slowly and are linked in an exchange of heat, water vapour, carbon dioxide and momentum with the atmosphere creating the underlying conditions that, over decades, control the broad patterns of rainfall, wind and atmospheric circulation. The upper layers of the sea store 1000 times the heat of the atmosphere and changes in sub-surface currents, temperature and salinity eventually change surface conditions. Argo will allow these changes to be followed and their prediction improved through better models. Argo will also allow sea surface height to be explained in terms of sub-surface temperature and salinity and allow height anomalies to be separately attributed to a real global sea level change, to the various periodic oceanic oscillations (above) and to changes due to differential heating and cooling, to advection of heat and fresh water and to the wind-driven redistribution of mass (Argo Science Team 1999). Argo will also establish what the global average temperature of the sea actually is, including the sea surface temperature, improving greatly on the past sparse, and not too accurate, measurements from ships (Roemmich & Owens 2000). Ocean Atmosphere models rely on information of the upper ocean to initialise them correctly (to get off to a good start). Argo will provide this. Remote sensing of the oceans (sea surface temperature and sea-level anomaly) can only measure surface condition (literally skin deep) and cannot provide what Argos will produce. Also, even the surface data measured from satellites need ground-truth to calibrate them correctly (see Chapter 19). Argo will also provide this. The role of the carbon cycle will also be clarified further by Argos data. Whereas weather forecasts, with radiosondes data, can be made for a few days in advance, Argo data, along with surface data from satellites and buoys, will also be able to make long-term seasonal forecasts of wet/dry, flood/drought conditions, while Argo data will provide operational oceanography with predictions of ocean conditions for commercial shipping as well as for planning research, the latter being the subject of the final section of this chapter.

Real-time data will be available (to all, without charge) within 12 h for forecasts and for operational use. Quality-controlled data will emerge a few months later for climatological and hydrographic use, after a check on salinity because of the sensor's liability to contamination.

Measuring ocean-atmosphere fluxes

Oceanography is researched by several major institutes around the world; for example, the Scripps and Wood Hole Institutes in the US, by the Southampton Oceanography Centre of Southampton University (SOC) in the UK and the Ocean Technology Group at the University of Sydney in Australia. Workers ply the seas from these centres on short or long voyages to make specialised observations for basic research – rather, that is, than making routine observations, although the techniques they develop may eventually become routine on VOS.

Such investigations are many and complex, but one interesting and important example is the measurement of the transfer of the wind's kinetic energy to the sea surface. As the wind blows over the surface, momentum is transferred from the air to the water, the so called *momentum flux*, which produces waves and currents, the ocean acting as a drag force on the atmosphere. Wind drag (or *stress*), expressed as a *drag coefficient*, slows the wind near the surface which in turn acts as a drag higher in the air, the effect extending up to a kilometre in strong winds (Bigg 1996). As discussed in Chapter 5, the vertical profile of windspeed is nearly logarithmic, the steepness of the curve being dependent on the roughness of the surface. The logarithmic profile results in wind shear as the layers slip across each other, generating the mechanism for momentum to be transferred down to the surface. Wind drag over the oceans is important in modelling ocean-atmosphere interactions and although it has been studied for at least 50 years, there is still incomplete understanding as to the physical processes involved over a complicated ocean surface.

Traditionally the drag coefficient has been assumed to be a function of the mean windspeed at a reference height, usually 10 m. Or it may be expressed in terms of the aerodynamic roughness of the sea surface. But observations and measurements have often pointed in the opposite direction to theoretical predictions; for example, most models have assumed that drag is caused mostly by the larger waves, while for the most part observations suggest that the shorter waves are more influential. Because of this uncertainty, a working group was set up in 1993 to consider 'The influence of the sea state on the atmospheric drag coefficient'. For readers wanting more information, the final report of the working group was published by Cambridge University Press in the book *Wind Stress Over the Ocean*, edited by Jones and Toba (2001).

Wind and waves have different velocities and it is assumed that airflow is accelerated above the crests and decelerates in the troughs. The momentum flux (or rate of transport of momentum) is mostly from the air to the sea, when the wind travels faster than the waves, generating currents and waves, but it can occasionally be in the reverse direction. The idea that swell running ahead of the wind returns momentum to the atmosphere was first suggested by Harris (1966); swell against the wind is less well investigated. The significance of wave age and wave direction relative to the wind coupled with the effects of surface tension of the water all play a part in what amounts to a complex and incompletely understood set of processes. Theoretical models have been developed of the processes, but measurements are required to resolve the many outstanding uncertainties. The work at SOC aims to develop a system that can operate on a ship, eventually routinely on a VOS, the development being carried out on research ships such as the RRS *Charles Darwin*.

Measurements are difficult to make. While over a land surface, stress can be measured fairly easily, for example by using a drag force anemometer (see Chapter 5), this is not possible over the sea because of the ever-changing topography. Instead, indirect measurements must be made from which stress can be inferred. Three methods are used: the Reynolds Stress method, the Inertial Dissipation (ID) method and the Profile method (a mast with an array of anemometers at different heights). The radar observations of sea-state that are now becoming available from satellites will add a further new dimension, but so far it is the ID method that has gathered most of the available data (Hicks & Dyer 1972, Pond *et al.* 1979).

Wind stress is defined as an average of the product of the horizontal and the vertical components of the wind speed (Yelland 2002), that is the stronger the wind and the more turbulent it is, the greater the momentum flux. The ID method requires measurements of the average horizontal windspeed and its vertical fluctuations, with means taken over minutes to an hour. Only the *inertial subrange* of turbulence, from about 1 Hz to 1000 Hz, needs to be measured. In this it differs from the eddy correlation method, which encompasses the full range of the turbulence spectrum. Early on (1970s and 1980s), this had to be measured using mechanical sensors such as small, light weight cup and fan anemometers, but recently sonic anemometers have become more widely available at reasonable cost and it is these that now dominate (see also 'The eddy correlation method', Chapter 7). The subrange of frequencies concerned are (fortunately) not greatly affected by the ship's presence or by its motion, but the long-term horizontal means are, and much development effort is going into quantifying this, through measurement on ships in detailed field studies (Yelland *et al.* 1994, 1998) and through modelling, using simplified ship shapes. Yelland estimates that wind stress can be estimated to about ±5%. But the ID method, and its derivation of the dissipation rate from the measurements, depends on the validity of several critical assumptions and it is not entirely clear just how valid all of these are; Janssen (1999) challenges some of them. It has also been shown, at low windspeeds, that when the ID method is compared with the eddy correlation method, disagreements are exposed. This must reflect the fact that we do not yet fully understand the complex mechanisms occurring at the air–sea interface. (As someone with deep-sea fishing experience in small boats, I can fully believe this.) The eventual aim, however, is to develop instruments and algorithms that can be operated on a routine, automatic, basis on voluntary ships.

As well as an input of momentum to the sea, there is also an input of solar energy, producing upward fluxes of heat and water vapour (evaporation) away from the surface (thereby increasing surface salinity). There is also an interchange in both directions of carbon dioxide. How these fluxes are measured over land, by eddy correlation instruments, has already been described in Chapter 7, but over the sea it is much more difficult to operate these complex instruments, especially on a moving ship, in amongst its superstructure tens of metres above the sea. SOC (in co-operation with four other organisations in Denmark, Sweden, France and The Netherlands and one commercial company in the UK) are attempting to develop an instrument (the 'Autoflux') to measure all of these fluxes at sea.

In addition, there is an input of rainfall to the sea, which freshens the top surface (perhaps we should call it the *rain flux*). But rainfall is also difficult to measure over the sea, the best hope at present being the detection of the sounds produced by the rain (see above), but also now through remote sensing (Chapter 19).

The problems of measuring such fundamental exchanges at the ocean surface, as well as our limited knowledge of the processes involved, illustrate just how far we have still to go in understanding the oceans, even at a most basic level. It is a field ripe for future research. In view of the fact that nearly three-quarters of the globe is covered by the oceans and consequently has little human presence on it and with the great difficulty (and expense) of making in situ measurements, this would seem an ideal application for remote sensing (Chapter 19). But, as will be shown, ground truth measurements will also be essential, and that is where the Autoflux, its successors and its equivalents will help.

References

Argo Science Team (1999) The global array of profiling floats. *Proceedings of OCEANOBS99 International Conference on the Ocean Observation System for Climate.* St. Raphael, France, October 1999, pp. 18–22.

Banfield, E. (1976) *Antique Barometers.* Baros Books, Trowbridge, UK.

Barlow, D. (1994) From wind stars to weather forecasts: the last voyage of Admiral Robert FitzRoy. *Weather,* **49**, 123–8.

Barlow, D. (1997) The devil within: evolution of a tragedy. *Weather,* **52**, 337–41.

Bigg, G. R. (1996) *The Oceans and Climate.* Cambridge University Press, Cambridge, UK.

Birch, K. G. & Pascal, R. W. (1987) A meteorological system for research applications – MultiMet. In *Proceedings of the 5th International Conference on Electronics for Ocean Technology.* Edinburgh, pp. 7–12. Institute of Electrical and Radio Engineers (IERE) Publication 72.

Bryden, H., Roemmich, D. & Church, J. (1991) Ocean heat transport across 24 °N in the Pacific. *Deep-Sea Res.,* **38**, 297–324 (Special issue).

Chenoweth, M. (1996) Nineteenth-century marine temperature data: comments on observing practices and potential biases in marine datasets. *Weather,* **51**, 280–5.

Clarke, C. S. & Painting, D. J. (1983) A humidity sensor for automatic weather stations. In AMS *Proceedings of the 5th Symposium on Meteorological Observation and Instrumentation,* Toronto.

Davis (1998) Preliminary results from directly measuring mid-depth circulation in the tropical and South Pacific. *J. Geophy. Res.,* **103**, 24619–39.

EGOS (1996) European Group on ocean stations. EGOS brochure, CMR 05/1996, WMO/IOC.

EUMETSAT (1995) Data collection system, user guide. EUM UG 02, pp. 29–34.

Feder, T. (2000) Argo begins systematic global probing. *Physics Today,* July 2000.

Folland, C. K. (1988) Numerical models of the raingauge exposure problem, field experiments and an improved collector design. *Q. J. R. Meteorol. Soc.,* **114**, 1485–516.

Folland, C. K. & Parker, D. E. (1995) Correction of instrumental biases in historical sea surface temperature data. *Q. J. R. Meteorol. Soc.,* **121**, 319–67.

Gilson, J., Roemmich, D. & Cornuelle, B. (1998) Relationship of TOPEX/POSEIDON altimetric height to the steric height of the sea surface. *J. Geophys. Res.,* **103**, 27947–65.

Harris, D. L. (1966) The wave-driven wind. *J. Atmos. Sci.,* **23**, 688–93.

Hasse, L., Grossklaus, M., Uhlig, K. & Timm, P. (1997) A ship rain gauge for use in high wind speeds. *J. Atmos. Ocean. Tech.*, **15**, 380–6.

Hicks, B. B. & Dyer, A. J. (1972) The spectral density technique for the determination of eddy fluxes. *Q. J. R. Meteorol Soc.*, **98**, 838–44.

Houghton, J. (1997) *Global Warming.* Cambridge University Press, Cambridge, UK.

Janssen, P. A. E. M. (1999) Note on the effect of ocean waves on the kinetic energy balance and consequences for the inertial dissipation technique. *J. Phys. Oceanogr.*, **29**, 530–4.

Jones, I. S. F. & Toba, Y. (2001) *Wind Stress Over the Ocean.* Cambridge University Press, Cambridge.

Kenworthy, J. M. & Walker, J. M. (1997) Colonial observatories and observations: meteorology and geophysics. In *Proceedings of the Conference of the Royal Meteorological Society*, University of Durham, April 1994. ISBN 0307 0913.

Lebedev, I. & Tomczak, M. (1999) Rainfall measurements with navigational radar. *J. Geophys. Res.*, **104**, 13697–708.

Met. Office (1995) *Marine Observers Handbook.* HMSO, London.

Met. Office (1997) *UK Meteorological Office Moored Data Buoy: Technical Description.* Meteorological Office Technical Description K-27-TD, Issue 1, August 1997.

Nystuen, J. A. & Selsor, H. D. (1997) Weather classification using passive acoustic drifters. *J. Atmos. Ocean. Technol.*, **14**, 656–66.

Parker, D. E., Folland, C. K. & Jackson, M. (1995) Marine surface temperature: observed variations and data requirements. *Clim. Change*, **31**, 559–600.

Patullo, J., Munk, W., Revelle, R. & Strong, E. (1955) The seasonal oscillation in sea level. *J. Mar. Res.*, **14**, 88–155.

Pearce, F. (2000) Tidal warming. *New Scientist*, **2232**, 12 [April 2000].

Pond, S., Large, W. G., Miyake, M. & Burling, R. W. (1979) A Gill twin propeller vane anemometer for flux measurement during moderate and strong winds. *Boundary-Layer Meteorol.*, **16**, 351–64.

Quartly, G. D., Guymer, T. H. & Birch, K. G. (2002a) Back to basics: measuring rainfall at sea: Part 1 – in situ sensors. *Weather*, **57**, 315–20.

Quartly, G. D., Guymer, T. H. & Srokosz, M. A. (2002b) Back to basics: measuring rainfall at sea: Part 2 – space-borne sensors. *Weather*, **57**, 363–6.

Roemmich, D. & Owens, W. B. (2000) The Argo project: global ocean observations for understanding and prediction of climate variability. *Oceanography*, **13**, 45–50.

Stammer, D. & Chassignet, E. (2000) Ocean state estimation and prediction in support of oceanographic research. *Oceanography*, **13**.

Stommel, H. (1955) Direct measurement of subsurface currents. *Deep-Sea Res.*, **2(4)**, 284–5.

Strangeways, I. C. (1996) Back to basics: The 'met. enclosure': Part 2(b) – raingauges, their errors. *Weather*, **51**, 298–303.

Swallow, J. C. (1955) A neutral-buoyancy float for measuring deep currents. *Deep-Sea Res.*, **3(1)**, 93–104.

Turton, J. (2002) Argo 2002. Progress towards a global array of profiling floats. *Paper presented at International Conference on Oceanology International 2000, ExCel, London.* (See also www.metoffice.gov.uk/research/ocean/argo/argo2002.pdf.)

Turton, J. & Cattle, H. (2000) The UK contribution to Argo. *NWP Gazette*, March 2000, pp. 6–7.

Warner, T. B. (1972) Ion-selective electrodes. Properties and uses in sea water. *J. Mar. Technol. Soc.*, **6(2)**, 24.

Yelland, M. J. (2002) Private communication.

Yelland, M. J., Taylor, P. K., Consterdine, I. E. & Smith, M. H. (1994) The use of the inertial dissipation technique for shipboard wind stress determination. *J. Atmos. Oceanic Technol.*, **11**, 1093–108.

Yelland, M. J., Moat, B. I., Taylor, P. K., Pascal, R. W., Hutchings, J. & Cornell, V. C. (1998) Wind stress measurements of the open ocean drag coefficient corrected for air flow disturbance by the ship. *J. Phys. Oceanogr.*, **28**, 1511–26.

18

Cold regions

And now there came both mist and snow,
And it grew wondrous cold:
And ice, mast-high, came floating by,
As green as emerald.

And through the drifts the snowy clifts
Did send a dismal sheen:
Nor shapes of men nor beasts we ken –
The ice was all between.

The ice was here, the ice was there,
The ice was all around:
It crack'd and growl'd and roar'd and howl'd
Like noises in a swound!

Samuel Taylor Coleridge *The Rime of the Ancient Mariner.*

Just as the oceans cover a large fraction of the Earth's surface, so too do the cold regions, and they are even more poorly monitored. In the past, the main reason for little data being collected here was simply that few people lived in these regions or travelled to them, data coming in the last 150 years mostly from the occasional expedition to Antarctica or from a special laboratory such as on Ben Nevis in Scotland. But modern automatic instrument systems do not require frequent attention and it is now possible to site them anywhere. The main difficulty now in cold regions is not the need for operators but the problems of the environment itself.

Observation problems at low temperatures

Low temperature in itself is not a major problem for instruments. Electronic and mechanical components can operate in temperatures down to $-40\,°C$ or lower, and

Measuring the Natural Environment, second edition, Ian Strangeways. Published by Cambridge University Press. © Ian Strangeways 2003.

batteries continue to function at these extremes, although perhaps with a reduced capacity. The main problems arise from the effects of snow and ice adhering to (or filling) the sensors, often accompanied by strong winds. The question considered here is how to protect sensors from their damaging and disabling effects. (Techniques for measuring the amount of snow falling, lying and melting are dealt with in Chapter 8.)

Several types of ice can form. Hoar frost is produced directly from water vapour in the air, but as the amount of vapour is small at sub-zero temperatures, this is not a serious problem and it lessens as temperatures fall. A particularly hazardous form of ice, rime, occurs when supercooled cloud droplets meet a surface and freeze, the ice building mostly into the wind, but also in the lee owing to eddies. Rime can be extremely thick and damaging (Fig. 18.1). Glaze ice can form when the cloud-drops, or raindrops, have time to form a thin film before freezing. The amount and type of cloud-induced ice that forms depends on the wind speed, air temperature, water droplet size and drop density. Even though the atmosphere holds less and less water vapour as the air becomes colder, rime can still be a problem down to the lowest temperatures of Antarctica. The worst rime ice is not, however, found in polar conditions (as might have been imagined), but on mountains in the mid-latitudes, such as in the Scottish highlands, just below freezing. Snow causes problems either

Figure 18.1. Ice can be a serious hazard to instruments (as well as to shipping and aircraft). Rime ice is seen here encasing an AWS on Cairn Gorm in Scotland. One cup of the anemometer remains nearly clear of ice (left); the net radiometer (lower right) is completely covered, as are the other, barely recognisable, sensors.

when wet, by adhering to sensors and then freezing, or as dry crystals, blowing like sand (Fig. 8.14) and able to enter the smallest of holes and completely fill cavities such as temperature screens.

De-icing manned stations

A brief history

Automatic and manual stations alike suffer from the snow and ice problem. There have been a number of manned observatories in cold places, a classic example being that operated on the summit of the Scottish mountain Ben Nevis, from 1883 to 1904 (Paton 1954, Begg 1984, McConnell 1988a, b). Its main purpose was not to study mountain climate but the vertical structure of the atmosphere in depressions and anticyclones (Roy 1983), similar observations being made at an observatory at Fort William at the foot of the mountain to provide data for comparison with those from the summit. Today, however, the mountain-climate aspects of the data are of equal interest.

Operators lived on the summit and made observations every hour during the day and the night for 21 years. Chart-recording instruments could not operate in the temperature screen under the conditions that prevail there for much of the year, being in supercooled cloud for a high proportion of the time. On many occasions the screen became completely ice-covered or ice-filled and often the thermometers had to have ice removed from them before readings could be taken. Conditions sometimes prevented the operators from getting to the instruments at all. Because of the exposed conditions and high winds, precipitation measurements were also prone to high errors and anemometers could not be kept continually ice-free. Even with people present all the time, it was an impossible task to keep the instruments fully operational, and there were inevitable gaps in the records.

At about the same time as the Ben Nevis operation, a similar station was being operated in northern Siberia at the mouth of the Lena River (Schutze 1883). But these examples were unusual and relatively short-lived, although very important in amassing information never before collected. During the second world war, Germany and Norway operated manned weather stations on Greenland, radioing back their measurements.

Modern Antarctic stations

The nearest modern equivalent to the pioneering installations in Scotland and Siberia are today's semi-automatic stations operated at bases in Antarctica. Typical are those operated by the British Antarctic Survey (BAS), which has several bases on the continent – at Rothera on the peninsula (and until recently also at Faraday) as well as at islands further north and at Halley on the mainland. At such bases, measurements are made by electronic or electrical sensors, of the types covered in earlier chapters; they are exposed in much the same way as manual instruments, with cables running some distance into the laboratories of the base. Here, mains power being available from diesel generators, the measurements can be processed by a PC, operating continuously and connected to a DCP, allowing the measurements to be processed into SYNOP for transmission over the GTS (see Chapter 12 for details of telemetry). In addition to inputs from the sensors, the stations accept manual inputs of non-instrumental observations such as cloud cover, state of precipitation, snow cover and similar variables, which are difficult or impossible to measure automatically at any location.

Ice and snow prevention and removal at these stations are entirely manual. A sharp hit with an ice axe is used to clear wind sensors, while accretions are removed from radiation sensors by a gentle brush by hand. The Stevenson temperature screen is of conventional wooden construction, but has fold-up doors on all sides, which can be closed on the side(s) exposed to the wind when there is snow or the possibility of rime formation (Fig. 18.2). Rainfall is not measured and snow is monitored by snow poles.

Figure 18.2. At Antarctic bases, the British Antarctic Survey prevents snow and rime entering or adhering to Stevenson screens by adding drop-down doors that can be closed when such hazards are occurring or are expected. Here the left-hand cover (at the back of the screen) is open and the other three are closed; the front is covered in adhering snow. Although this procedure reduces the air flow considerably, it is only necessary when radiation levels are fairly low, and so does not affect the temperature readings unduly; but clearly it is a compromise.

Figure 18.3. The cup anemometers and wind vane at the British Antarctic Survey base at Faraday, on top of a 10-m tower, are prone to icing and snow cover. However, since for much of the year the snow is dry and rime is not frequent, protection against them is not essential, although it is necessary to be aware of when sensor performance is impaired and to flag the data accordingly.

Even though the stations are manned continuously, the manual de-icing methods obviously have their logistical problems and limitations, just as they did on Ben Nevis. No manual system can ensure continuous observations. If snow starts to fall, or rime to build up, it is necessary to go out, lift the appropriate door on the screen and climb the 10-m-high wind-sensor mast (Fig. 18.3) in order to remove the accretions manually; this can be difficult or impossible in high winds or at night, if for no other reason than safety. Working in this way, the sensors (temperature apart perhaps) cannot be kept continuously ice-free and although for much of the year these methods are adequate or are not required at all, inevitably there will be periods when the data are not correct. However, it will generally be clear when the data are suspect and these can be flagged as doubtful. Users of such data should, however, be aware of these facts.

De-icing automatic stations

In Antarctica

Even in the case of fully automatic weather stations (AWSs), operated remotely from bases in Antarctica, there is no widely used automatic ice and snow prevention or removal. However, in the remoter, colder regions where these stations are usually deployed, there is less of a problem since the lower temperatures reduce the frequency of severe icing or snowfalls. Rime can form, nevertheless, and snow does blow as spindrift, and so the problem still exists. But just as it is possible to infer from the data from manned stations when measurements are likely to be affected by ice and snow, so too is it possible with AWSs; the data give clues as to when to flag the measurements as suspect, for example when the wind is indicated as from a fixed direction for an extended period. But in marginal conditions, where some movement of the vane continues, errors may go undetected. Again, users of such data should be cautions.

What this illustrates is that one solution is to do nothing to prevent snow or ice forming on the sensors, but to be aware of when it is occurring and to quality-control and flag the data appropriately. But this does mean that the data will have gaps in them, and/or be inaccurate, and it is only feasible at locations where rime formation and wet or blowing snow do not occur for long periods. Having gaps in the data from these locations (as indeed from any location) means that long-term means and totals are not fully reliable.

By the use of heat

Techniques have been developed to combat the snow and ice problem automatically, the normal strategy being to use heat, mostly electrically generated. But large amounts of energy are necessary for this approach, owing to the size of the sensors, the high winds prevailing and the energy required to melt ice. Thus a cup anemometer (Hartley 1970) used 1.2 kW of heat, while a Pitot-tube wind sensor required 750 watts (Gerger 1972) to be kept ice-free. Gerger also describes a solarimeter heated by air blown over the dome. If each sensor is heated individually in this way, the power demands increase pro rata, and the large amount of power required is not usually available at remote sites, although it would be at Antarctic bases; it is surprising that more prevention has not been done where power is available.

Gerger (1972) reported on the technique of putting sensors in a housing that opens periodically to expose them for a short time, the housing being heated only at its opening point and internally (sufficient to keep the sensors just above freezing), thereby using less power. Alexeiev *et al.* (1974) noted the use of this technique for anemometers on ships, where adequate power is available, although earlier

Figure 18.4. The 'opening box' AWS, designed by Heriot Watt University, is heated electrically (internally and at its opening point). Here it is seen open while one of its half-hourly readings is being taken; this exposes the anemometer (top plate), wind direction and temperature sensors (central base plate). The high altitude of the mountain (1245 m) makes it possible to telemeter its data to Edinburgh, 133 km distant, in one single UHF radio link.

comments about the problems of using anemometers on ships, even in non-icing conditions, must raise doubts about the usefulness of this.

A more recent example of a heated enclosure is an AWS on Cairn Gorm in Scotland, developed by Heriot Watt University (Curran *et al*. 1977, Barton & Borthwick 1982, Barton & Roy 1983). In this AWS, sensors for measuring wind speed and direction, temperature and humidity are housed together in an insulated cylinder (Fig. 18.4), heated at its opening point and internally, and opened by an electric motor every 30 min for 3.5 min. This gives sufficient time to measure the mean wind speed and direction and to detect peak gusts as well as allowing the temperature sensor time to attain equilibrium with the atmosphere before being read, yet it is a sufficiently brief period to avoid ice and snow accretion. Above freezing point, the heating is disabled. To keep all four sensors ice-free requires an average power of only about 400 watts, a considerable reduction on earlier heated sensors exposed continuously and individually. But even 400 watts is beyond the resources of most remote sites. The Heriot Watt AWS is, however, quite large (1 m high by 0.75 m diameter) and power requirements could be reduced, perhaps considerably, if the design was miniaturised.

Indirect methods of de-icing

Non-heating techniques for ice prevention have been used in various applications, such as helicopter blades, aircraft wings, radar domes and ship's superstructures, but few have been applied to meteorological sensors. They have the advantage of requiring much less power to operate them than those based on heating. Jellinek (1957) and Jellinek *et al.* (1978) observed that ice adheres less well to hydrophobic materials, such as plastics, than to metals, but the usefulness of this for ice prevention is, in practice, minimal since although the strength of adhesion is certainly about ten times less, it is still very large (about 2 kg cm^{-2}) (Sayward 1979). However, while ice adheres well to all rigid materials, irrespective of composition, in most cases it sheds easily from all that are flexed – perhaps just slightly more easily from hydrophobic materials. But if the ice has formed as numerous, unconnected small individual crystals, bending the surface may not remove them; it must have formed as a sheet. These facts were used by the Institute of Hydrology in the design of a cold regions AWS (CRAWS; Strangeways 1985a, b).

A CRAWS consists of an aluminium plate 45 cm square, sandwiched between 1-cm-thick foam rubber sheet, the whole being sealed in a polythene envelope (Fig. 18.5). Cavities in the foam-rubber house sensors measure temperature (by PRT), humidity (by ion-exchange sensor), solar radiation and reflected solar radiation (by light-sensitive diode). Each cavity is protected by a material to suit its function, (reflective) mylar for temperature (water-vapour-permeable) PTFE for humidity (as used also on Met. Office buoys – Chapter 17) and (diffusive) white polythene for radiation, all plastic and all flexible. On top of the plate, a drag-force anemometer (Chapter 5), encased in a flexible silicone-rubber cover, senses wind speed and direction. The whole structure is deployed on a pneumatic cylinder containing a piston and protected by rubber bellows. Air from an aqualung, controlled by a timer, periodically fires the piston, causing a sudden short vertical movement and thereby inducing a mechanical shock; the vented air is then directed so as to flex all the covers, the combined effect being to remove any ice that has formed or snow that has accumulated or adhered.

Nevertheless, both the Heriot Watt and Institute of Hydrology designs remain specialised and expensive. For wider application, it is possible to visualise further developments of these ideas, perhaps miniaturised and perhaps combining heating and pneumatics. These are open fields for further research.

Observations of sea ice and icebergs

This chapter has so far viewed cold regions from the point of view of being a hazard to making observations and to operating instruments, but cold regions are

Figure 18.5. This pneumatically de-iced AWS was developed at the Institute of Hydrology and tested alongside the opening box station (Fig. 18.4), as well as in Antarctica for 12 months (on the right in Fig. 18.2). It measures wind speed and direction with a drag anemometer (the top cylinder, see also Fig. 5.10), and solar radiation (the sensor on top of the anemometer) and reflected solar radiation, temperature and humidity by means of sensors in the cavities on the underside of the plate.

also something that need to be observed and measured in their own right. This applies particularly to the extent of ice cover over the sea since it is a hazard to shipping and of climatological significance. As is shown in Chapter 19, ice cover is one of the variables ideally suited to observation from satellites because of its easily detected yes/no nature, but there is also a history of its observation and reporting from ships and these remain essential. Its importance to travellers is historically embedded in our memories in such events as the *Titanic* (Lawrence 2000) and Shackelton's ill fated but heroic journey to Antarctica in 1915 and final rescue in South Georgia (Alexander 1998).

This all became very clear to me as I travelled from the Falkland Islands to the Antarctic base at Faraday aboard the BAS ship RRS *Bransfield* on a mission to set up and test the cold regions weather station seen in Figs. 18.2 and 18.5. (Which also telemetered its measurements for 12 months via Meteosat to the UK from Antarctica, demonstrating that even at this relatively high latitude of 65°S, geostationary satellites can be reliably used.) This journey, through different kinds of ice, and the photographs I was able to take help in understanding and describing them.

Ice on the sea takes two forms: ice formed when the sea itself freezes, and the ice produced when parts of glaciers and ice shelves break off and float away. Being able to recognise the various types of ice is important in observing and reporting ice cover and so definitions and descriptions are the key to accurate observation.

Sea ice

The amount of sea ice varies considerably from year to year and as its presence affects the weather over a large area, knowledge of its distribution is important. Broad-scale observations have been greatly improved through satellite imagery but observation from shore stations, ships and aircraft are still of great importance for, as usual, ground truth is necessary to calibrate remotely sensed measurements. Nor can satellites observe and report the small-scale details that only on-the-spot observers can see. Even today, estimates of floating ice still rely completely on visual observation. The only instruments used are conventional radar and sideways-looking radar from aircraft (Fig. 19.5). But icebergs are relatively poor reflectors of radar and cannot always be seen in this way.

First-year ice

Sea ice develops slowly, the first stage being small *spicules* known as *frazil ice* that form in large amounts, making the sea look 'oily'. As cooling continues the spicules join up making the sea look 'greasy' and matt in appearance. When snow falls on this it turns into a slush and the wind and waves form what is termed *shuga* (Fig. 18.6). As cooling progresses, sheets of *rind* or *nilas ice* form, at rates depending on the temperature and water salinity, rind forming when lower salinity water freezes as a thin layer of easily broken ice nearly free of salt. Nilas forms when higher salinity water freezes when the freezing is rapid and is not disturbed by wind; this ice has a more flexible property. Nilas is divided according to its thickness, dark nilas being the thinner, while white nilas is the more advanced and thicker (10 cm). Wind and waves can then break up rind and nilas to produce *pancake ice*. As cooling continues further, pancake ice can thicken into grey or grey-white ice, the latter reaching up to 30 cm thickness. The wind can break this ice up into ice cakes and floes of a range of sizes. All of this ice is known generically as *young ice*. As the winter advances, the floes thicken, becoming known as *thin*

Figure 18.6. Shuga ice, with its greasy texture, forming in the Gerlache Straits, photographed from the BAS ship RRS *Bransfield* at the onset of winter on the return journey from Faraday to the Falkland Islands.

(30–70 cm), *medium* (70–120 cm) and *thick first-year ice* (up to 2 m) by the end of the winter.

Old ice

If thick first-year ice survives the summer, it is then called *old ice*, which is progressively named *second-year ice* and then *multi-year ice* depending on the number of seasons it has survived through. Old ice can range in thickness from 1 to 3 m or more and can be recognised by its bluish colour, in contrast to the green of first-year ice.

Snow falling on the ice in winter insulates it from the cold air above and this slows its growth. The depth of lying snow varies considerably depending on its geographical location, wind and topography, just as it does on the land. While the snow is present, about 90% of any solar radiation is reflected, preventing melting, but when the air temperature rises above freezing when the summer arrives, melting starts and puddles of fresh water form, and these absorb most of the solar radiation, further accelerating melting. Eventually the puddles reach through to the bottom of the ice floes forming *thaw holes*. If the ice is free to move into warmer water, melting is further accelerated, through wave action and because of the warmer water and air.

Figure 18.7. Pack ice between Anvers Island and the Peninsula mainland, seen from the deck of the *Bransfield.*

Sea ice movement

Sea ice is further subdivided into whether it can move under the influence of wind and currents – *drift ice*; or if it is attached to the coast – *fast ice*. If drift ice covers more than seven-tenths of the sea surface it is known as *pack ice* (Fig. 18.7). Wind stress causes the pack ice to move downwind, although the Coriolis force also causes it to veer to the right of the surface wind direction in the northern hemisphere, just as the surface wind is deflected. This is a complex matter that will not be pursued further here (but see Chapter 5 for references).

The rate of drift is affected not just by the wind strength but also by the amount of ice cover. Between 10% and 30% cover, there is enough open sea to allow the ice to move more freely than if more tightly packed. Very closely packed ice will move at only about 2% of the windspeed on average. If the ice has become deformed with ridges (see below) resulting in underwater unevenness, movement is reduced due to increased drag. Because the ice is floating, currents in the surface layers also act on it and cause it to move. When there is wind and current, the direction of movement is the vector sum of the two forces, although, offshore, the wind usually dominates. (In the case of Shackleton's journey to Antarctica in 1915, one of the problems was that a northeast wind forced the pack ice up against the ice shelves and the land for a long period, preventing it from moving, closing up any free sea between the floes. Eventually the jammed pack ice crushed their ship the *Endurance*.)

As the ice moves under the pressure of wind and currents, young ice floes may override each other, while in thicker ice the pressures lead to ridges and humps forming, which rise above the mean level. This causes large amounts of ice to be forced downwards to support the weight of the ridges. When this occurs the draught below the ridges can be up to five times their height and such ice can be a major hindrance to shipping and also slows its movement under the influence of wind stress.

Icebergs

Icebergs, or *bergs*, are floating ice originating from glaciers or ice shelves (Fig. 18.8). The depth of a berg under the water compared with its height above is a familiar point of discussion. The ratio depends on the shape and origin of the ice. If the top 10–20 m is of snow the ratio is about 1 part above with 5 below, but if it is all ice, the ratio is 1 to 8.

Icebergs reduce in size by melting, by wave erosion and by breaking up (Fig. 18.9), a process known as *calving*, whereupon the balance will be disturbed

Figure 18.8. An iceberg, probably about 30 m high, in the South Atlantic near Signy in the South Orkney Islands. The brilliant white flat top surface causes the mist or clouds above to be illuminated.

Figure 18.9. As the larger bergs break up they form smaller ones, as here near to Deception Island. It will have broken up many times and is showing the effects of sculpturing by the sea.

and the berg may capsize or lie at a different angle. Melting occurs mostly at the water line in cold water, but if the water is warm, it will take place mostly underwater, which makes it unstable and unsafe to approach as it may capsize at any time. When melting in this way, the break-up leads to many smaller pieces of ice, known as *bergy bits* (Fig. 18.10) and *growlers* and these are a hazard to ships. Unfortunately, bergs are poor reflectors of radar energy and so are not always detectable in this way. The fragments breaking off, particularly the growlers, are even more difficult to detect with radar not only because they are smaller, but also because their echoes are lost in the noise of radar scatter off the waves (which in other contexts can be useful – see Chapter 19). These (relatively) smaller pieces of ice can be a danger to shipping.

Changes in sea-ice cover with time

The latest (2001) IPCC Climate Change reports conclude that the systematic decrease in Spring and Summer Arctic sea-ice extent in recent decades is broadly consistent with increases of temperature over most of the adjacent land and ocean. A large reduction in the thickness of summer and early autumn sea-ice over the last 30–40 years is consistent with this decrease in spatial extent, but there is uncertainty as to what extent poor temperature sampling and multi-decadal variability are affecting the conclusions. However, the IPCC unequivocally states that the Antarctic sea-ice extent has shown no reduction over the last 25 years (Folland *et al.* 2002).

Figure 18.10. This *bergy bit* may look small, but its scale can be judged by the penguins standing on top (possibly too small to be seen in the printed picture). Even a small part of this, should it break off forming a growler, would be a danger to ships striking it at cruising speed.

Observing and reporting sea-ice

As in the case of clouds, visual observation is the principal method of identifying sea-ice. In both cases, much depends on the experience of the observer and their ability to differentiate between the many types of ice or a mix of several types. No instrument can as yet do this. A WMO handbook (1970) describes and illustrates sea ice for mariners.

There are four characteristics of sea ice that need to be reported: thickness (new first-year to old-ice, with observations on deformation), amount (in tenths of the sea covered by ice), movement (its speed and direction) and the number of ice-bergs, particularly important at night and in poor visibility. (The RRS *Bransfield*, for example, has a searchlight looking forward at night. Standing on deck watching the bergs appear out of the dark is a remarkable experience.) All of these characteristics can be reported in plain language or in special WMO codes, the simplest being the ICE group at the end of the SHIP code. The ICEAN code is for specialised use for the dissemination of sea ice analysis with a prediction for the coming hours. WMO have also produced symbols for use on maps that are internationally disseminated by radiofax. But as noted in Chapter 12, the old WMO codes such as SYNOP are being replaced by a new general code called the *buffer code*.

In making these observations, from ship or shore, two rules should be observed: a high vantage point should be used, such as the top of a lighthouse or the crow's nest of a ship, and only conditions up to a radius of half the distance to the horizon should be reported.

Ice accretion

Much has already been said earlier in this chapter about the nature and problems of ice forming on instruments. The build up of ice on ships can also represent a danger, particularly to smaller ships of less than 1000 gross tonnage. Even on large ships it can encase the radar or radio antennas, and visibility on the bridge can be affected. (Experience in the Scottish Highlands has shown me just how dangerous reduced visibility can be in a white-out if conditions also need goggles to be worn because of spindrift. If the goggles then ice up, you are virtually blind.) WMO report formats include provision to report ship icing, again in plain language or in code. My all too brief experience of the Antarctic was one of the most impressive of all my many travels to remote places. It is like nowhere else.

References

Alexander, C. (1998) *The Endurance*. Bloomsbury Publishing, London.

Alexeiev, Jo. K., Dalrymple, P. C. & Gerger, M. (1974) Instruments and observing problems in cold climates. *WMO Technical Note No. 135.*

Barton, J. S. & Borthwick, A. S. (1982) August weather on a mountain summit. *Weather*, **37**, 228–40.

Barton, J. S. & Roy, M. G. (1983) A monument to mountain meteorology. *New Scientist*, 16 June, 1983.

Begg, J. S. (1984) At Ben Nevis observatory. *J. Scot. Mountain. Club*, **33**, 43–51.

Curran, J. C., Peckham, G. E., Smith, S. D., Thom, A. S., McCulloch, J. S. G. & Strangeways, I. C. (1977) Cairn Gorm summit automatic weather station. *Weather*, **32**, 60–3.

Folland, C. K., Karl, T. R. & Salinger, M. J. (2002) Observed climate variability and change. *Weather*, **57**, 269–78.

Gerger, H. (1972) Report on the methods used to minimise, prevent and remove ice accretion on meteorological surface instruments. (WMO) *CIMO-VI*/ Doc. 35, Appendix A.

Hartley, G. E. W. (1970) A heated anemometer. *Met. Mag.*, **99**, 270–4.

Jellinek, H. H. G. (1957) Adhesive properties of ice. *US Army Snow, Ice and Permafrost Research Establishment*. Corps of Engineers, Research Report No. 36.

Jellinek, H. H. G., Kachi, H., Kittaka, S., Loe, M. & Yokota, R. (1978) Ice-releasing block-copolymer coatings. *Colloid Polym. Sci.*, **256**, 544–51.

Lawrence, E. N. (2000) The *Titanic* disaster – a meteorologist's perspective. *Weather*, **55**, 66–78.

McConnell, D. (1988a) The Ben Nevis observatory log-books, part 1. *Weather*, **43**, 356–62.

McConnell, D. (1988b) The Ben Nevis observatory log-books, part 2. *Weather*, **43**, 396–401.

Paton, J. (1954) Ben Nevis observatory. *Weather*, **9**, 291–308.

Roy, M. G. (1983) The Ben Nevis meteorological observatory 1883–1904, Part 1. Historical background, methods of observation and published data. *Met. Mag.*, **112**, 218–329.

Sayward, J. M. (1979) Seeking low ice adhesion. *Cold Regions Research and Engineering Laboratory, Special Report:* No. 79–11.

Schutze (1883) The arctic meteorological station on the Lena (River*). Nature*, **28** (59).

Strangeways, I. C. (1985a) The development of an automatic weather station for use in arctic conditions. PhD Thesis, Reading University, Department of Meteorology.

Strangeways, I. C. (1985b) A cold regions automatic weather station. *J. Hydrol.*, **79**, 323–32.

World Meteorological Organisation (1970) *Sea-ice Nomenclature*, Vol. I & III. WMO-No 259, Geneva.

19

Remote sensing

... it seemed unto me no other than a huge Mathematicall Globe, leasurely turned
before me, wherein successively, all the Countries of our earthly world within the
compasse of 24 howres were represented to my sight.
Francis Godwin – Bishop of Hereford 1617–1633. *The Man in the Moone, 1638*
(Domingo Gonsales, flying towards the moon on a framework drawn by swans,
looks back towards the Earth).

This is but a brief review of a complex subject that is constantly developing. It is
intended to give an overall impression rather than a detailed account. The aim is
to outline the sensors used on spacecraft, what wavelengths are being measured
and to investigate what variables can be sensed remotely and how this can be done.
Putting remote sensing (RS) into context with ground-based measurements, the
main topic of this book, is also an important consideration for this chapter. For those
wanting more detail on any one topic, the web pages of NASA and EUMETSAT
are invaluable. Start with the home pages (www.nasa.gov and www.eumetsat.de)
and select the appropriate subject.

Satellite orbits

These have already been touched on in Chapter 12 on Telemetry, but some additional
points are worth introducing here that are relevant to RS.

Atmospheric drag limits orbital height to 300 km or more. Satellites at altitudes
of 500–2000 km are usually operated in nearly circular polar orbits (giving orbital
times of 1–2 h), 880 km being typical for meteorological satellites. Because the
Earth has a slightly greater diameter at the equator than through the poles, a small
change occurs in the satellite's orbital period, causing precession of the orbital
plane. For a given height, an inclination from the purely north–south path can be

Measuring the Natural Environment, second edition, Ian Strangeways. Published by
Cambridge University Press. © Ian Strangeways 2003.

calculated that causes the plane to precess one rotation per year. This results in the plane staying fixed relative to the Sun (as the Earth orbits the Sun) and so this is termed a Sun-synchronous orbit, the satellite passing over the equator at the same time at each pass through the year. Most remote sensing satellites are in this type of orbit. The same mechanisms are also at work in a geostationary orbit, the satellite moving slightly, in a figure-of-eight, each day. So the orbit should strictly be called *geosynchronous* rather than geostationary (King-Hele 1994), although the latter will be used hereafter in accordance with tradition.

The electromagnetic spectrum

Figure 19.1 illustrates the electromagnetic spectrum, with the extreme ends (low radio frequencies and gamma rays) omitted to save space and also because these extremes of frequency are not used in RS from satellites as yet. The figure brings together information regarding the wavelength, frequency and photon energy, the names of the radiation, the spectrum of the Sun and the Earth's long-wave infrared spectrum (see also Fig. 2.1), along with an indication of the types of instrument used to measure the various bands. Approximate indications of the bands of radiation that are absorbed and transmitted by the Earth's atmosphere are also shown. These 'windows' are important, since RS of the surface is only possible from space in the bands that are not absorbed by the atmosphere, in particular the visible and the two infrared windows A and B, where transmittance is high, shown in Fig. 19.1. The bands that are absorbed, however, are also important because they also radiate, and this can be used to obtain profiles of temperature and other variables down through the depth of the atmosphere (see subsection 'Sounders' below).

The division of radiation into ultraviolet, infrared, microwave, etc. does not, of course, indicate a sudden change in the nature of the radiation at the point where the name changes. Except in the case of visible light, the division is partly for convenience, and various break-points will be found in different writings; the ones used here are typical and as good as any other. The division of radar into K, X, C, S, L and P bands, as shown in Fig. 19.1, is also just one of several possible divisions.

The most-used wavebands

The largest concentration of measurements from satellites is in the visible region (0.4–0.75 μm) with only slightly fewer being made in the near infrared (0.75–1.1 μm). The next most common band is in the thermal infrared window marked B in the transmittance graph of Fig. 19.1 (10.5–12.5 μm). Fewer sensors operate in the window marked A (3.5–4.0 μm). In the region between 6 and 7 μm water vapour radiates strongly, and a few satellites (such as Meteosat) make measurements of humidity by sensing this frequency band. There is no atmospheric window in this band, since water vapour not only radiates at these frequencies but inevitably also strongly absorbs. Meteosat (which is typical of its type) measures in the three bands 0.45–1.0 μm (visible and near IR), 5.7–7.1 μm (water vapour IR) and 10.5–12.5 μm (thermal IR).

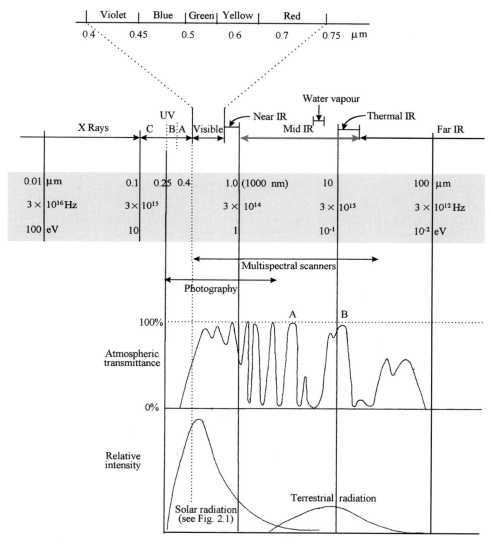

Figure 19.1. The electromagnetic (em) spectrum, from X-rays to HF radio, showing the subdivisions by name, frequency and wavelength, with approximate curves of atmospheric transmittance and the radiation spectra of the Sun and the Earth. $\lambda = c/f$, where λ is wavelength in m, f is frequency in Hz and c is the speed of light in a vacuum, 3×10^8 m s^{-1}. In each part of the em spectrum, the photon energy is given in electronvolts, eV; 1 eV is equal to 1.6×10^{-19} J.

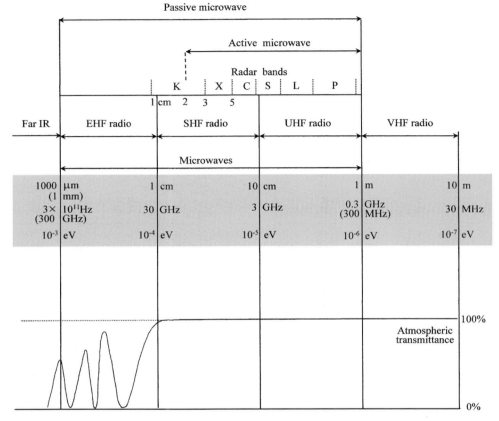

Figure 19.1. (*Cont.*)

In the microwave region, there is a concentration of satellites operating in the C band using Synthetic Aperture Radar (SAR) active sensing techniques (see later), with another group in the X and K bands acting as passive scanning radiometers. But there is continuous development and change with the launch of each new satellite.

Measuring the radiation

Sensors and electronics are subject to different problems in space to those on the surface, in particular to wider temperature swings which need to be controlled. Heat generated internally by the equipment and received as solar radiation has to be balanced by long-wave radiation outwards into space (since there is no cooling by an atmosphere). Control is achieved by varying the attitude of the spacecraft relative to the Sun and by shutters that vary the area radiating the IR radiation.

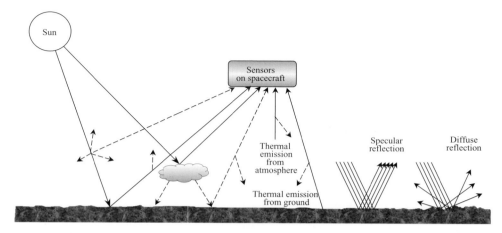

Figure 19.2. Because of the presence of the atmosphere, solar radiation may have followed a complex path of multiple reflections, scattering, diffusion and absorption before arriving at a sensor, such as a radiometer on a satellite. The solid lines show the direct and reflected solar radiation; the dotted lines show diffuse radiation from a cloud and radiation scattered by the atmosphere. Natural thermal emission originating from the warmed ground and atmosphere is also indicated. The general absorption of solar radiation as it passes through the atmosphere is shown by the large arrow on the left. How radiation is reflected at the ground depends on the nature of the surface and can be *specular* when the surface is smooth or *diffuse* when it is rough, the processes varying with wavelength and angle of incidence. At water surfaces, ultraviolet is reflected and near-infrared absorbed, while the blue–green spectrum achieves maximum penetration, although water waves complicate the situation. All these factors make remote sensing a difficult and complex task.

Most RS sensors measure the natural radiation from the Sun reflected from the Earth and its atmosphere and also the long-wave infrared and microwaves originating directly from the Earth and the atmosphere by virtue of their constituting a 'black body' at a temperature above absolute zero. Systems measuring radiation from these sources are referred to as passive. Figure 19.2 illustrates the various routes taken by, and sources of, the radiation. It is also possible for a satellite to have its own source of radiation, in particular microwaves generated by a radar transmitter or infrared generated by laser. Sensors on board the platform then detected the reflected radiation. These methods are called *active* (Blyth 1981).

Film cameras

For RS in daylight from manned platforms (including spacecraft) in the visible and near-infrared bands, direct photography on film offers many advantages by providing high-image resolution (10 million bits on a 70 by 70 mm film) with modern optics and photographic emulsions, at relatively low cost.

The spectrum from 0.3 to 0.9 µm can be recorded by film, the shorter-wavelength threshold being limited by the transmittance of glass lenses and the longer-wavelength threshold by the sensitivity of the film emulsion. Quartz lenses can extend the spectral range, and mirrors can be used in place of lenses to avoid the losses that occur through glass.

Filters can be used to separate the image into several discrete bands; typically four cameras are operated together, fixed to a frame to ensure good image alignment, each sensitive to a specific band. The usefulness of splitting images into several wavebands is that it is often possible to identify materials (water, vegetation and soil for example), or their condition, by their different spectral reflectance. Film cameras are widely used from aircraft and from manned spacecraft.

Vidicon tubes

Vidicon cameras are well adapted to automatic, remote use, having the capability to produce electrical signals that can be digitised, stored and telemetered back to ground. However, while these were used widely in the 1970s and 1980s, they have been largely superseded by scanning radiometers and so will not be described here. Charge coupled devices (CCDs) are also now finding a place as image sensors, for example in the detection of lightning (Chapter 15).

Radiometers

Visible wavelengths

Visible-light radiometers are fairly simple and measure the radiation within a fixed angle of view, either over a wide angle, averaging the whole scene (as in camera exposure meters), or in a highly focused way, using lenses for spot measurements. The light is directed onto a sensor such as a photoresistive (photoconductive) cell of cadmium sulphide (sensitivity 0.5–0.9 km) or a silicon photodiode of the type used to measure solar radiation (Chapter 2). Semiconductor radiation sensors respond to individual incoming photons and are thus known as quantum sensors, responding rapidly to variations in radiation level – although they do not have a response to very high frequencies because of the relatively slow decay of the charge carriers.

Infrared wavelengths

There is a range of pure and doped semiconductor photoresistors covering the spectrum from 0.5 to 20 µm. For example, in place of cadmium sulphide a sensor of indium antimonide can be used to detect the shorter IR wavelengths while mercury cadmium telluride (kept at a low temperature) can be used for the longer-IR wavelengths in an infrared radiometer. In some radiometers, a rotating shutter passes the incoming radiation directly onto the sensor, alternating it with radiation reflected from an internal calibration reference cavity at a controlled temperature (see 'scanning radiometers' below). The difference in reading between the two sources is proportional to the temperature of the incoming radiation; the internal reference also counteracts any drift in sensor and electronic characteristics.

Microwaves

Although visible and shorter IR radiation can be collected by optical methods, in the microwave bands an antenna is used, focusing the radiation onto a horn which then passes it to a diode detector. Very low levels of natural radiation emanate from the ground in the microwave region and, by using a parabolic dish, it is possible to boost the signal by increasing the collecting area. The size of the dish also influences the spatial resolution obtained, this being dependent on the distance from the target and the beam width. The beam width is proportional to the wavelength and inversely proportional to the diameter of the dish. In practical terms this means that if the radiometer is on board a satellite at a height of 500 km and the dish is 1 m in diameter, then at a wavelength of 3 cm the resolution will be a spot 18 km in diameter (Blyth 1981). While this might be useful for obtaining average measurements over large areas, it is clearly no good for finer observations. (If flying in an aircraft at 5 km altitude, the resolution of the same dish is about 180 m.) Radars with one dish working in this way are known as real-aperture systems, to differentiate them from synthetic-aperture radars, which have better resolutions; see 'Active microwave sensing' below.

A microwave radiometer measures what is called the *brightness temperature* (see later), which is a function involving both the temperature of the object and its emissivity, the latter being largely dependent on its dielectric constant (see the subsection on the capacitance probe in Chapter 9); surface texture also affects the emissivity (Fig. 19.2). There is, therefore, no direct and simple equivalence between the measured radiation and its temperature. More on this later, in 'Brightness temperature'.

Spectroscopy

By inserting a filter into the radiation path of a visible or IR radiometer, the beam can be split into discrete spectral bands and the intensity of each measured. By using a variable interference filter it is possible to divide up a wide spectrum into very narrow (5-μm wide) bands, and if a reference source of radiation is also included in the instrument, precise calibration of the intensity of each band is possible.

But radiometers only look at one point at a time, or encompass very wide angles of view, and so are limited in what they can achieve. If they can be made to scan a complete scene, they become much more useful.

Scanning radiometers

To combine the potentially high resolution of a visible or IR radiometer with wide areal coverage, it is necessary for the radiometer to scan the view, both vertically and horizontally, as in a TV picture.

Scanning radiometers in polar orbits

In a geostationary satellite such as Meteosat, both the X and Y axes have to be scanned, but if the satellite is moving relative to the ground, such as one in a polar orbit, the forward (along-track) motion of the satellite provides the X scan, only the Y axis having to be scanned by the instrument. If the along-track motion of the satellite is used, the frequency of the across-track scan and the width of the scan strip can be chosen so as to take account of the altitude and speed of the platform, to ensure that adjacent individual strips neither overlap each other nor have gaps between them.

The most usual way of scanning the scene is to rotate a mirror at $45°$ to the horizontal at several hundred revolutions per minute by means of a motor, the radiation being reflected into the system's optics, typically a small Cassegrain telescope (Fig. 19.3). When the mirror is pointing downwards it scans the Earth; as it looks up it reflects radiation from an on-board calibration source, in a similar way to that in a fixed radiometer. It can also look at the low temperature of deep space

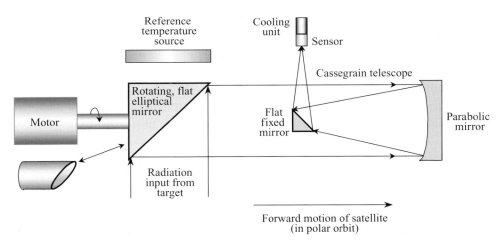

Figure 19.3. A scanning radiometer aboard a satellite in a polar orbit uses the forward motion of the satellite to generate the X scan, a rotating mirror generating the Y scan. The collected light is directed to a small Cassegrain telescope, which focuses the radiation onto the sensor. When facing upwards, the rotating mirror receives radiation from a body of known temperature for calibrating the system. This body can be deep space.

as an additional reference at a few degrees above absolute zero. In some designs the mirror vibrates rather than rotates, but rotation is more usual.

The need to use a large dish in the microwave region to get sufficient resolution was alluded to above, and this problem is exaggerated further if the dish has to move, as in a scanning system. From an aircraft the problems are less, in part because the resolution is better due to the low altitude, and in part because planes are larger than RS satellites. On aircraft it is also possible to use three dishes rotating on a common axis, switching so as to receive from just the forward-looking dish – it is usual to look slightly forward, rather than straight down, in the case of microwaves as this helps to identify the nature of the surface better, through polarisation effects. It is also possible to use two dishes spaced, say, n metres apart and rotated in synchronism, which, with suitable computer processing, can make the combination look like a dish n metres in diameter, thereby greatly improving resolution. However, on small spacecraft this arrangement is too large to be practical and special aerials known as phased array antennas, which can reproduce the properties of large antennas, have been developed, allowing higher resolution. They have no moving parts, and this allows fast scanning and alterations to the shape and width of the beam as well as control of the polarisation sensitivity, all these characteristics being controlled electronically. But they are expensive and (it is understood) none are yet in operation on spacecraft.

Scanning radiometers in geostationary orbits

Aboard a geostationary satellite such as Meteosat or GOES, where there is no motion of the satellite relative to the ground, radiometers have to scan in both the X and Y directions. To achieve this, the whole satellite rotates on its axis at 100 rpm, the spin creating the X scan, while a mirror steps down by a very small angle at each rotation, giving the Y scan. In this way, over 25 min, the satellite scans the entire visible globe, producing 2500 scan lines. (To compensate for rotation of the whole satellite, its many antennas have to be electronically switched, continually – so that radio communication is maintained with the Earth even though the satellite is rotating.)

Multispectral scanners

Multispectral scanners (MSSs) represent the next step-up in complexity, combining a scanning radiometer with spectroscopy. Instead of radiation from the Cassegrain telescope being directed to a single radiation sensor, it is passed through several dichroic mirrors or through a mechanically switched filter. The radiation from each is focused onto a separate sensor, sensitive to a particular part of the spectrum, possibly with an additional filter to modify the sensor's range of spectral response (Fig. 19.4).

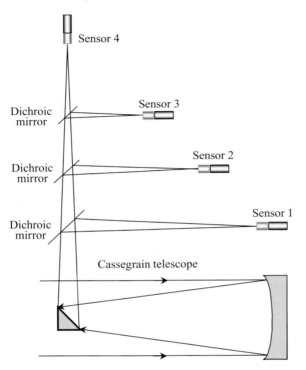

Figure 19.4. In principle, a multispectral scanner is a scanning radiometer in which the incoming radiation is split into several separate beams by dichroic mirrors, with a sensor associated with each that responds to a different wavelength, perhaps with its frequency discrimination further enhanced by filters.

The best known form of multispectral scanner in a polar orbit is the Advanced Very High Resolution Radiometer (AVHRR) operated by NOAA. The satellite flies at an altitude of 833 km giving a resolution of 1.1 km at the nadir point covering a swath of 2600 km width. Five channels are scanned, one in the visible (0.58–0.68 μm) range for daytime cloud mapping and one in the near IR (0.725–1.00 μm) range for surface water and ice, with three in the thermal IR (3.55–3.93, 10.30–11.30, 11.50–12.50 μm) used for sea surface temperature measurements and night time cloud mapping.

Typical multispectral scanners in geostationary orbit are those aboard Meteosat and GOES, the latter being familiar as the Visible and Infrared Spin-Scan Radiometer (VISSR). The Meteosat (second generation) Spinning Enhanced Visible and Infrared Imager (SEVIRI) will be operational in 2003, until which time the earlier Meteosat satellites will continue to be used. At present there is a Meteosat Transition Programme in operation to handle the changeover to the improved new satellites. The new satellites will make observations in 12 channels from the visible (one visible channel having a resolution of 1 km – very good from a geostationary orbit – the remainder are 3 km) through the near IR to the far IR at 12.5 μm. The various spectral bands will be used for cloud detection, precipitation monitoring, winds from cloud motion, cloud height detection, surface temperature, winds from water vapour channels, convection and the detection of various hazards such as fog, thunder and intense depressions.

Sounders

Another class of instrument aboard satellites, in both types of orbit, are known as *sounders* or *atmospheric sounders*. Their job is to obtain measurements of air temperature and water vapour content at different heights down through the depth of the atmosphere from the surface to 40 km altitude, by making observations in both the IR and microwave bands at different wavelengths. The principle is similar to that already discussed in Chapter 16, 'The upper atmosphere', in which microwave radiometers on the ground look upwards to achieve the same ends. The technique involves a complex mathematical inversion process which there is no need to describe here.

Typical of sounders in polar orbits is one operated alongside the AVHRR, known as the High Resolution IR Sounder (HIRS), in which a single Cassegrain telescope with an elliptical mirror providing across-track scanning is combined with a rotating wheel of 20 filters. The mirror steps, holds position while each filter is moved into the optical path (taking 100 ms) and a reading is taken. The field of view is relatively wide at 1.4° which, from an altitude of 833 km, gives a ground resolution at the nadir point of 20 km. The same IR sensors are used as in the higher definition scanning and fixed radiometers. Calibration is provided by two views, one of a warm target of known temperature within the craft and one of space. The bands (central wavelengths) measured are: 0.70 μm for cloud detection, 3.70 and 4.00 μm for surface temperature, 4.25, 4.40, 4.46, 4.52, 4.57, 13.40, 13.70, 14.00, 14.20, 14.50, 14.70 and 15.00 μm, all for temperature soundings at various heights. In addition water vapour is measured at three heights at wavelengths of 6.70, 7.30 and 8.30 μm.

A Microwave Sounding Unit (MSU) is operated alongside the HIRS, with still lower resolution (50 km at nadir or sub-satellite point). It consists of two separate units measuring a total of 15 frequencies, giving profiles of temperature from the ground to 45 km (3 hPa), the frequencies ranging from 23.8 to 89 GHz, with most between 50 and 60 GHz.

There are also similar IR multi-frequency sounders aboard geostationary satellites, for example alongside the VISSR imager on GOES-7, measuring in the 3.95–14.74 μm band. Two bands are in the atmospheric windows (A and B in Fig. 19.1) allowing the surface to be measured, seven observe in the CO_2 absorption/radiation bands and three in the H_2O bands, giving observations of radiation from varying altitudes and thus estimates of temperature up to the stratosphere.

There are variations in the exact details of wavelength, number of channels and the resolutions achieved in all of the above imagers and sounders, depending on the literature consulted. This no doubt reflects the rapidly changing situation. However, the important point to get across here is that both polar- and equatorial-orbit satellites produce two types of image. Firstly high definition images with resolutions of about 1 km (at the sub-satellite point) in both the visible and IR bands, allowing specific tasks to be accomplished, such as cloud detection and temperature measurement. Secondly there are the sounders that measure similar IR, as well as microwave, frequencies, but with much less resolution (25–50 km), these being used to obtain profile measurements of temperature and water vapour down through the depth of the atmosphere. The exact details of what is in orbit at any one time are less important than these general principles, which do not change. For current information, consult the web sites.

Calibration of radiometers

Visible channels

Radiometers measuring visible wavelengths, such as the AVHRR, are calibrated before launch. This has the disadvantage that any drift occurring over the years of the satellite's lifetime cannot be corrected or even measured. It is probable that future systems will include on-board calibration.

Infrared channels

Radiometers working in this part of the spectrum are calibrated continuously on board the satellite. Two references are scanned at the end of each line, one being a view of space acting as a black body at 3 K, or virtually absolute zero, the second is a warm target aboard the craft, maintained at a known temperature. A linear relationship between the radiance and the sensor output is derived from these two references, with just a small correction required for non-linearity.

Sounders (IR and microwave)

In the (HIRS) sounders, calibration is done every 40 lines, occupying three lines, which are lost from the Earth scan. In the MSU sounder, calibration is at the end of each line as with the AVHRR imager, the same two references also being used in each case.

Digitisation

The analogue signals from the radiometers are converted to digital form for teleme-tering to ground. In the case of the AVHRR the signal is resolved into 10 bits, giving 1024 steps (see Chapter 11), while the IR channels of Meteosat are digitised to 8-bit resolution.

Active microwave sensing

All the above instruments measure radiation that is of natural origin. It can be very useful to emit radiation actively from the platform and measure the reflected signal. This is most commonly done in the microwave bands using a radar transmitter, although lidar is becoming available (as from the ground – see Chapter 16). These techniques can be used to measure altitude, particularly useful for sea surface height (see Chapter 17) and to obtain information of sea surface state and thus of winds (see 'Scatterometer sea surface winds' later). More recently radar has also been used to measure precipitation from space (see later in this chapter). The frequency used is usually fixed on any one platform, but can be anywhere in the active microwave range, as shown in Fig. 19.1. There are advantages, however, in using multiple radar frequencies – for the same reasons that it is useful to employ a number of visible or IR bands.

In the most commonly used radar technique, a radar pulse is emitted sideways and downwards from the platform, at right angles to the direction of travel, at a pulse repetition rate of from 1000 to 10 000 pulses per second. A fan-shaped volume of radiation is generated as the pulse travels outwards from the antenna (Fig. 19.5), the pulse arriving at the ground at increasingly greater distances and at increasingly later times, as it propagates outwards at the speed of light. At the moment shown, the region of ground returning the signal is the shaded area; the receiver on the platform thus detects returned radiation from each such area sequentially, using the same antenna as that which transmitted the pulse, switched to the receiving mode. The amplitude, phase, polarity and scatter of the returned signals all give information on the topography, texture and nature of the surface. The *angular spread* of the transmitter, β, and the *depression angle*, θ, determine the area of ground illuminated. The width of the ground g across the swath area depends on the width of the pulse, τ, and on θ, which is usually set at between $30°$ and $60°$ to obtain optimum resolution.

But the resolution of the ground area covered in the direction of flight, W, is limited by the antenna beam width, which is a function of antenna size and the frequency used. In most real-aperture systems (see earlier in the chapter, in the sub-section 'Scanning radiometers'), the across-track resolution g is ten times

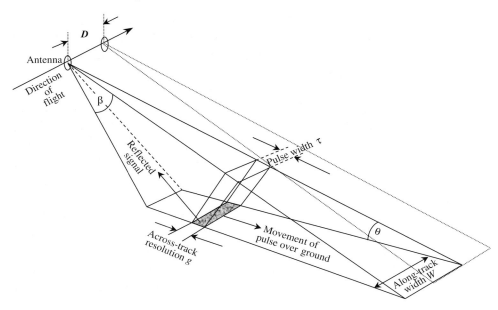

Figure 19.5. In a common form of active remote sensing, a radar pulse is emitted sideways and downwards, the reflected signal being returned from the shaded area at increasingly distant points as the pulse travels outwards. To make the along-track resolution width, W, as good as that across the track, g, a technique known as *synthetic aperture radar* is used. By timing pulse emissions (relative to the speed of travel), the scanned areas can be made to partly overlap. Then, by receiving echoes from the same area but originating from two positions, the effective diameter of the radar antenna can be artificially increased (through complex processing) to D, thereby effectively narrowing the beam width and so increasing the resolution.

better than W. As already discussed (in the sub-section 'Microwaves') it is possible to simulate a larger antenna by using two dishes spaced some distance apart. In the case of active radar instruments, this 'second' antenna is the same antenna, having moved further down the flight-path (Fig. 19.5) separated by time. In this way, the same effect is produced as having two separate antennas spaced by the distance D. This is referred to as *synthetic aperture radar* (SAR). How the information is merged to produce this artificially narrower beam is mathematically complex, but the end product is a resolution in the order of 25 to 50 m from a polar orbit.

Image reception and dissemination

For retrospective use, images may be obtained from the satellite operators, either as printed images or as digital data. For use in real time, however, the images must be received directly from the satellites. Something has already been said in Chapter 12 about the latter, with regard to Meteosat, and how image reception is associated with the reception of data telemetered from DCPs. PDUSs and SDUSs (primary and secondary data user stations) receive image data directly from Meteosat, the former as high-quality digital data, the latter as lower-quality analogue Wefax data (known as APT, automatic picture transmission – from satellites in polar orbits – soon to be changed in new satellites to a digital format). Figure 12.16 shows a small dish for the reception of both DCP data and Wefax images from Meteosat. Meteosat images are divided into nine square sections showing different parts of the Earth's disc as well as the whole disc. In addition there are more frequent repeat-images of North Africa and Europe. The other geostationary satellites produce similar images of their area (Chapter 12).

The dissemination of satellite images is explained fully in several EUMETSAT publications (see the reference list for Chapter 12), and how images can be received by the amateur directly from the various satellites is well documented by Harris (1996). Many of these images are now available on the appropriate web pages, such as those for EUMETSAT and the UK Met. Office, although not necessarily in real time.

Interpreting satellite images and soundings

Having received the digitised data from the satellites, what use can be made of them; how can geophysical quantities be retrieved from measurements of radiance? Satellites measure radiance in a number of spectral bands and these measurements are converted either to brightness or reflectance in the visible spectrum or to *brightness temperature* in the IR wavelengths (see below). But there are limits to the accuracy of what can be extracted from these measurements, in part because many factors affect the manner in which the radiation is emitted by or reflected from the objects of interest, or how the radiation is subsequently absorbed or scattered by the intervening atmosphere and its clouds and aerosols. The WMO (1996) list many such factors, a few being: the elevation of the Sun, the satellite–Sun angle, the satellite viewing angle, the transparency of objects through which the radiation must pass, the reflectivity of the underlying surface (which can vary depending on conditions), sun-shadowing by higher objects, the shape of the objects giving shades to the reflected light and variations of the emissivity of the surface and of atmosphere. We are familiar with all of these things in our everyday life for they also affect how things look to our eyes, but our brains make sense of them and allow for the angle of viewing and the brightness of the light. The task of the researcher is to find ways around all of these problems in the form of equations (Blyth 1981, Barrett & Curtis 1982, Rango *et al.* 1983, Farnsworth *et al.* 1984, Barrett & Herschy 1986, Schultz & Barrett 1989).

It is useful for our purposes to divide the techniques used to extract the required quantities from the radiation received into those that are qualitative, not requiring precise quantification as a number, which also includes those cases where the image can be used to give a simple yes/no indication (the edge of an ice field for example), and those that demand a precise figure, such as the temperature of the ground. This latter group extends into more difficult requirements, where not only is a precise value required, but where the remotely sensed measurement cannot give it directly. Examples of this are the air temperature a metre or so above the ground or the water temperature a metre or so below the sea surface.

Brightness temperature

Satellite images are composed of numerous individual pixels that, when displayed on a screen, give the impression to the eye of a recognisable continuous two-dimensional view of the Earth. To quantify the data, however, in order to derive numerical information, each pixel is treated as a point value (even if they do represent an area as large as 1 km^2).

The radiometers measure the radiance in specific bands of wavelength. In the visible spectrum, the radiance is converted to *brightness*, as recognised by the eye or to the reflectance of the surface. Readings in the IR or microwave bands are converted to temperature using the concept of *brightness temperature*. An ideal 'black body' absorbs all the radiation falling on it. Real objects, however, only absorb a fraction of the radiation, known as the absorptivity. The efficiency of radiation, the emissivity, of the same body is the same as its absorptivity at the same wavelength. *Radiance* is the power radiated per unit area per steradian per unit wavelength and is related to the temperature of the body radiating it, at a given wavelength, via a complex equation involving the speed of light and the Planck and Boltzmann constants, in what is known as the Planck function. If radiance in a narrow wavelength band is measured, the temperature of the black body that radiated it can be calculated using the Planck function. This is known as the *brightness temperature* and for most RS qualitative observations the radiance is converted to this quantity.

Algorithms

Some variables need not be expressed as a precise quantity, a yes/no being sufficient (cloud/no cloud, snow/cloud, sea/land or desert/forest, for example) but other variables require numerical values to be derived (temperature or rain amount, for example). Such variables are less amenable to remote sensing because the electromagnetic spectrum reflected from, or emitted by, the Earth's surface or the atmosphere and measured by the satellite instruments is rarely directly related to such variables.

To overcome this, intermediate models or *algorithms* have to be developed to interpret the remote measurements, and these algorithms can be quite complex and may not be very precise, or they may only be applicable to a particular location, or to a particular set of circumstances, or to one season. Another difficulty is the way in which the atmosphere, its clouds and particulate content affect the passage of the radiation through it. The largest effort being put into RS by individual research groups is concerned with developing algorithms. Algorithms are also required for the simpler yes/no situations, so that data extraction can be done automatically.

Clouds

The most obvious information from satellites are features such as clouds, snow and ice, geological formations and vegetation. The majority of RS applications are of this type, although our interest here is more inclined to the less obvious quantitative requirements, dealt with later.

Images of clouds are perhaps one of the greatest benefits and most obvious features of satellite RS, giving information never before available. In fact EUMETSAT state that the *main* mission of Meteosat is to generate 48 cloud images a day for use in short-period forecasting, the visible channel producing images in daytime and the IR during both day and night. The International Satellite Cloud Climatology Programme, part of a WMO research project, produces global maps of monthly averages of cloud amounts (see also Chapter 14, on clouds).

The methods used to extract cloud information automatically from the images are typified by those developed by the UK Met. Office, in which a series of tests are applied to the individual pixels. Any colder than a specified brightness temperature are deemed to contain cloud. A second test checks adjacent pixels, a high variance suggesting either a mix of clear and cloudy pixels or clouds at different levels. Small variations and low temperature indicate full cloud cover. By comparing the brightness temperatures at different wavelengths further deductions can be made. For example, by comparing the 3.7, 11 and 12 μm bands of the AVHRR, thin cirrus cloud can be differentiated from thick low cloud or fog. During daytime, reflected solar radiation (with allowance for the elevation of the Sun) can be used to separate bright cloud from dark ground. By comparing the visible with the near-IR images, clouds can be differentiated from water and both distinguished from the land and vegetation. There is nothing particularly clever or surprising about this. Our eyes are effectively CCD multispectral cameras and we differentiate between things to a great extent by colour and brightness. We can detect small and quite subtle differences of wavelength when comparing the colour of two similar paints, for example, and how a white wall will take on the colour of the radiation falling on it. All we are doing in RS is copying this process. In a similar way, we are intuitively familiar with atmospheric windows, for we are unable to see through fog while radar can. Viewed in this way, RS is not complicated, it is intuitive, although the maths may be quite complex. Our brains do the same maths without our being aware of it.

The height of clouds can be estimated in several ways, the simplest being to measure the temperature of their tops, from which height can be inferred. While this gives good results for stratus and large cumulus clouds, it is not particularly effective for semi-transparent cirrus or small cumulus clouds. By similar techniques of comparing radiances of the same pixel at different wavelengths, it is possible to distinguish between snow and cloud, between cumulonimbus, nimbostratus, altocumulus, cumulus and cirrus clouds over the land. Sunglint can also be detected and land differentiated from the sea. There is, however, some considerable overlap in many of these cases and thus some uncertainty. There is, on the other hand, clear differentiation between cloud and snow and cumulonimbus and nimbostratus are also well differentiated. The data obtained by these techniques may be used as input to numerical models, but much information can also be obtained simply by viewing them as images. In effect they are extending the capabilities of our eyes to see at different wavelengths.

Open water

Open water such as lakes can be easily detected using the near-IR bands which are strongly absorbed by water but highly reflected by plants. This difference can be detected in a yes/no way, allowing images to be analysed automatically. It is not so simple, however, in the case of rivers because their width is usually much less than that of a lake, and will normally be less than one pixel wide, even from a polar orbit. Here, indirect evidence is often sought instead, such as the tell-tale signs of changes in vegetation in the proximity of water (see below), but these can only be identified by eye, not automatically. (It might be wise of me to add 'at present' for developments are rapid.) The ability to detect the geographical extent of water also gives the ability to monitor flooding.

Snow and ice

The presence and extent of snow and ice is another of the few hydrological variables that can be measured from satellites operationally, because, like open water, snow and ice can be recognised on a yes/no basis. Several methods have now been developed of differentiating snow from cloud, such as that of Ebert (1987), which uses an algorithm based on albedo thresholds. Dozier (1987) describes a method using the 1.55 to 1.75 μm channel of Landsat, in which snow exhibits a much lower reflectance than cloud. But manual methods, looking at changes in clouds and the ability to identify ground features such as roads and forests, are also used.

However, at IR wavelengths, although snow and clouds can be differentiated from each other, the presence of cloud prevents detection of what the snow and ice cover is below the cloud. And snow-covered areas are rather prone to being cloudy. While microwave frequencies offer an all-weather technique for detecting sea ice, based on the high contrast between the unfrozen and frozen sea at these wavelengths, microwave radiometers have only a coarse resolution compared with IR.

Temperature of the surface

As an instance of lack of direct equivalence between the satellite radiation measurement and the variable, consider the same lake whose areal limits were simple to define. Radiation measured in the thermal infrared gives an indication of the temperature of the lake, but only of the water surface – the top millimetre or so, literally the skin – with no direct way of measuring the temperature even just a few millimetres below the surface. The situation is further complicated by the state of the surface, whether it is placid or wave-covered. The same problem exists in making sea surface temperature measurements. Just the skin temperature is sensed and, despite the name, this is not exactly what is required (Harris 1998). The in situ sea surface temperature, measured by a ship or a buoy, is that of the water about 1 m below the surface – the bulk temperature (Chapter 17). The same is true of soil temperature; it is the temperature of the very top skin of the soil (or plants) that is sensed by satellite, the temperature just beneath the surface of the ground is unknowable directly from RS measurements. Equally, there is no direct way of knowing what the temperature of the air a metre above the ground is (as determined by an in situ thermometer in a Stevenson screen, for example), because the ground temperature is rarely the same as that of the air above it and can differ quite considerably at times. These other temperatures have to be inferred indirectly through a knowledge or estimate of other variables, processes or conditions, and the result is inevitably not as accurate as a primary measurement of the temperature at the surface.

The surface temperature measurements themselves are also likely to be in error owing to the passage of the radiation through the atmosphere – even when the measurement is made through one of the two main mid-IR atmospheric windows (A and B, Fig. 19.1). Methods have been developed to compensate for atmospheric effects (see below), but they can never be completely successful.

Sea surface temperature is of particular importance because it acts as a proxy for land-air temperature when calculating global mean temperatures. Even though only the skin temperature is measured, satellite data do compare reasonably well with data collected in situ and have the great advantage of global cover on a frequent basis.

Measurements of SST are usually made using IR scanning radiometers (such as the AVHRR), the first vital step being to remove any pixels that contain clouds. For this purpose special algorithms have been developed specifically for SST data retrieval (Saunders & Kriebel 1988). Measurements are made in the 10.5–12.5 μm window, but it is still necessary to correct the brightness temperature for the attenuating effects of water vapour which can introduce errors up to 10 K. This is tackled in two ways.

The *differing path length (multilook)* approach makes measurements of the same sea area at different angles. Because attenuation is proportional to path length, the two measurements can correct for the attenuation. The European Space Agency satellite ERS-1 carries an IR radiometer known as the *along track scanning radiometer* (ATSR) which has a dual angle view of the sea, allowing SST measurements to be corrected for atmospheric interference.

In the *split window* method, corrections for atmospheric interference are made by measuring the SST at different wavelengths, taking advantage of the fact that water vapour absorption is very wavelength dependent. The difference in attenuation between a pair of wavelengths is proportional to the difference between a second pair. By looking at two pairs of frequencies in the 10–12 μm band, with one frequency being common to both pairs, the attenuation can be calculated and corrected for. Algorithms using this principle have been developed for both day- and night-time measurements (McClain *et al.* 1985). Different day/night algorithms are necessary because the errors in SST measurement from satellite include factors such as heating of the top layers by sunlight, this being amongst the more problematic. Other factors, however, include evaporation, rainfall, currents, upwelling and downwelling and the mixing of the top layers by wind, all affecting the skin temperature. Night-time readings are often preferred because the problem of solar heating is eliminated, although it does not help with the other problems. Rao (1990) has shown that, in a comparison with ground-truth measurements, the errors of satellite SST measurements are in the order of half a degree in the IR, while in the microwave errors vary from a quarter to three-quarters of a degree. Problems in merging datasets from different sources, such as surface observations and those from satellites, can be minimised by tuning the algorithms so that they produce results consistent with a reference such as a buoy observation. But some error will undoubtedly remain. At microwave frequencies the attenuating effects of clouds are much smaller, but the resolution of the radiometers is in the tens of kilometres compared with the 1–5 km in the IR bands.

Temperature and humidity profiles through the atmosphere

Sounders such as the HIRS and MSU receive an intricate mix of IR and microwave radiation emerging from the top of the atmosphere. Analysis of the sounder observations is done using a complex equation known as the *radiative transfer equation* (RTE), an expression involving the Planck function (see 'Brightness temperature', above). The equation is in two parts, the first representing the radiation originating from the surface; the second, radiation from the atmosphere. The latter incorporates a *weighting function*, which gives profile information, the function peaking at different altitudes allowing temperature and humidity soundings to be made by selecting wavelengths that correspond to radiation originating (largely) from specific levels. The solution of the RTE is complex, in part because the weighting functions overlap, making it difficult to sense the radiation from a precisely specified level. Comparisons with ground-truth from radiosondes have helped check the accuracy of satellite retrievals, but sondes have their own errors and there are errors caused by differences in both the timing of the measurements and differences in position.

If a profile of the temperature through the depth of the atmosphere is required, one technique uses a microwave sounder to measure the temperature-dependent thermal emissions of oxygen. The atmospheric concentration of oxygen is relatively constant in space and time and, since it both absorbs and emits at frequencies of around 60 GHz, it provides a stable temperature tracer. Measurements made at 53.74 GHz give an indication of temperature in the middle troposphere and those at 67.95 GHz, the lower stratosphere. These can be combined to give an indication of the average temperature of the lower atmosphere, weighted mostly in the region of 1.5–5 km altitude. These measurements have been used in the debate about global warming (Jones & Wigley 1990, Spencer & Christy 1990), but I do wonder if they are accurate enough for this.

Various retrieval methods have been developed to measure humidity profiles using the RTE (Smith 1985), a common approach being to obtain two temperature profiles by measuring radiances in the bands at which carbon dioxide and water vapour emit, with an assumed vertical distribution of water vapour. Since the concentration of CO_2 is relatively constant, while that of water vapour varies considerably, the difference between the temperature profiles is due to the difference between the assumed and the actual humidity profile, allowing the latter to be deduced.

Precipitation

No method of remote sensing of rainfall from satellites measures the individual drops, simply the bulk properties of the rain system. Techniques fall into three categories: measurement of the rain as it falls, estimation of how much rain may be falling and observation of the resultant ground condition after rain has fallen.

Thermal-infrared and visible wavelengths

Many indirect methods of estimating how much rain is likely to be falling have been developed over the last 20 years. These involve looking at the clouds themselves from above, rather than at the rain falling from beneath them. The techniques use the satellite measurements of cloud-top brightness and/or temperature and/or texture, with use being made of the visible, infrared or both bands. These are known generically as *cloud indexing* methods and may be coupled with other 'predictors' such as ground-truth raingauge surface measurements.

There are many such techniques, a well-known one being the operational method developed by Reading University in Niger using Meteosat images, which estimates 10-day and 30-day rainfall totals (Dugdale & Milford 1986). The method relates the duration and extent of cold cloud-tops, as measured by the satellite's thermal infrared channel, to the precipitation, temperatures around −40 °C being typical. It is applicable only to convective cloud systems and is most useful for the production of longer-term rainfall summaries, rather than brief events. However, a large part of the rainfall over Central Africa originates from such deep convective clouds. Figure 19.6 shows a section of a whole-disc Meteosat thermal infrared image (10.5–12.5 μm), in which such cloud formations can be seen over Africa.

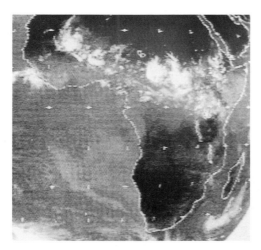

Figure 19.6. Cloud-top temperatures can be used to assess the possibility of and probable intensity of rainfall below convective clouds of the type that occur in many equatorial regions, as seen here in a Wefax image of central Africa from Meteosat.

However, the more detailed images produced by Meteosat are, in practice, used in preparing rainfall estimates by this method. Correlation is only moderate for individual pixels but if several adjoining areas are included, results are improved (Ebert & Manton 1998). The threshold temperature selected to indicate the probability of rain varies with latitude, being 15° lower in mid-latitudes (Arkin & Meisner 1987). By looking also at the visible wavelengths, the cloud albedo can be estimated and this helps identify the cloud type and thus aids in inferring the likely rainfall rate. These same techniques can be used with polar orbiting satellites which give better spatial resolution, with cover up to the poles, but with much less frequent cover in time; a snapshot of possible rainfall only every three hours is of limited value.

A related approach, known as the *life-history* method, uses a series of consecutive images from a geostationary satellite to estimate the stage of development of the cloud. Clouds with the same appearance may produce differing amounts of rain depending on whether they are growing or decaying, this being measured in terms of the rate of change of the cloud area. However, there has been evidence that this procedure may not improve prediction on all occasions. Nevertheless, developments continue (see 'Combination of techniques', later).

Passive microwave

Microwave frequencies are radiated naturally from the ground over a wide range of frequencies. Those in the 50–100 GHz band are scattered by the precipitation, especially by ice crystals (which tend to be more common over the land than the sea), resulting in a reduction of the measured 'brightness temperature' of the surface of the Earth. If the unimpeded level of radiation is also known (no cloud) the difference can be used to infer the rainfall intensity through coefficients determined empirically. The coefficients are different over land and sea and they also vary with season (Wilheit *et al.* 1991).

The lower frequency microwaves in the 10–30 GHz band, although radiated by the surface, also originate from the raindrops themselves, in fact they are much more efficient emitters than the surface of the sea, which has a low emissivity, producing radiation polarised in the vertical and horizontal directions equally. Emissions increase as the density of raindrops increases, but scattering by the drops higher up becomes influential so that eventually most of the radiation originates from the higher (and so cooler) parts of the rain column. This difficulty can be minimised by measuring at several different frequencies (Ferraro & Marks 1994). Because the ground emits microwaves much more strongly than the sea surface, the method is more useful for rain measurement over the oceans.

Active microwave

Weather radar, which is Earth-based RS, has already been described (Chapter 8). It is complex and expensive, even on the ground, but nevertheless the first radar system designed to measure the reflectivity of precipitation from space is now operational on the TRMM (Tropical Rainfall Measuring Mission) satellite, although only, of course, in the tropical regions. It was launched in 1997.

One radar method measures the time delay and strength of the reflected signal, giving the intensity of the rain (just as in ground-based weather radar). A rotating dish cannot be used, however (on a satellite), but rotation can be simulated electronically by phase adjustments (in a way similar to that used in sodar sounders; see Chapter 16). By this means, a path 220 km wide is scanned to a resolution of about 4 km (Kummerow *et al.* 1998), giving a vertical resolution of rain distribution to about 250 m. As with ground based weather radar the strongest reflection comes from the 'bright band' (of partly melted snow near the base of the clouds) and allowance has to be made for this.

A different radar technique measures the reflected signal from the ground, after it has passed through the rain during its downward journey and back through it on return, the level of attenuation indicating the quantity of rain in the path. This requires a measure of the reflectance of the surface, be it land or sea. This is done either by sensing the strength of the reflected signal from rain-free adjacent pixels or by using a second radar frequency (imaginatively known as the *dual-frequency surface reference technique* or DSRT) that is little affected by rain. Reflectance from the sea depends on the surface conditions which depend mostly on the wind (Chapter 17). Radar is also used to measure sea altitude (TOPEX, Chapter 17) and the effects of rain on this need to be known (Quartly *et al.* 1996). These techniques, as imperfect as they are, are particularly useful for measuring rainfall over the sea, where land-based instruments and techniques are impracticable and data are consequently scarce.

Comparisons

If the IR, visible, and active and passive microwave techniques are compared, differences of up to 30% are found (Kummerow *et al.* 2000), varying both with region and between convective and stratus rain types. Snow and hail are not measured as accurately as rain by some of the methods. The reasons for the various differences are being investigated. But even if the measurements are made with precision from a satellite in polar orbit, it is not possible to detect all rain events since rainfall varies rapidly with time and space and may well be missed completely or sampled just once every three hours. This will result in the system failing to see many of the small to moderate rain events although it should certainly catch the larger and longer-duration ones. So there is some way to go yet. Quartly *et al.* (2003) review the various methods.

A new project, known as Global Precipitation Measurement (GPM), funded by NASA, is developing both the passive and active microwave techniques (Gadian 2001). The project will develop over the next decade and is in its very early stages at present. It does indicate, however, that there is much effort being applied to the improvement of rain measurement from satellites and by 2010 we will be much further advanced. The same project also includes space-borne multi-frequency microwave radiometers working in the 10.7, 19, 22, 37, 85 and 150 GHz bands similar to those used from the ground at the Chilbolton site in the UK.

By the after-effects of rain

Remote measurement of the after-effects of rainfall is based on detecting the changes in reflectance that occur as the soil gains water and, to a lesser extent, the changes in reflectance from vegetation. This is most successful in arid countries, where the bare ground is exposed, but at best it is qualitative, giving little more information other than that it has rained. The method cannot be placed in the same class as those using algorithms to quantify the amount of precipitation.

Combination of techniques

At the end of Chapter 8, the FRONTIERS project was introduced (forecasting rain optimised using new techniques of interactively enhanced radar and satellite data). This involves combining ground-based radar measurements of rainfall with estimates of the probability of rainfall based on satellite images. In the UK, where it was developed, FRONTIERS uses Meteosat data, comparing measured radar precipitation intensities with the reflectance and radiance values of the clouds deduced from visible and infrared data. Not only does it allow the radar measurements to have checks made on them (along with the additional input from ground-truth telemetering raingauges) but it allows the rainfall estimates to be extrapolated somewhat beyond the immediate radar coverage. As the name indicates, it is an interactive method and needs the involvement of an experienced forecaster who exercises judgement on how to modify the radar data, based on the satellite images, and produce an estimate of present rainfall along with a forecast of likely rainfall amounts over the next few hours. This combination of radar, telemetering ground-based raingauges and satellite data, as well as the input from the forecaster, produces an end product that is more than the sum of the parts. Nevertheless, while the RS measurements do add useful extra information and will generally result in a better forecast, the satellite images could mislead and could have the opposite effect if wrongly interpreted.

Wind

Cloud-drift winds

By following cloud top movements for several consecutive half-hour images from geostationary satellites, the movement of clouds can be used to estimate the speed and direction of the wind, at cloud level. It must be remembered, however (Chapter 5), that clouds are not used to measure wind direction at ground level since direction varies with altitude. Target areas of about 20×20 pixels are chosen that have suitable ranges of brightness temperature for easy recognition, these 'targets' being followed automatically each half-hour. Of course, some areas do not have suitable clouds and so the method is not globally applicable. In the high latitudes, where the geostationary satellites cannot penetrate, polar orbiting satellites can be used.

Scatterometer sea surface winds

It has been known since radar was first developed in the 1940s that at low angles of radar beam elevation, sea waves produce large unwanted echoes. It was soon recognised that this backscatter increased with increasing windspeed and this opened the way to measure wind remotely. Radars designed to use this effect are known as scatterometers.

Such backscattering is caused by in-phase reflections from rough surfaces. First-order (Bragg, see Chapter 16) scattering at microwave frequencies originates from the smaller ripples, known as 'cat's paws'. The smaller waves are produced by the local wind, unlike the larger swell that reflects the effects of the more distant wind (distant in time or space). The level of backscatter ($\sigma°$) received by the radar from a large extended surface such as the sea, known as the *normalised radar cross-section*, is governed by the transmitter power, the slant angle of the beam, the radar wavelength, atmospheric attenuation and the antenna gain in an expression known as the *radar equation*. $\sigma°$ increases with surface windspeed, decreases with the angle of incidence of the beam and is dependent on the angle of the beam relative to the wind direction, being greater looking up- or down-wind and least at right angles to the wind. By measuring $\sigma°$ at two angles, both the speed and direction of the wind can be measured, although there can be a $180°$ ambiguity in the direction. The first scatterometer was flown on the SEASAT-A satellite in 1978, achieving accuracies of 2 m s^{-1} for speed and $20°$ for direction provided that a rough wind direction was already known so that the ambiguity could be resolved. The ERS-1 satellite was the next vehicle for experiment.

Microwave radiometer wind speeds

Microwave sounders (above) give a measure of brightness temperature. A number of algorithms have been developed to use these data to give an indication of windspeed over the sea, the roughness of the water affecting the brightness temperature. These systems are not yet operational, but they do show the extent of the work in progress and the potential of RS in the measurement of environmental variables. Given the same pace of development, RS has much to offer in the coming decades. We are just at the start.

Sea waves

The same radar altimetry as is used to measure sea surface height (see Argo in Chapter 17) is also used to measure *significant wave height* (SWH – the mean height of the one-third highest waves, Chapter 17) (Carter 1999). Using satellites, starting with SKYLAB (in 1973), followed by the ERS-1 (1991) and ERS-2 (1995), which have nadir-pointing altimeter radars working in the Ku-band at about 13 GHz, and the TOPEX/Poseidon satellite (1992) and more recently Jason (2001) (Chapter 17), the waveform of the returned radar pulse is analysed to derive SWH.

Typically, altimeter radars have a surface resolution (footprint) of about 10 km. Radar pulses arriving at the sea surface from the satellite are reflected off the horizontal surfaces of the waves within the whole of the area. The larger the waves, the more drawn-out, in time, are the returned pulses, since they are reflected from a greater range of distances. This has the effect of changing the slope of the returned pulse, a perfectly flat sea producing a near-vertical leading edge, a rough sea a sloping one. There are problems with noise on individual pulses, but means are taken producing an average slope every second. The best calibration strategy is to derive a regression between the radar derivation and measurements from nearby deep sea buoys (another example of the important symbiotic relationship between RS and ground measurements). The satellites are in near-Earth orbits that result in the footprint moving across the ground at about 7 km a second, the wave data being reflected from an area about 10 km wide by 17 km long. Accuracy is about 0.5 m or 10%. A good summary of the principles is given by Aage *et al.* (1998) and Lipa and Barrick (1981).

The same signals can also be used to derive windspeeds, since wave height is related to wind action. However, the larger waves are probably due to distant or old winds; it is the smaller waves that are generated by the local winds. There is some uncertainty over the accuracy of wind measurements derived from altimeters, algorithms being purely empirical (Witter & Chelton 1991). Nevertheless, as an indicator of significant wave height, altimeter techniques are well established and reliable.

Radiation

Solar radiation arriving at the Earth's surface evaporates water and heats the atmosphere and the ground. It is, therefore, a very important quantity, as Chapters 3 and 7 have shown. It is also one of the few variables that can be determined quantitatively and with reasonable accuracy by RS. A satellite radiometer measures the short-wave (solar) radiation being reflected back to space from the atmosphere and from the ground. Subtracting this measurement from the (precisely known) amount of incoming radiation just outside the atmosphere gives the amount reaching the Earth's surface. This value is usually expressed as the fraction of radiation that would have been received under clear-sky conditions, this being a function of Sun elevation and atmospheric turbidity. To take these factors into account, as well as the multiple reflections, absorption and scattering that occur throughout the depth of the atmosphere, Möser and Raschke (1983) derived a polynomial equation giving the incoming radiation under clear-sky conditions for varying values of Sun elevation, which incorporated coefficients based on atmospheric visibility data. Using this method, tests from a project in Mexico (Stewart *et al.* 1999) demonstrated that the error, averaged over a six-month period, was around 9% for daily values. If just clear days were included the error fell to 5%. This is a very good result for RS.

Evaporation

Even with the best of ground-based instruments, it is difficult to measure evaporation – even the eddy correlation method, one of the best methods available, having an accuracy of only 15% to 25% (for hourly values). Most ground-based methods, such as those developed by Penman, estimate evaporation indirectly by measuring net radiation, temperature, humidity and windspeed. However, a simplified form of the Penman method (Makkink 1957, de Bruin 1987) uses measurements of incoming solar radiation (which can be measured well by RS), combined with the saturation vapour pressure versus temperature curve (Chapter 4), using average monthly air temperatures for the area concerned. By using this abbreviated method, it is possible to estimate potential evaporation remotely from satellite data (Stewart *et al.* 1999). It is, however, difficult to say just how accurate these estimates are.

Total ozone and ozone profile

UV from the Sun is partly absorbed by the atmosphere and partly backscattered to space, ozone being the main scatterer. The measurement of ozone from the ground and from satellites is dealt with in Chapter 20 on atmospheric composition.

Volcanic activity

Volcanic ash is a hazard to aircraft; it is also a factor affecting the climate, but it is near-impossible to differentiate it from water/ice clouds using single IR or visible images because they both look very similar. However, their microphysics are not the same causing them to radiate differently in the two thermal IR bands 10.3–11.3 and 11.5–12.5 μm, the difference in brightness temperature being 10 K for ash and near-zero for water/ice clouds (Prata 1989, Potts 1993). But as little has been done in the way of in situ observations of ash clouds, the accuracy of the remote observations is not certain. Nor is the method without other problems; for example, if the ash clouds become diffused, underlying water clouds will give a false indication.

Vegetation

Satellite observations are very useful for differentiating between vegetation type, allowing crop monitoring, forest management and drought and flood assessment. It is an example of qualitative observations not requiring a precise figure. The methods used are based on green vegetation having a low reflectance at the visible wavelengths but very high values around 1 μm (near and mid IR) due to the interaction of the solar radiation with the chlorophyll. Reflectance of soil and water are, in contrast, low in the 1 μm band, allowing vegetation to be detected by observing reflectances in these two adjacent bands, this ratio being known as the *normalised difference vegetation index* (NDVI). The ratio is near to zero for clouds, snow, water and bare rock. The same corrections are required as must be applied when making many other measurements, such as satellite and Sun angles and the effects of the atmosphere. The fact that the leaves will be at many different angles and that there may be different types of plant intermixed, along with vegetation-free areas between or intermingled with the plants, also needs to be taken into account. As with sea surface temperature, cloud-free pixels have to be identified and excluded. Again the AVHRR is a foremost source of such data, along with Landsat's multispectral scanner and thematic mapper.

Fires

Because of the strong non-linearity of the Planck function and the peaking of radiation from hot objects at around 4 μm, the 3.7 μm band of the AVHRR is very sensitive to hot spots, regions where the brightness temperature reaches 500–1000 K. This allows fires to be detected automatically.

Other variables

The above does not exhaust the possibilities of RS and many other variables are now being measured using the satellite images and more will undoubtedly be added.

These will not be included since the aim of this chapter has been to illustrate basic RS principles, not to give an exhaustive and detailed account of each and every variable and the techniques used to measure them.

The strengths and weaknesses of remote sensing

Whereas in situ measurements are made by sensors in direct physical contact with the variables that they measure, at one single point, RS measurements are made entirely by sensing the electromagnetic radiation reflected from, or emitted by, the surface of the Earth and its atmosphere, without any contact with the variable being measured.

While a platform from which RS measurements are made can be simply a mast on the ground, an aircraft or a balloon, it is sensors aboard spacecraft in orbit around the Earth that are having the biggest impact on RS, and have, therefore, been the main concern of this chapter. However, many of the instruments and techniques used on spacecraft and described here can also be used on the other platforms.

Cost

Global weather forecasting requirements could be met by about 5000 conventional ground stations, regularly distributed around the globe, at a cost of about 10 billion US dollars (WMO 1996). Just a few satellites can provide similar cover at a cost of 500 million dollars. So a significant motivation for using satellites is financial.

Cost also enters the equation in another way. Although the expense of the design, development, launching into orbit and operational costs of the satellites are high, they are borne by the satellite operator, usually NASA or possibly EUMETSAT. The user, however, can often get the data that the satellites produce cheaply, or even free of charge. Even though a primary data user station (PDUS) is not as cheap as a Wefax receiver (see Chapter 12), compared with the price of installing and maintaining in situ instruments over large areas it is very cheap indeed. So individual users can benefit financially by using the free RS data instead of installing and operating a ground-based instrument network. The large price differential, however, tempts cut-backs in ground-based observations, and this is unfortunate and ill-advised, for RS cannot replace ground observations, indeed it relies on them.

Spatial and temporal restraints

Satellites have the well-known advantage that they can monitor the whole globe and access areas where it would be difficult to site ground stations, in particular the oceans, although this is less of a problem with automatic stations and satellite telemetry. But there is a downside to this. While satellites operated in geostationary orbits can produce repeat images (of the whole disc) every half-hour, their spatial resolution is 2.5 × 2.5 km (improving to 1 km in the new generation of Meteosat in the visible spectrum). But this is only achievable at the sub-satellite point and as the field of view extends away from the nadir point, the pixels become increasingly larger, elongated and distorted, so that at the limits of view the surface is seen at a grazing angle. The polar regions cannot be seen at all.

Polar orbits may be true polar (at 90° to the equator) or skewed to obtain sun-synchronous operation. Satellites in polar orbits are low-flying and so collect images with a much higher spatial resolution than those from geostationary satellites. In the case of the Spot satellite, for example, the pixels are as small as 10 m square, while for Landsat the resolution is from 30 to 120 m square, depending on the instrument used – multispectral scanner (MSS) or thematic mapper (TM). But while polar orbits give much better spatial resolution than geostationary orbits, the price paid here is reduced time coverage. In the case of satellites in true polar orbits, the poles are in view at each pass (14 times a day in the case of Argos). As the satellite moves from the poles towards the equator, however, overpasses of any given section of ground naturally become increasingly less frequent, and at the equator (although there are still 14 passes a day somewhere over the equator) there may be many hours without an image of any particular area. With some satellites, repeat images may be days apart, even up to two or more weeks. Three-hourly spot observations are not adequate for many quantitative applications.

Clouds and night

If clouds are the subject of interest, then their presence is to be welcomed, but if it is some aspect of the surface beneath the clouds that is of concern, clouds are a problem in the visible and the infrared spectrum. In mid-latitudes and in the tropics, clouds can obscure the view frequently and for long periods. If the measurements are also made by an infrequently passing polar satellite, opportunities to observe the ground might be weeks or months apart. In the visible and non-thermal IR spectrum, night-time also restricts observations to half the day. With rapidly changing variables, such as many of those of concern to meteorology and hydrology, both these limitations are obviously a major problem. For example, it would be impossible to reproduce the performance of a conventional meteorological site using remote sensing if measurements were only available every few days. There are, of course, some things that do not need to be measured frequently because they change relatively slowly, such as ice- and snow-cover and land-use. But if high-resolution observations are required at frequent intervals in the visible and IR spectrum, then cloud, night-time and infrequent overpasses would make them impossible. These factors alone limit greatly the usefulness of RS in many applications.

Scaling

There is an additional problem, referred to as *scaling up*, which concerns the compatibility of a single point measurement made on the ground with the often much larger areal average measurement produced by many RS instruments, perhaps over a 25 square kilometre area; and there is also the problem of being certain of the precise geographical location of the RS pixel relative to the ground observation. In addition there is the question of the degree of compatibility between the instantaneous readings made by the satellite and the long-term means measured by the in situ ground instruments, or of ensuring that the RS measurements apply to the same instant that the ground-truth measurement was made. Furthermore, there is the question of the scaling-up of models (Stewart *et al.* 1998) – can models developed for use over a few square kilometres be expected to be correct on the normally much larger scale of RS?

Accuracy and ground-truth

RS far out-performs ground-based measurements in many applications, particularly in cloud observing, the detection of the extent of ice and snow and of fires, crop development and many other yes/no or qualitative observations.

But where precise analogue values are needed, for example of temperature, rainfall, evaporation or soil moisture, accuracy is limited by the skill of the algorithm and by the physical complexity that needs to be modelled. However, it is possible to ameliorate both the lack of any direct correlation between the measured radiation and the variable itself, as well as the effects of atmospheric interference on the passage of the radiation from ground to satellite, by incorporating in situ ground-truth measurements. These, combined with the RS quantitative data via the algorithm, greatly increase the accuracy of the RS observations. Indeed it is this combination of ground-truth and remotely sensed data that gives some of the best results in RS. Ground-truth combined with RS also allows the precise in situ measurements from the ground-based instruments to be extrapolated to give better estimates of the variation between point observations and thus a better areal average.

It is thus not an either/or but a 'both' choice. The combination of in situ with remote data is widely used for most of the variables that hydrologists and meteorologists need to measure. Ground-truth is also essential for the verification of algorithms. That NASA has gone to the great lengths and expense of soft-landing instruments on other planets to make in situ measurements, of the weather on Mars for example, bears out the importance of ground-truth.

References

Aage, C., Allan, T. D., Carter, D. J. T., Lindgren, G. & Olagnon, M. (1998) *Oceans from Space: a Textbook for Offshore Engineers and Naval Architects.* Repères Océans 16. IFREMER, Brest. ISBN 2-84433-010-X.

Arkin, P. A. & Meisner, B. N. (1987) The relationship between large-scale convective rainfall and cold cloud over the western hemisphere during 1982–1984. *Monthly Weather Rev.*, **115**, 51–74.

Barrett, E. C. & Curtis, L. F. (1982) *Introduction to Environmental Remote Sensing,* second edition. Chapman & Hall, London, 352p.

Barrett, E. C. & Herschy, R. W. (1986) A European perspective on satellite remote sensing for hydrology and water management. *Hydrologic Applications of Space Technology.* IAHS Publication No. 160, pp. 3–12.

Blyth, K. (1981) Remote sensing in hydrology. International Hydrological Report No. 74.

de Bruin, H. A. R. (1987) From Penman to Makkink. Evaporation and weather. In Volume 39, Proceedings and Information, TNO Committee on Hydrological Research, Technical Meeting 44.

Carter, D. J. T. (1999) Variability and trends in the wave climate of the North Atlantic. *Proceedings of the 9th International Offshore and Polar Engineering Conference.* Brest, France, Volume III, pp. 12–18.

Dozier, J. (1987) Remote sensing of snow characteristics in southern Sierra Nevada. In *Proceedings of the Vancouver Symposium on Large Scale Effects of Seasonal Snow Cover*, pp. 305–14. IAHS Publication No. 166.

Dugdale, G. & Milford, J. R. (1986) Rainfall estimation over the Sahel using Meteosat thermal infrared data. In *Proceedings of Conference ISLSCP. ESA SP-248.* European Space Agency, Paris, pp. 315–19.

Ebert, E. E. (1987) A pattern recognition technique for distinguishing surface and cloud types in polar regions. *J. Clim. Appl. Meteorol.*, **26**, 1412–27.

Ebert, E. E. & Manton, M. J. (1998) Performance of satellite rainfall estimation algorithms during TOGA COARE. *J. Atmos. Sci.*, **55**, 1537–57.

Farnsworth, R. K., Barrett, E. C. & Dhanju, M. S. (1984) Applications of remote sensing to hydrology including ground water. *International Hydrological Programme,.* UNESCO, 122p.

Ferraro, R. R. & Marks, G. F. (1994) The development of SSM/I rain-rate retrieval algorithms using ground-based radar measurements. *J. Atmos. Ocean. Technol.*, **12**, 755–70.

Gadian, A. (2001) Global precipitation measurement. *Book of Abstracts.* Royal Meteorological Society, Second National Conference, Manchester, September 2001. p. 63.

Harris, A. R. (1998) Satellite retrievals of sea surface temperatures for long-term climate monitoring. In *Proceedings of the American Meteorological Society,* UNESCO 25–29 May 1998, Volume 1, pp. 162–5.

Harris, L. (1996) *Satellite Projects Handbook.* Newnes (Reed Elsevier), London, ISBN 0 7506 2406 X.

Jones, P. D. & Wigley, T. M. L. (1990) Satellite data under scrutiny. *Nature*, **344**, 711.

King-Hele, D. (1994) *Theory of Satellite Orbits in an Atmosphere.* Butterworth, London.

Kummerow, C., Barnes, W., Kozu, T., Shiue, J. & Simpson, J. (1998) The Tropical Rainfall Measuring Mission (TRMM) sensor package. *J. Atmos. Ocean. Technol.*, **15**, 809–17.

Kummerow, C., Simpson, J., Thiele, O., Barnes, W. & Chang, A. T. C. (2000) The status of the Tropical Rainfall Measuring mission (TRMM) after two years in orbit. *J. Appl. Meteor.*, **39**, 1965–82.

Lipa, B. J. & Barrick, D. E. (1981) Ocean surface height-shape probability density function from SEASAT altimeter echo. *J. Geophys. Res.*, **86**, 10921–30.

Makkink, G. F. (1957) Testing the Penman formula by means of lysimeters. *J. Inst. Water Eng.*, **11**, 277–88.

McClain, E. P., Pichel, W. G. & Walton, C. C. (1985) Comparative performance of AVHRR-based multi-channel sea surface temperatures. *J. Geophys. Res.*, **90**, 11587–601.

Möser, W. & Raschke, E. (1983) Mapping of solar radiation and of cloudiness from meteosat image data. *Meteorologische Rundsch.*, **36**, 33–41.

Potts, R. J. (1993) Satellite observations of Mt. Pinatubo as clouds. *Aust. Met. Mag.*, **42**, 59–68.

Prata, A. J. (1989) Observations of volcanic ash clouds in the 10–12 micron window using AVHRR/R. *Int. J. Remote Sensing*, **10**, 751–61.

Quartly, G. D., Guymer, T. H. & Srokosz, M. A. (1996) The effects of rain on Topex radar altimeter data. *J. Atmos. Ocean. Technol.*, **13**, 1209–29.

Quartly, G. D., Guymer, T. H. & Srokosz, M. A. (2002) Back to basics: measuring rainfall at sea: Part 2 – Space-borne sensors. *Weather*, **57**, 363–6.

Rango, A., Martinec, J., Foster, J. & Marks, D. (1983) Resolution on operational remote sensing of snow cover. In *Proceedings of Hydrological Applications of Remote Sensing and Remote Data Transmission,*. IAHS Publication No. 145.

Rao, P. K. (1990) *Weather Satellites: Systems, Data, and Environmental Applications.* American Meteorological Society, Boston, Mass.

Saunders, R. W. & Kriebel, K. T. (1988) An improved method for detecting clear sky and cloudy irradiances from AVHRR data. *Int. J. Remote Sens.*, **9**, 123–50.

Schultz, G. A. & Barrett, E. C. (1989) Advances in remote sensing for hydrology and water resources management. UNESCO Publication IHP-III Project 5.1.

Smith, W. L. (1985) Satellites. In: Haughton, D. D. (ed.) *Handbook of Applied Meteorology.* Wiley, New York, pp. 380–472.

Spencer, R. W. & Christy, J. R. (1990) Precise monitoring of global temperature trends from satellites. *Science*, **247**, 1558–62.

Stewart, J. B., Engman, E. T., Feddes, R. A. & Kerr, Y. H. (1998) Scaling up in hydrology using remote sensing: summary of a workshop. *Int. J. Remote Sens.*, **19** (1), 181–94.

Stewart, J. B., Watts, C. J., Rodriguez, J. C., de Bruin, H. A. R., van den Berg, A. R. & Garatuza-Payán (1999) Use of satellite data to estimate rainfall, radiation and evaporation for Northwest Mexico. *Agric. Water Manage.*, **38**, 181–93.

Wilheit, T. T., Chang, A.T. C. & Chiu, L. S. (1991) Retrieval of monthly rainfall indices from microwave radiometric measurements using probability distribution functions. *J. Atmos. Ocean. Technol.*, **8**, 118–36.

Witter, D. L. & Chelton, D. B. (1991) A Geosat altimeter wind speed algorithm and a method for altimeter wind speed algorithm development. *J. Geophys. Res.*, **96**, 8853–60.

World Meteorological Organisation (1996) *Guide to Meteorological Instrumentation and Methods of Observation*, sixth edition. WMO, No. 8

20

Atmospheric composition

Generally the atmosphere is hazy: and this is caused by the falling of impalpably
fine dust, which we found to have slightly injured the astronomical instruments.
The morning before we anchored at Porto Praya, I collected a little packet of this
brown-coloured fine dust. . . . The dust falls in such quantities as to dirty
everything, and to hurt people's eyes. . . . Professor Ehrenberg finds that this dust
consists in greater part of infusoria with siliceous shields . . . and has ascertained
no less than sixty-seven different organic forms!

Charles Darwin *Voyage of the Beagle* (Cape de Verd Island).

This chapter is not as detailed as the previous ones. In part this is because most
of the instruments used are largely of the industrial type, pressed into the service
of environmental monitoring and, being very specialised, would require too much
space to do them justice; nor is such a description really necessary. Also in the
sections on total ozone and ozone profiles, the techniques have already been covered
in Chapters 16 and 19 and it would be tedious to repeat it all. Readers should refer
to these chapters for the detail of the methods.

The variables and their measurement

Earth's atmosphere is composed of 78% nitrogen, 21% oxygen and 0.9% argon. The
remaining 0.1% is made up of carbon dioxide (0.036%), hydrogen, ozone, carbon
monoxide, helium, neon, krypton and xenon in more or less fixed quantities, and
water vapour in amounts that vary greatly. In addition, there are very small amounts
of other, trace, gases measured in parts per million (ppmv, etc., where v indicates
that the ratio is by volume) and parts per billion or trillion. It is these trace gases, and
atmospheric particulate matter, that are the topic of this chapter, more specifically
those that are generated by, or are affected by, human activities. Because of this,

Measuring the Natural Environment, second edition, Ian Strangeways. Published by
Cambridge University Press. © Ian Strangeways 2003.

water vapour is not included in the WMO repertoire of trace gases, even though it accounts for up to 2% of the atmosphere (at 20 °C and 100% RH) and is the major greenhouse gas. We should be measuring it just as carefully as the other, lesser, greenhouse gases. It could well be a key element in climate change (Folland *et al.* 2002).

The instruments and methods used to measure these substances are complex and need specialised operators. They were designed and made for monitoring processes in the chemical, food and manufacturing industries; they are simply being pressed into the service of environmental monitoring. The instruments are used in laboratories, powered by mains electricity with the gases piped to them, and not generally able to operate at remote, unattended sites. The measurements are difficult to make because of the very low concentrations of the trace gases and because of the need for careful calibration. Sensor exposure, in the sense of from where the air sample is collected, is also important to avoid contamination or biases, especially since the concentrations are so low.

Greenhouse gases

The WMO set up the World Data Centre for Greenhouse Gases (in Tokyo) in 1990 to archive data for all greenhouse gases (except water vapour) (WMO 1995).

Carbon dioxide (CO_2)

That CO_2 is a naturally occurring gas is general knowledge, as is the fact that its concentration has fallen and risen naturally and widely over the entire period of geological time. This has been detected by such proxy methods as air trapped in ice cores from Greenland and Antarctica and revealed by the even longer, but less direct, records to be found as shells in the sediment at the bottom of the seas and oceans. That burning fossil fuel puts more of it into the atmosphere is also now familiar to all. There is a massive natural cycling of the gas (the carbon cycle) from the atmosphere into and out of the sinks and sources of the water of the oceans and of the biomass of the land and oceans. Our industrial products simply join in this ancient and huge cycle.

The level of CO_2 in the atmosphere is measured using non-dispersive infrared (NDIR) industrial gas analysers, which measure the attenuation of an IR beam passing through the air sample (see also 'The eddy correlation method', Chapter 7). With care, a precision of ± 0.1 ppmv is achievable (currently the CO_2 concentration is about 360 ppmv or 0.036%). A reference calibration cell, in which the concentration of the gas is known, is measured at least once a week, thereby checking the system's calibration, giving an absolute measure of concentration. An alternative is to collect samples in a specially designed glass or stainless steel flask for return to a central laboratory where the same type of gas analyser measures the concentration (Komhyr 1989). Such methods can also be used for most trace gases. To ensure global compatibility, a three-level calibration system has been developed consisting of a primary, secondary and working reference gas, the latter being exchanged between the various national measurement centres and the central calibration laboratory in the USA.

Chlorofluorocarbons (CFCs)

Chlorofluorocarbons CFCs, in particular $CFCl_3$ (CFC 11) with a concentration of 268 pptv (and CF_2Cl_2, CFC 12), are a family of man-made compounds that have been in use since the 1930s as refrigerant gases and aerosol-can propellants. They contribute to the greenhouse effect and also destroy the ozone layer.

The principal method of measuring CFCs is to pass the air sample through an industrial gas chromatograph (based on electron capture detectors). Periodically, gas of known concentration is passed through the system, an absolute value of concentration being obtained by the ratio of the two readings. Again samples can be collected in flasks (Prinn 1983).

Nitrous oxide (N_2O)

There are numerous small sources of N_2O both natural (about twice the size of anthropogenic sources) and anthropogenic (fossil fuel and biomass burning and fertilisers) sources, contributing about 6% to the (enhanced) greenhouse effect. Levels are currently around 312 ppbv (parts per billion by volume) with a growth rate of about 0.7 ppbv per year. It has a lifetime of around 150 years, being broken down slowly in the stratosphere by sunlight. It is essentially inert in the troposphere.

As with CFCs and several other trace gases, N_2O concentrations are measured in an electron capture gas chromatograph, with a gas of known concentration being measured periodically to give absolute values (Elkins 1996). Yet again samples can be collected for central analysis.

Methane (CH_4)

Methane occurs naturally, about 20% coming from wetlands, as well as due to human activities, in particular agriculture and waste disposal. In the mid 1990s its concentration was about 1720 ppbv with a growth rate of about 8 ppbv per year (causes still not clear), although the growth rate has been falling since around 1980. The main removal process is reaction with the hydroxyl radical OH in the troposphere where it is oxidised to produce water, with a removal time of about 12 years. It also reacts with chlorine in the destruction of ozone. Absorbing strongly at 7.66 μm (where CO_2 and H_2O absorb weakly), it is one of the important greenhouse gases.

Flame ionisation detection (FID) gas chromatographs are used to measure methane concentration, this method being very reliable and less difficult to operate than other methods. The gas is usually separated-out first with a molecular sieve column. The measurements are made relative to a gas sample of known concentration. An alternative method of measurement is by IR laser, the attenuation of the beam being dependent on the concentration of the methane (a method described earlier for other gases) (Dlugokencky 1995).

Reactive gases

Although these gases (CO, SO_2, and NO_x in particular) do not have any direct greenhouse effect, they can influence the chemistry of the gases that do, through interaction with OH in the atmosphere. Some are also of importance as acid rain precursors and in determining ozone concentrations at ground level.

Carbon monoxide (CO)

CO is not a greenhouse gas itself, but it reacts with OH and thereby affects some of the greenhouse gases indirectly. There are several methods of measuring CO at atmospheric levels of concentration, one using the same FID gas chromatographs as are used for methane, in which an accuracy of 5–10% is attainable. In another method the CO is passed over hot mercuric oxide, releasing mercury vapour which is detected by UV absorption methods. IR absorption using diode lasers as the light source is an alternative technique, but the costs are higher and the method more complex. Reference standards of dry air with precisely known CO mixing ratios are essential in all of the methods (Weeks 1989). If samples are collected in flasks for later analysis, there are contamination problems and CO values can rise or fall in a few days. The container materials need to be tested for contamination before any test programme.

Sulphur dioxide (SO_2)

Sources of atmospheric SO_2 are both natural and man-made, including the sea, volcanoes, biomass decay and anthropogenic emissions. Its concentration varies from 0.05 ppbv in remote country to 10 ppbv in urban areas, although power plants and volcanoes can emit concentrations as high as 1000 ppbv. SO_2 is a greenhouse gas, as it absorbs IR, but because of its low concentration it is not a significant one.

However, it is climatologically active, reacting photochemically with particulate matter to produce sulphate particles which act as nuclei for cloud condensation, and thus affect the energy input to the ground (Charlson 1987). The dry particles, or the resulting rain, both transfer acidity to the ground and to the vegetation, which can be damaging.

Concentrations of SO_2 can be measured continuously using a pulsed-fluorescence technique. This is the reverse of the technique that measures the absorption of radiation by a gas at certain well-defined wavelengths. Instead the gas is induced to emit radiation, which is characteristic of its structure, and it does so at an intensity proportional to its concentration. It is a reliable and accurate method and simple to calibrate (Luke 1997). Gas chromatograph methods can also be used but they need more frequent attention and technical expertise. Filters can also be used to collect samples and these give potentially greater accuracy, but the method needs more frequent attention. (See below under 'Wet deposition' and 'Dry deposition'.) Because SO_2 has a short lifetime in the atmosphere, sampling frequencies of under an hour are used, the pulsed-fluorescence technique allowing such high sampling rates in real time. By combining this method with filter samples, exposed periodically to obtain an integrated measure, the two measurements can be compared and accuracy improved. SO_2, being a reactive gas, can combine with water drops in the tubes used to convey the sample to the instrument and can also react with some tube materials. Thus materials such as Teflon or stainless steel have to be used and kept as short, clean and dry as possible.

Nitrogen oxides (NO_x)

There is a large family of nitrogen oxides that permeate the atmosphere, originating from biomass burning, lightning, soil bacterial action and from anthropogenic combustion. These oxides are important determiners of ozone concentration as well as being important contributors to acid rain and dry deposition. Their concentration, outside of urban areas, is very low and little studied, in part because of the difficulty of measuring such low concentrations – parts per trillion (1 in 10^{12}). The most important, and first-formed, compounds are nitric oxide (NO) and nitrogen oxide (NO_2), the sum of these often being shown as NO_x. The others – nitric acid (HNO_3), aerosol nitrate and peroxy-acetyl-nitrate (PAN) – are produced from them through conversion once they are in the atmosphere. Conversion from one to another is often rapid and the best way of expressing the concentration of these gases is as the sum of them all (but not the inert N_2O). This total is denoted as the *total reactive nitrogen* or NO_y.

In measuring these gases, care is required to avoid contamination, especially from cars and trucks, since levels are otherwise very low even in less remote areas.

Nitric oxide (NO) and nitrogen oxide (NO₂)

NO can be measured most conveniently and accurately using chemiluminescence, in which the NO is mixed with ozone (O_3) in a reaction chamber. The air sample is drawn into the chamber, at a controlled rate, with a vacuum pump where it is mixed with ozone, generated within the instrument by exposing oxygen to a high voltage. As the NO reacts with the O_3 to produce NO_2, some of the NO_2 is formed in an electronically excited state. As it relaxes back to its ground state, it emits red photons, which are detected by a photomultiplier tube, giving a measure of the NO content. The NO_2 is measured the same way, after first being converted to NO by passing it over heated molybdenum or gold. This process converts all reactive nitrogen species to NO. NO_2 can be measured directly by a similar chemiluminescence method, in this case using luminal in place of ozone, with blue photons emitted, but it becomes non-linear in response at low concentrations (2–3 ppbv) (Luke & Valigura 1997).

Peroxy-acetyl-nitrate (PAN)

This particular oxide of nitrogen is abundant in urban air, but it is little studied. Roberts (1990) reviews its sources, sinks and chemistry. It can be measured by gas chromatographs with electron capture detection or by the luminal chemiluminescence method, after conversion to NO_2 by heat. The largest problem with its measurement is obtaining reliable calibration standards, not the least because of their short lifetime.

Nitric acid (HNO₃) and nitrate aerosols

The most common, initial, nitrogen oxide is NO, but this is rapidly converted to NO_2 and subsequently to HNO_3. The most usual method of measuring nitric acid and nitrate aerosols is to collect samples in filters. Air is drawn at a known rate (several litres per minute) for several hours through, firstly, a filter of Teflon to separate out the aerosol particles followed by a nylon filter to collect the acid gases. In a laboratory the materials are extracted and analysed by ion chromatography for nitrate ions (Luke & Valigura 1997). Other methods are available and readers needing more information should refer to Klemm (1994) and Anlauf (1985).

Total reactive nitrogen (NOy)

NO_y is comprised of NO_x (itself composed of NO and NO_2) and a number of other lesser nitrogen oxides. They are all measured by converting them to NO or NO_2 by passing them through a converter containing gold or molybdenum heated to about 300 °C together with a small amount of CO or hydrogen (Luke & Valigura 1997).

Deposition from the atmosphere

Material from the atmosphere is deposited on the ground, either periodically as wet deposition in the form of precipitation or as dry deposition as a continuous slow flux at all other times. Wet deposition is a rapid delivery of concentrated materials while dry deposition is a slower, more continuous exchange of materials both to and from the surface.

Wet deposition

Measurements of the chemistry of precipitation are important in order to understand the exchange of materials between the atmosphere and both the land and the oceans. In making any such measurements it is important to make certain that local contaminants are not collected in error (Bigelow 1987). Measurements are typically made by collecting samples of the precipitation, be it rain or snow, followed by laboratory analysis. The amount of rain must also be measured to give the total input. The gauges used to collect the deposition must be covered when there is no precipitation to exclude dry deposition. This can be done automatically by gauges with lids that open when it rains, or manually by removing a cover during precipitation, although this would have to be done at night as well. Plastic collectors are preferred to metal ones to avoid chemical interactions. Daily collections are better than weekly since some of the ions can change during storage. Also by frequent collection, several rain events (which might be quite different chemically) are not all pooled into one gross sample. Some measurements may be made on-site in real time, such as pH and conductivity, but care is required since sensors for these two variables can become contaminated if left for any length of time and their readings can also drift with time (Gillett & Ayers 1991). Another point requiring care is that any pH electrodes used, be they in the field or the laboratory, must be of a type that operates in low ionic strength solutions, rather than of the more usual laboratory types that are designed for the much higher solution strengths found in typical laboratory work (Fig. 20.1) (Rodda *et al.* 1985). Many 'acid rain' tests may be in error because of this. It is also to be noted that samples stored in bottles even for a day can change in pH quite considerably. Keeping samples for a week and shipping them over distances will probably result in unreliable data. Chemistry of this type cannot be done easily and casually. Ideally a laboratory should be set up specifically for the purpose, with samples being measured immediately after collection. Although analysis techniques vary a lot from place to place, ion chromatography and absorption spectrophotometry are the preferred methods. The ions measured are typically sulphate (SO_4), chloride (Cl), nitrate (NO_3), hydrogen (pH), calcium (Ca), magnesium (Mg), sodium (Na), ammonia (NH_4) and potassium (K). Also acids such as formic and acetic may be measured in pristine areas.

Figure 20.1. The manual measurement of the pH of rainfall requires care. Firstly it is preferable to collect the rain in a plastic gauge to prevent any chemical reaction that might occur with metal funnels. Equally important is the choice of pH electrode (right, in plastic container). Normal laboratory sensors are not suitable and in their place a sensor designed for use in low ionic strength solutions is necessary. The recent 'history' of the sensor is also important since different indications will result from storing the probe prior to the measurement in distilled water, in water near the pH being measured or in a buffer solution. (See also how to measure pH in Chapter 10.)

Dry deposition

Deposition is a misleading term here because some materials can be transported both to and from the surface. It is a complex process, the rate of transfer being determined by such factors as wind strength and turbulence, the characteristics of the material, gradients of the materials within the atmosphere and the chemical and physical nature of the surfaces. Dry deposition tends to be a more local process than wet.

Eddy correlation methods have been used to measure the flux of some of the materials in much the same way as already described in other chapters for water vapour and CO_2 fluxes, different sensors being substituted to measure the rapid fluctuations of ions such as SO_2, O_3 and NO_x. A summary of the sensors required for this is given by Baldocchi *et al.* (1998). A simpler alternative is to use filters similar to those described for nitric acid, except that here a third filter of cellulose is added impregnated with, for example, potassium carbonate to collect sulphur. In addition to the measurement of the specific variables, meteorological measurements are also made of all the usual variables, to give information, for example, on wind direction. Hicks (1991) gives a full description of these methods.

Particulate matter

Particles of course are also part of dry (and contained in wet) deposition and so are measured in a similar way, that is with filters. The filters can collect all the particles on one filter or the filter can be cascaded to collect increasingly small particles. Water-soluble matter can then be dissolved from the filter and measured in an ion chromatograph. Insoluble matter is usually analysed using neutron activation analysis, proton-induced X-ray emission or other very sophisticated and complex laboratory instruments that need not be considered here. Harris and Kahl (1990) give a description of the procedures used at the Mauna Loa observatory.

Ozone

Ozone makes up about 0.5 ppmv of the atmosphere and acts as a greenhouse gas. It absorbs in the 0.1–0.34 and 0.4–0.65 μm bands in the solar spectrum and in the 4.7, 9.6, 10.5 and 14.1 μm bands in the terrestrial infrared band. It also absorbs in the microwave region. Most of the gas is in the stratosphere where it causes warming, despite the altitude, because of its greenhouse effect. It also reacts photochemically with the other trace gases. Measurements are made of ozone at ground level, of total ozone through the depth of the atmosphere and as a vertical profile up through the atmosphere. At the ground, ozone is of interest because it is a man-made pollutant, while in the stratosphere its significance is because it is destroyed by CFCs, thereby weakening the shield it provides against UV radiation.

Units and terms

Surface ozone is that occurring up to about 10 m above the ground, measurements being given in units of barometric partial pressure or possibly as a mixing ratio. *Total ozone* is the total amount of the gas in a vertical column of the atmosphere above a point on the ground, expressed as the equivalent layer-thickness if all the gas was at one point and at standard temperature and pressure (STP). The *vertical profile* is a measure of O_3 concentration at intervals up through the atmosphere, expressed as partial pressure, mixing ratios or local concentration. By integrating these profile measurements it is also possible to derive the total column amount. In recognition of J.M.B. Dobson's early work on ozone, the unit of the vertical column of ozone is called the *Dobson unit (DU)*, one unit being defined as the amount of ozone in the vertical column which would occupy a depth of 10^{-5} meters at $0\,°C$ and standard pressure.

Measuring surface ozone

The usual method of measuring surface ozone is by drawing air through a cell and measuring the UV absorption at the $0.254\,\mu m$ emission line of a mercury vapour light. Ozone absorbs strongly at this wavelength, while other gases do not. For calibration, the cell alternately measures the air sample and then air sent through a tube containing manganese dioxide, which converts the ozone catalytically to oxygen while leaving all the other gases unchanged. Knowing the ozone absorption coefficient at the wavelength used and the length of the path through which the light passes, the ratio of the two readings gives the O_3 content. Considerable efforts have been made to measure the spectrum of ozone absorption in the laboratory over a range of temperatures because the coefficient is strongly temperature dependent. In 1992 the International Ozone Commission recommended a standard ozone absorption spectrum based on the work of Bass and Paur (1985). The absolute accuracy of the standard is between ±3–5% and this affects all ozone measurements.

The air sample is pumped along a Teflon tube to an indoor instrument, the intake being sited so as to be representative of the air to be measured. This should be at least 3 m above the ground, most typically at rooftop height. There are many requirements that have to be met about exposure and siting, criteria having been agreed by the Global Atmosphere Watch (WMO 1988). The main source of error is the loss of ozone along the inlet tube, minimised by keeping it dry and clean and by changing it regularly. Because the method compares two samples, one with and one without ozone, small drifts in light output or detector sensitivity do not affect the measurement. Nor do the other trace gases which may absorb small amounts of the UV light because they are the same in both samples. The absolute calibration of the instrument is achieved by measuring the absorption in a cell containing a known amount of ozone, the amount being determined by a process that gives the number of ozone molecules in the cell.

Measuring total ozone

Total ozone is measured on a routine basis from the ground using the Sun and Moon as light source and from satellites measuring the solar UV backscattered from the atmosphere.

Direct Sun and Moon measurements

The best and most accurate method of measuring the total amount of ozone in the full depth of the atmosphere is to measure the level of radiation received directly from the Sun at the surface in the waveband between 0.305 and 0.330 μm. If the atmosphere was composed of just ozone, measurement at one wavelength at different times of the day (giving a ratio of actual path length to the vertical path, μ) would be adequate. However, using just one wavelength can give errors because other atmospheric constituents, such as aerosols, affect the reading. This problem is reduced by comparing measurements made at several wavelengths.

Of course, this method only works during the day and when the Sun is not obscured by clouds. The Sun must also be at least 15° above the horizon. The most usual ground-based instruments used to make these measurements are those by Dobson (1957) and Brewer (Evans 1987).

The Moon can also be used as the light source, but its much lower radiation levels make for a less accurate observation. The method can also only be used five days either side of the full Moon and, as with the Sun, must have an angle of at least 15° above the horizon. Despite its limitations, the Moon method is invaluable in the polar nights, when it is the only applicable ground-based method.

Zenith sky measurements

Methods have been developed for use when the Sun is obscured by clouds. It has been shown empirically that UV scattered from the sky at the zenith point (immediately overhead) gives enough information to enable a fair estimate of total ozone to be obtained. The relative magnitude of measurements of UV at two wavelengths is dependent on the total ozone and the value of μ. Empirical charts are obtained by making near-simultaneous readings of the zenith sky and direct sunlight. Each station needs to make its own chart.

Satellite measurements

The longest ozone records from satellites come from NIMBUS-7 since 1978 using the total ozone mapping spectrometer (TOMS) and the solar backscatter ultraviolet (SBUV) instruments (Heath *et al.* 1978). UV from the Sun penetrates the atmosphere, being gradually absorbed in its progress downwards. But it is also scattered back to space from the various depths that it reaches in its downward journey. During its downward journey and again during its return, it is attenuated by absorption by ozone at various wavelengths. Thus an instrument looking downward at selected wavelengths will be able to detect how much UV radiation is being scattered back. If the same instrument periodically measures the input of the UV, by looking directly at the Sun, the ratio of input to output will give a measure of the total amount of ozone. In practice, measurements are made at five wavelengths between 0.312 and 0.380 µm. The longest wavelength is not absorbed by ozone and so gives an indication of ground and cloud reflectivity at the time of the observation. The other four wavelengths are absorbed by ozone to differing extents. There is also a radiation band at 9.3 µm in the thermal IR allowing similar observations (Ma *et al.* 1984). The amount of absorption depends not just on the amount of ozone present, but on solar angle, angle of view of the instrument, the level from which the radiation is being scattered and the ozone absorption coefficient. Multiple scattering up and down at all depths clearly results in a complex calculation to derive the ozone content, suitable models (inversion algorithms – see 'Sounders' in both Chapters 16 and 19) being required to accomplish the task. These models rely greatly on ground-truth from direct-Sun measurements made at the surface, and different algorithms are necessary for different seasons and latitudes. Such observations require the Sun to be at least 2° above the horizon and consequently cannot be used during polar nights. TOMS is an example of such an instrument and, as with other types of scanning radiometers (Chapter 19), TOMS scans across-track and has a resolution of 50 km at the nadir point and 250 km near the horizon. The SBUV instrument uses a similar detection system but its purpose is to measure a vertical profile of ozone, in a way similar to radiometers already described for measuring temperature profiles (Chapters 16 and 19). During the 1990s new instruments measuring the backscatter from ozone, and from other atmospheric minor constituents, have become operational on satellites such as NASA's upper atmosphere research satellite (UARS), along with similar instruments on Spot-3 (launched 1993) with a range of instruments launched on the Earth observation system (EOS) in polar orbit.

Measuring the ozone profile

The ozone profile is measured in three ways, all quite familiar by now if you have read the rest of the book. If not, reading Chapters 16 and 19 gives a useful background so that this section can be kept short and so avoid repetition.

Ozonesondes

This is simply another sensor aboard a radiosonde, rising to 30–40 km (Chapter 16). The sensor carried aloft is based on the reaction of ozone with potassium iodide producing free iodine, every molecule of ozone resulting in the generation of two electrons. To achieve this reaction, the air is bubbled through the solution by means of a small pump, the electrical output of the cell being telemetered along with the other sonde signals. An accuracy of about ±5% is possible (Komhyr 1986).

Umkehr measurements

The differential absorption of solar UV at two wavelengths as it passes through the atmosphere and is scattered to the Earth's surface depends on the vertical distribution of ozone. In this method, measurements are made, at two different wavelengths, of the level of solar UV scattered from clear blue sky directly overhead and received at ground level using, typically, a Dobson spectrophotometer, with Sun elevation angles from 0 to 30°. If the log of the ratio of the two signals is plotted against the Sun elevation angle, the slope of the curve changes sign at a particular Sun angle (*Umkehr* is German, meaning 'turn around'). The plot is then compared with the results from a multiple-scattering radiation model (algorithm) and, using a separate measurement of total ozone, a profile can be derived.

Lidar ozone measurements

This general method of measurement has already been dealt with in Chapter 16 and so little more needs to be added. As in all such systems, a laser emits a pulse vertically, in this case at a suitable UV wavelength, and the backscattered radiance is received from different heights. From this the amount of ozone can be estimated by integrating many shots over a period of hours. It only operates in cloud-free conditions and is interfered with by aerosols.

Satellite measurements

The same satellite instrument (SBUV) that is used to measure total ozone from space also measures the ozone profile. The effective height of UV scatter from the ozone in the atmosphere, back to space, is dependent on the amount of ozone above that height and the absorption coefficient. Because the coefficient varies greatly between 0.25 and 0.40 μm the scattering height is a function of wavelength. The measurements made by the SBUV instrument at five different wavelengths are sufficient to obtain an ozone profile, using an algorithm of a scattering atmosphere. The vertical resolution is about the same as the Umkehr method, that is nine distinct levels from 0.98 to 500 hPa.

Total water vapour

Although water vapour is not listed by the WMO as an atmospheric component to be measured under the Global Atmospheric Watch (GAW) (because it is not a gas directly affected by human activity) it is nevertheless an important component, being a powerful, natural, greenhouse gas. Its measurement, therefore, would seem vital.

While humidity profiles are already measured by radiosondes, and humidity is measured at the surface by most AWSs, as well as at traditional meteorological enclosures, a new technique is currently being developed to measure *total water vapour* through the depth of the atmosphere using GPS transmissions.

The GPS network consists of 24 satellites in orbit at an altitude of about 18 000 km, with atomic clocks that transmit signals giving the exact time. The speed of transmission of a pulse from a GPS satellite to the ground is not quite the speed of light, but slightly slower due to the combined effects of the ionosphere, the atmosphere and its water vapour content. The effect of the ionosphere can be removed by measuring the delay at the two different frequencies used by the receivers and the atmospheric effect is constant, the changing delay being due to changes in the amount of water vapour. Since the orbital position of the satellite and the height of the receiver are known accurately, the slant path distance to the satellite is also known accurately, and so the expected time of arrival of the pulses can be determined. From this, the delay in propagation can be measured. Between 8 and 12 satellites are usually in sight from any one receiver, giving measurements along paths of different directions and angles of elevation. If these measurements are made from a network of GPS receivers, the zenith (vertical) delay can be calculated and converted to total water vapour, using complex software (provided the surface pressure and temperature are known). The units used to express the total amount of water vapour in a column of air is the weight of the water if it was all concentrated together at ground level, expressed in kg m^{-2}. Currently there are eight sites operated in the UK, mostly at radiosonde or sounder stations and while the method is still experimental, it will undoubtedly be a most useful new tool (Jones 2002, Nash 2002).

References

Anlauf, K. G. (1985) A comparison of three methods for measurement of atmospheric nitric acid and aerosol nitrate and ammonium. *Atmos. Env.*, **19**, 325–33.

Baldocchi, D. D., Hicks, B. B. & Meyers, T. P. (1998) Measuring biosphere atmosphere exchanges of biologically related gases with micrometeorological methods. *Ecology*, **69** (No 5), 1331–40.

Bass, A. M. & Paur, R. J. (1985) The ultraviolet cross-sections of ozone, I: The measurements. In: Zerefos, C. S. & Ghazi, A. (eds) *Atmospheric Ozone*. Reidel, Dordrecht, pp. 606–10.

Bigelow, D. S. (1987) Siting and sampling characteristics of the national atmospheric deposition programme and the national trends network: regional influences. *EMEP Workshop on Data Analysis and Presentation*. Cologne, Germany, 15–17 June 1987, *EMEP/CCC-Report 7/87*, Norwegian Institute For Air Research, Lillestrom, Norway, December 1987, pp. 149–60.

Charlson, R. J. (1987) Oceanic phytoplankton atmospheric sulphur, cloud albedo and climate. *Nature*, **326**, 655–61.

Dlugokencky, E. J. (1995) Atmospheric methane at the Mauna Loa and Barrow observatories: presentation and analysis of in situ measurements. *J. Geophys. Res.*, **100** (Number D11:23), 103–13.

Dobson, G. M. B. (1957) Observer's handbook for the ozone spectrophotometer. *Ann. Int. Geophys. Year*, **5**, 46–114.

Elkins, J. W. (1996) Airborne gas chromatograph for in situ measurements of long lived species in the upper troposphere and lower stratosphere. *Geophys. Res. Lett.*, **23** (Number 4), 347–50.

Evans, W. F. J. (1987) Stratospheric ozone science in Canada: an agenda for research and monitoring. *International Report ARD-87–3*. Atmospheric Processes Research Branch, Atmospheric Environment Service, Canada.

Folland, C. F., Karl, T. R. & Salinger, M. J. (2002) Observed climate variability and change. *Weather*, **57**, 269–78.

Gillett, R. W. & Ayers, G. P. (1991) The use of thymol as a biocide in rainwater samples. *Atmos. Env.*, **25** (No. 12), 2677–81.

Harris, J. M. & Kahl, J. D. (1990) A descriptive atmospheric transport climatology for the Mauna Loa observatory using clustered trajectories. *J. Atmos. Res.*, **95** (No. D9), 13651–67.

Heath, D., Krueger, A. J. & Park, H. (1978) The solar backscatter ultraviolet (SBUV) and total ozone mapping spectrometer (TOMS) experiment. *The NIMBUS 7 User's Guide*. NASA Goddard Space Flight Centre, Greenbelt, Maruland, pp. 175–211.

Hicks, B. B. (1991) Dry deposition inferential measurement techniques – 1: Design and test of a prototype meteorological and chemical system for determining dry deposition. *Atmos. Env.*, **25A** (No. 10), 2345–59.

Jones, J. (2002) Real time observations from the UK GPS network. In: *Proceedings of COST Action 716. Exploitation of Ground Based GPS for Climate and Numerical Weather Prediction Applications*. Bern, 14 October 2002.

Klemm, O. (1994) Low to middle tropospheric profiles and biosphere/troposphere fluxes of acidic gases in the summertime Canadian Taiga. *J. Geophys. Res.*, **99**, 1687–98.

Komhyr, W. D. (1986) *Operations Handbook – Ozone Measurements at 40-km Altitude with Model 4A Electrochemical Concentration Cell (ECC) Ozonesondes*. NOAA Technical Memorandum ERL ARL-149.

Komhyr, W. D. (1989) Atmospheric carbon dioxide at the Mauna Loa observatory, 1. NOAA global monitoring for climatic change measurements with nondispersive infrared analyser, 1974–1985. *J. Geophys. Res.*, **94**, 8533–47.

Luke, W. T. (1997) Evaluation of a commercial pulsed fluorescence for the measurement of low level SO_2 concentrations during the gas-phase sulphur intercomparison experiment. *J. Geophys. Res.*, **12** (No. D13), 16255–65.

Luke, W. T. & Valigura, R. A. (1997) Methodologies to estimate the air-surface exchange of atmospheric nitrogen compounds. In: Baker, J. E. (ed.) Atmospheric deposition of contaminants to the Great Lakes and coastal waters. *Proceedings from a Session at the SETAC Fifteenth Annual Meeting*. 30 Oct to 3 Nov 1994, Denver, Colorado. SETAC Press, Pencacola, Florida, pp. 347–377.

Ma, X. L., Smith, W. L. & Woolf, H. M. (1984) Total ozone from NOAA satellites: a physical model for obtaining measurements with high spatial resolution. *J. Clim. Appl. Met.*, **23**, 1309–14.

Nash, J. (2002) GPS total water vapour – potential for future operational observations. In: *Proceedings of the WMO Technical Conference TECO-2000*, Bratislava, 23–25 September 2002.

Prinn, R. G. (1983) The atmospheric lifetime experiment, 1: Introduction, instrumentation and overview. *J. Geophys. Res.*, **88** (No. C13), 8353–67.

Roberts, J. M. (1990) The atmospheric chemistry of organic nitrates. *Atmos. Env.*, **24A**, 243–87.

Rodda, J. C., Smith, S. W. & Strangeways, I. C. (1985) On more realistic measurements of rainfall and their impact on assessing acid deposition. *Proceedings of the ETH/WMO/IAHS Workshop on Correction of Precipitation Measurement*, 1–3 April 1985, Zurich.

Weeks, I. A. (1989) Comparison of the carbon monoxide standards used at Cape Grim and Aspendale. In: Forgan, B. W. & Ayres, G. P. (eds.) *Baseline Atmospheric Programme 1987.* Australian Government Department of Science and Technology, Canberra, pp. 21–5.

World Meteorological Organisation (1988) *Technical Regulations*, **1** (WMO-No 49), Geneva.

World Meteorological Organisation (1995) Report of the Meeting of Experts on the WMO World Data Centres (E.W. Hare). *Global Atmosphere Watch Report No. 103*, WMO/TD-No. 679, Geneva.

21

Forward look

Most advances in measuring the natural environment that have taken place over the last three to four decades have come about through developments in microelectronic chip technology. This has led to the now ubiquitous PC and to an abundance of other devices including (relevant to us) data loggers and satellites able to both generate images and relay data from ground stations to a distant base.

Developments in sensor technology for in situ measurements of the environment over the same period have been relatively modest in comparison; when they have occurred, they have usually been driven by the same microelectronic technology. Much of the actual 'front-end' of the sensor technology has, however, changed remarkably little. We are still using cups and vanes to measure wind, but with electrical sensors attached (although sonic anemometers may slowly replace them if costs can be brought down further). Temperature is still measured by electrical resistance thermometers in small screens. Despite being able to measure solar radiation with precision, sunshine duration measurements are still in demand, although sensed electronically. A most useful microelectronic development has been the introduction of general purpose photo diodes which can also be used as cheap, but slightly less precise, solar radiation sensors. Thermal solarimeters, the most accurate, have remained virtually unchanged since 1965, as have net radiometers. Humidity is still widely measured by the wet and dry method although thin-film capacitive sensors are now replacing it (developed originally for radiosondes). While they are not as accurate over the full range, the capacitive sensors are much more practical; they, too, are produced by microelectronic technology. River level is still widely measured by staff gauges and by floats (the floats now turning a digital shaft encoder instead of driving a chart recorder) or by gas bubbled through a tube to near the river bed, the pressure being measured by the new flexing diaphragm sensors, fabricated using microelectronic techniques. These same sensors can also be directly

Measuring the Natural Environment, second edition, Ian Strangeways. Published by Cambridge University Press. © Ian Strangeways 2003.

submerged in the water to measure the depth, including the oceans. The same principle is now also used to measure barometric pressure, replacing aneroid capsules. Rainfall is still predominantly measured with a funnel collecting the rain with a tipping bucket to measure the water, representing little advance on Sir Christopher Wren in the seventeenth century. Weighing raingauges come second with electronic load cells. Optical raingauges are now available but they remain somewhat specialised. We still cannot measure snowfall very well because of the wind problem. Evaporation pans, despite their primitiveness and indifferent accuracy, continue to be the main 'sensor' for evaporation. Eddy correlation and similar methods remain too specialised and expensive to replace pans. Soil moisture measurement is little advanced on the 1970s when the neutron probe was developed, although methods measuring soil dielectric have appeared since, but are less precise. Radiosondes are a slightly different case and new, very miniature, sensors have been specially developed for them. But they are a breed apart; a self-contained world.

In looking to the future of environmental monitoring we should consider three separate tasks that need instrumentation: the local or national operational projects, the global long-term view and specialised research.

As in the past, so it will be in the future that the majority of measurements of the natural environment will be made by individual organisations, small and large, commercial and governmental – research institutes, universities, national weather services, water resources agencies and a host of others. These organisations will continue to buy, develop, operate and maintain in situ equipment of their own choice, to their own standards and to suit their own budgets and purposes, just as they have in the past. The instruments now available are generally more than adequate for these tasks, and in many cases the highest accuracy is not required.

Until very recently, climate was a rather obscure, academic field of study, with little practical application. It is only of late, with global warming so up-front and politically charged, that this has changed, highlighting the lack of good data. We must assume that climate change will continue to hold everyone's attention and so we need to think about how to improve its measurement.

In situ instruments are less satisfactory for land-based global measurement than for the small, local, self-contained applications mentioned above. This is not because modern in situ instruments are unsuited to the task, they meet requirements well, but because there is too great a variation of skills, commitment, instruments and procedures in each individual country worldwide to produce consistent, intercomparable high quality data. There are certainly many pockets of excellence around the world, but there are also tracts of un-excellence, and I have seen them both. I also fear that there is a gradual reduction in ground-based, in situ measurements in many of the unstable or poorer countries. If you have a civil or tribal war going on, or live in any dysfunctional country, the will or ability to keep networks of instruments going dwindles. I need not list the countries in order for the extent of the problem to be appreciated.

An example of the limited cover of ground-based measurements comes from the IPCC assessment of stations reliable enough to be used for calculating mean annual global temperature anomalies. Many of the world's records had to be rejected as uncorrectable, uncheckable, or too short in duration, leaving 1584 out of 2666 land-surface air temperature stations in the northern hemisphere, mostly clumped together in the US, Europe and the former USSR (Jones *et al.* 1985, Parker 1994). Only 293 out of 610 in the southern hemisphere were acceptable, concentrated in the southern countries of South America and in South Africa, Australia and New Zealand. Another problem is that these older instruments were not designed and were never intended for precise climate research but for the less demanding task of weather forecasting. Such instruments are usually those seen on conventional meteorological enclosures (Strangeways 1995).

In situ instruments are also a problem when it comes to the oceans, for they can only be sited on moving ships, on expensive buoys and on a few small islands, resulting in extensive geographical gaps.

On the other hand, remote sensing (RS) from satellites has many advantages over in situ measurement, just one being unfettered global access. An additional advantage of this is the ability to use one type of instrument everywhere, with the same processing of the data, thereby aiding consistency and so providing comparable accuracy around the globe. (The same freedom of access applies to the oceans, allowing instruments such as Argo floats to be used worldwide.) However, as has been shown in Chapter 19, RS also has limitations. The same point-precision cannot be obtained by RS as it can by in situ instruments, nor is a continuous record of quickly changing variables usually possible. RS is snapshot and broad brush, in situ measurements are continuous and point-specific.

One appeal of RS is that it is based simply on the writing of algorithms. This makes it easy in one sense, since it can all be done in an office with nothing more than a PC and access (often free) to satellite image data, perhaps obtained (free) over the internet, making it a very low cost option. It could almost be done from home. Contrast this with the high cost of developing and testing even simple new in situ instruments, requiring a laboratory and workshops with test and construction equipment and machinery. There then follows the high cost of installing the instruments, often in remote places, and of subsequently operating and maintaining them. The considerable price differential between RS and in situ measurements should not be allowed to result in a falling off of in situ measurements or a decline in the development of new instruments.

Assuming algorithm developments continue at their present rate, I would imagine that in 10 years' time it will be possible to extract much more information from the various discrete bands of radiation measured by satellite sensors than is possible at present. The on-board satellite instruments will also continue to be developed and improved by the space agencies, such as NASA and ESA. It is to these organisations that RS is really indebted, for that is where most of the investment has been – in complex radiometer and radar instrument designs and in the launching of the satellites that carry the instruments into orbit. It is also where all the data come from. Without NASA there would be little RS.

As I have said several times, RS and in situ measurements both have their limitations but, when combined, much more is achievable than through each separately. Such a symbiosis allows one set of data to correct for the weaknesses of the other, for they are, in a way, opposites. This allows frequent adjustment of the satellite readings to conform to ground-truth, an extrapolation of the in situ readings between ground stations and, gradually, a refinement of the algorithms. Five hundred ground stations, judiciously sited around the globe, would be sufficient to allow this symbiotic blending. The ground stations would also provide data in their own right, for use quite independently of their satellite data-correction function. The same satellites that produce the images also relay data transmitted by DCPs from ground stations, allowing the two sets of measurements to be correlated and processed together. By this means, RS and in situ measurements, upper atmosphere radiosonde data and Argo/Jason ocean observations will all combine to give an integrated dataset of considerable power.

In addition, there is much specialised research in progress. I have described some of that going on in the UK in earlier chapters; for example, the measurement of the water vapour and carbon dioxide fluxes using eddy correlation instruments (which have been used over the Amazon to measure the flux of carbon dioxide between forest and atmosphere), a study of the transfer of wind energy (the momentum flux) between atmosphere and ocean, and the work using ground-based radars, lidars, sounders and microwave radiometers, to investigate the many processes occurring up through the atmosphere and within clouds.

So it has been since Galileo and Torricelli started their simple but revolutionary experiments in the seventeenth century – to understand the natural environment we must measure it.

References

Jones, P. D., Raper, S. C. B., Santer, B., Cherry, B. S. G., Goodes, C., Kelly, P. M., Wigley, T. M. L., Bradley, R. S. & Diaz, H. F. (1985) The grid point surface air temperature data set for the Northern Hemisphere. *US Department of Energy, Report TRO22*.

Parker, D. E. (1994) Effects of changing exposure of thermometers at land stations. *Int. J. Climatol.*, **14**, 1–31.

Strangeways, I. C. (1995) Back to basics: the 'met. enclosure': part 1 – its background. *Weather*, **50**, 182–8.

For readers wishing to keep up with progress in the topics dealt with in this book, I have set up a web site www.ianstrangeways.org.uk on which to post interesting news as it emerges, corrections to the book if and when I find them or am told of them, extra images of environments and of instruments and useful links to other web sites.

Appendix: abbreviations and acronyms

ACCLAIM	Antarctic circumpolar current level from altimetry and island measurements
ADC	analogue-to-digital conversion
ADEO	advanced Earth observing satellite (Japan)
AMDAR	aircraft meteorological data relay
APT	automatic picture transmission
ARGO	array for real-time geostrophic oceanography
ATD	arrival-time-difference
AVHRR	advanced very high resolution radiometer
AWS	automatic weather station
BAS	British Antarctic Survey
BCD	binary coded decimal
BOD	biochemical oxygen demand
BSI	British Standards Institute
BT	brightness temperature
CCD	charge coupled device (image sensor)
CEH	Centre for Ecology and Hydrology (formerly IH)
CFC	chlorofluorocarbon
CLIVAR	climate variability and predictability experiment
CLS	Collecte Localisation Satellite (Argos)
CMOS	complementary metal-oxide semiconductor
CNES	Centre National d'Études Spatiales (French Space Agency)
CRAWS	cold regions automatic weather station
CRDF	cathode ray direction finder
DCP	data collection platform
DF	direction finder (lightning)
DFIR	double-fence international reference (snow fence)
DO	dissolved oxygen
DSRT	dual-frequency surface reference technique
EEPROM	electrically erasable and programmable read only memory
EGOS	European Group on Ocean Stations
ELF	extra low frequency (radio and lightning)
ENSO	El Niño southern oscillation

Measuring the Natural Environment, second edition, Ian Strangeways. Published by Cambridge University Press. © Ian Strangeways 2003.

EOS	Earth observing system
EPROM	electrically programmable read only memory
ESA	European Space Agency
FID	flame ionisation detection (gas chromatography)
FRONTIERS	forecasting rain optimised using new techniques of interactively enhanced radar and satellite data
GAW	global atmospheric watch
GCM	global circulation model
GCOS	global climate observing system
GMS	geostationary meteorological satellite
GODAE	global ocean data assimilation experiment
GOES	geostationary operational environmental satellite
GOMS	geostationary operational meteorological satellite
GOSAMOR	global ocean salinity monitoring
GPM	global precipitation measurement
GPS	global positioning system
GTS	global telecommunication system
HF	high frequency
HIRS	high resolution IR sounder
IC	integrated circuit
ICE	(a WMO code)
ICEAN	(a WMO code)
ID	inertial dissipation
IH	Institute of Hydrology (now CEH)
IOC	Intergovernmental Oceanographic Commission
IPCC	Intergovernmental Panel on Climate Change
IPTS	international practical temperature scale
IR	infrared
JMA	Japan Meteorological Agency
LED	light emitting diode
LES	land earth station
LIDAR	light detection and ranging
LIS	lightning imaging sensor
LORAN	long-range navigation
LPATS	lightning position and tracking system
LRIT	low-rate information transmission
LRPT	low-rate picture transmission
LSB	least significant bit
MDWF	moisture dry weight fraction
MOR	meteorological optical range
MOS	metal-oxide semiconductor
MRF	met. flight research
MSB	most significant bit
MSS	multispectral scanner
MSU	microwave sounding unit
MVF	moisture volume fraction
MVP	moisture volume percentage
NASA	National Aeronautical and Space Administration
NDIR	non-dispersive IR (gas analyser)
NDVI	normalised difference vegetation index

NEDIS	National Environmental Satellite, Data and Information Service
NOAA	National Oceanographic and Atmospheric Administration
NPL	National Physical Laboratory
NWS	national weather service
ORP	oxidation–reduction potential
OTD	optical transient detector (lightning)
OWS	ocean weather ships
PAN	peroxy-acetyl-nitrate
PAR	photosynthetically active radiation
PC	personal computer
PDUS	primary data user system
PRT	platinum resistance thermometer
PSTN	public switched telephone network
PTFE	polytetrafluoroethylene
RAM	random access memory
RASS	radio-acoustic sounding system
RH	relative humidity
RMS	Royal Meteorological Society
RS	remote sensing
SAFIR	(a French lightning direction finding system)
SAR	synthetic aperture radar
SAT	static air temperature (aircraft measurements)
SBUV	solar backscatter UV (instrument)
SCADA	system control and data acquisition
SDUS	secondary data user system
SHIP	(a WMO code)
SOC	Southampton oceanography centre
SODAR	sound detection and ranging
SST	sea surface temperature
STP	standard temperature and pressure
SVP	saturation vapour pressure
SWH	significant wave height
SYNOP	(a WMO code)
TAT	total air temperature (aircraft measurements)
TDR	time domain reflectrometry
TM	thematic mapper
TOA	time of arrival
TOGA	tropical ocean global atmosphere
TOMS	total ozone mapping spectrometer
TOPEX	ocean topography experiment
TRMM	tropical rainfall measuring mission
UARS	upper atmosphere research satellite
UHF	ultra-high frequency
UV	ultraviolet
VHF	very high frequency
VISSR	visible and IR spin-scan radiometer
VLF	very low frequency (lightning)
VMO	voluntary marine observer
VOF	voluntary observing fleet
VP	vapour pressure

VWC	volumetric water content
WCRP	world climate research programme
WHYCOS	world hydrological cycle observing system
WMO	World Meteorological Organisation
WOCE	world ocean circulation experiment
WWW	world weather watch
XBT	expendable bathythermograph
ZFP	zero-flux plane

There are some groups of letters that look like acronyms (e.g. Meteosat, Inmarsat, Wefax, Landsat) but they are in fact abbreviations, although often wrongly printed in upper-case. Their meaning can usually be inferred from the context in which they are quoted and so they are not included above. However, a useful list of related acronyms and similar abbreviations has been compiled by Padgham (1992), amounting to around 3000 items.

Reference

Padgham, R. C. (1992) *A Directory of Acronyms, Abbreviations and Initialisms.* Natural Environment Research Council, Polaris House, Swindon, UK.

Index